A Coast of Scenic Wonders

Coastal Geology and Ecology of the Outer Coast of Oregon and Washington and the Strait of Juan de Fuca

Miles O. Hayes and Jacqueline Michel

Joseph M. Holmes

Illustrator

a Division of Research Planning, Inc.
Columbia, South Carolina

PANDION BOOKS
a division of Research Planning, Inc.
P.O. Box 328
Columbia, South Carolina USA 29202

Email: permissions@researchplanning.com

A CIP catalog record for this book has been applied for from the Library of Congress.

ISBN 978-0-9816618-5-8

All photographs by Miles O. Hayes and Jacqueline Michel unless otherwise indicated.
Illustration and design by Joseph M. Holmes.

The paper used in this book meets the minimum requirements of the American National Standard for Information Sciences – Permanence of Paper for Printed Library Materials, ANSI/NISO Z39.48-1992

Printed in the USA through
Four Colour Print Group, Louisville, KY

Front cover: View looking north at a large sea stack called Proposal Rock at the outlet of Neskowin Creek. The village of Neskowin, Oregon is visible in the distance. Photo taken 28 September 2018.

Back cover: Sand and gravel beach named Sporthaven, south of the jetty at Brookings, Oregon. Photo taken on 1 August 2014 by Walter J. Sexton.

DEDICATION

TO THE ONES WHO DID THE WORK

While writing this book, we relied extensively on the incredible amount of research conducted along the Oregon and Washington Coasts by coastal geologists and oceanographers. In particular, we want to acknowledge the work of Paul D. Komar and Peter Ruggiero and their students and colleagues at Oregon State University; and Jonathan C. Allan at the Coastal Field Office, Oregon Department of Geology and Mineral Industries. Their work has greatly enhanced the understanding of the geological history, coastal processes, sediment dynamics and budgets, and coastal hazards of the region. We also want to acknowledge the contributions of David Alt and Donald Hyndman, both who were at the University of Montana, who's books on the "Roadside Geology of..." different states inspired us to write our series of books on the coastal geology and ecology of different states. This book on Oregon and Washington is the fifth in our series.

TABLE OF CONTENTS

LIST OF FIGURES

PREFACE

One might rightfully wonder why a duo of scientists native to and permanent residents of the Carolinas would write a book on the coastal geology and ecology of the coasts of Oregon and Washington. We are coastal geologists by training and practice, having carried out projects on coasts all over the World. However, because we have been conducting scientific response to major oil spills since 1975, we also have a keen interest in coastal ecology. In 1976, we invented a coastal mapping system called the Environmental Sensitivity Index (ESI), and since that time, our group has mapped in detail the coastal environments and biological resources of all the coastal states in the U.S. plus several international areas (e.g., Gaza Strip, Kuwait, El Salvador, Niger Delta, etc.) using the ESI technique multiple times. Of course, those mapping projects included the coasts of Oregon and Washington, the results of which are presented herein.

Having such extensive environmental data on the coasts of the different states inspired us to write a book series on the coastal geology and ecology of some of them. This is the fifth such book in that series, that started with the coast of South Carolina in 2008. Other books in the series include ones on the coasts of Central California (2010), Georgia (2013), and Southern Alaska (2017). Two additional books are in preparation: Cape Cod and Florida.

Whereas we won't be taking you on any boat rides, beach walks, or overflights in a small airplane, which is usually a part of our field mapping projects, we hope the descriptions and illustrations in this book will convey a little of the excitement that we always feel when we are out in the field along the fascinating coasts of Oregon and Washington.

Miles O. Hayes and Jacqueline Michel, Columbia, S.C., 24 February 2019.

ACKNOWLEDGEMENTS

First and foremost, we would like to thank Paul Komar for his review of an earlier draft of this book and providing copies of key research papers. The ecology sections were reviewed by Chris Boring, Lauren Szathmary, and Jennifer Weaver, the RPI biologists who worked on the Environmental Sensitivity Index (ESI) atlas for Oregon and Washington. All of their suggestions are much appreciated. Walter J. Sexton shared his photographs of the Oregon and Washington Coasts.

And for all of the great quotes and details on the geology of the region that we have lifted from the writings of David Alt, the Earnest Hemingway of coastal geology writers.

However, we should add that we take full responsibility for any misinterpretations we might have made in our conclusions, with respect to both the suggestions of the reviewers, and our interpretation of the impressive volume of scientific literature that now exists for the coasts of Oregon and Washington (on a variety of ecological and geological topics).

At Research Planning, Inc. (RPI), our science technology company located in Columbia, South Carolina, Joe Holmes created all the graphics in this book, obtained permissions for the figures and quotes, and prepared the manuscript for printing. Wendy S. Early assisted in preparation of the manuscript. Mark White, Lee Diveley, and Katy Beckham synthesized the ESI shoreline data. Pandion Books, a Division of RPI, provided funding as well as staff support.

That land is a community
is the basic concept of **ecology**,
But that the land is to be loved and respected
is an extension of **ethics**.

Aldo Leopold - SAND COUNTY ALMANAC

My words are like the stars –
they do not set.
How can you buy or sell the sky –
the warmth of the land?
The idea is strange to us.
Yet we do not own the freshness of the air
or the sparkle of the water.
How can you buy them from us?
We will decide in time.
Every part of this earth
is sacred to my people.

Chief Seattle - IN LETTER TO THE PRESIDENT IN 1885.

1 INTRODUCTION

The authors of this book have been studying coasts for many years, with co-author Hayes' analysis of the coasts of the World for the U.S. Navy in 1960-65 being a starting point (Hayes, 1964; 1967a). Since that time, the authors have worked on the coasts of over 50 countries, including almost all of those in North America, the Middle East, parts of Central and South America, and West Africa. Hayes also supervised Masters and Doctoral research of 72 graduate students, most of which had to do with coasts, while a Professor at the Universities of Massachusetts and South Carolina. Michel's specialty of oil-spill response has enabled her to carry out research in a wide variety of coastal settings.

Our goal for this book is to introduce you to the natural setting of the coasts of Oregon and Washington, without a doubt two of the most beautiful and most geologically complex coasts in all the "lower 48" states. Also, those are coasts you can enjoy visiting during all seasons of the year (see satellite image of the entire area under discussion in Figure 1).

There are two ways you may want to consider using this book:

1) If you wish to gain an introductory understanding of the coastal processes and landforms, and general ecological trends, you should spend some time right away reading Sections I and II.
2) On the other hand, if you are visiting the coast and you want to learn about a specific area, you can look up that area in Section III. Then, if you have questions about some aspect of that particular area, such as why the beach is eroding or what principal bird groups are present, you can read up on those topics in Sections I and II.

The overall approach and underlying theme of this book are based first on a description of the physical characteristics of the coasts, because that physical framework is the basic skeleton upon which other relevant features, such as coastal wetlands, sand beaches, and rocky shore complexes, rest. For example, certain fundamental questions, such as why do some segments of the coast erode dramatically while others do not, cannot be answered without a firm understanding of the physical processes that cause the erosion and the geological history of the landscape being eroded. However, none of the major coastal habitats, including their biological components, are neglected as we explore this coastal area from the California border in the south to the Olympic Peninsula in the north.

Before getting further into this discussion, we need to introduce the term **geomorphology** (*geo* = earth; *morph* = shape, form; *ology* = to study), a scientific discipline devoted to understanding the origin and three-dimensional shape of the landforms of the Earth. The discipline of **coastal geomorphology** is focused on the understanding of how the different coasts of the World originated, as well as their resulting forms and organization.

Both of the authors of this book are coastal geomorphologists. When we started our research in coastal geomorphology, understanding how the coast was made was a young field of scientific research with but a few practitioners, such as Francis P. Shepard, Willard Bascom, and Douglas Inman, who produced many of the first ideas on the coastal processes and geomorphology of the Pacific Coast of North America.

With regard to our early work, hypotheses on how barrier islands evolve, how the tides and waves shape the coast, and how coastal storms change things were there for the alert observer to deduce, assuming enough examples had been seen and that the global processes had been properly accounted for. We were using what

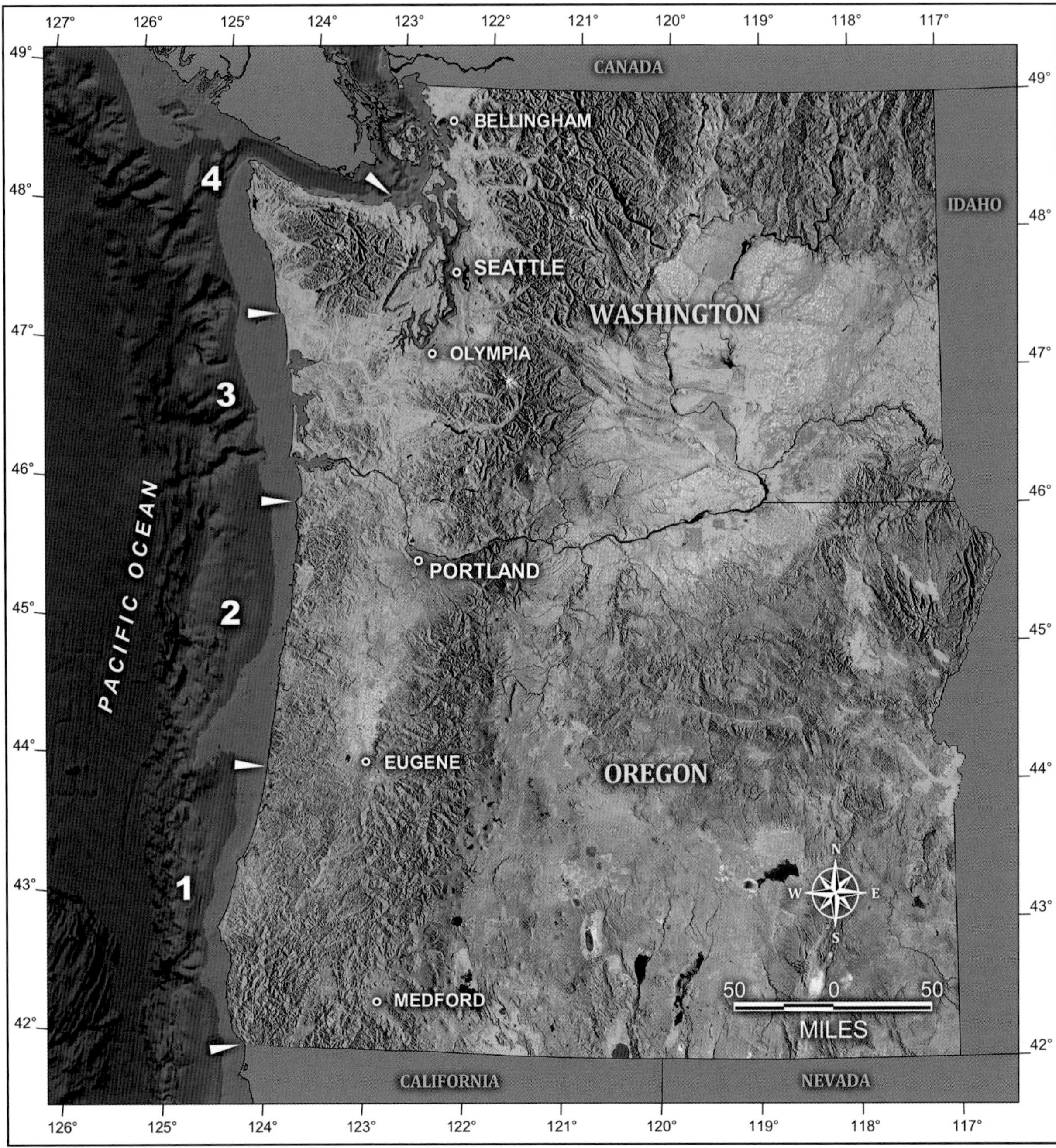

FIGURE 1. Satellite image of the states of Oregon and Washington. The boundaries of the four major morphological Compartments of the coast [Southwest Coast (1); Central Upland (2); Central Estuaries (3); and Olympic Peninsula (4)], that are discussed in detail in the text, are also shown. Image courtesy of USGS and NOAA NGDC.

Comet (1996) referred to as "one of Aristotle's two forms of logical inference" – namely inductive reasoning from the observation of a generalized pattern or distribution in order to develop a principle or law.

Although we call ourselves coastal geomorphologists (or sometimes coastal geologists), our interests and professional activities have provided for us a broad experience in coastal ecology. We are ardent birders and the work of our company, Research Planning, Inc. (RPI) in Columbia, South Carolina, in areas such as beach erosion, oil-spill response, and natural resource mapping, has broadened our understanding of coastal ecosystems. We assisted the RPI geologists and biologists as they compiled a summary of the coastal ecosystems of the two coasts under discussion in this book in 2015 under contract to the National Oceanic and Atmospheric Administration (NOAA) that included detailed maps of the coastal habitats and biological and human-use resources (NOAA/RPI, 2015). The data gathered during that project provided necessary information on many of the sites discussed in Section III.

This book begins in Section I with a brief comparison of this coast with other coasts around the World, with emphasis on those coasts on the western boundaries of North and South America. Then we discuss the origin of the coasts of Oregon and Washington, emphasizing the current theories on their geological evolution. The rest of Section I includes a description of the dynamic physical processes that shape the present shore, namely waves, tides, major storms, earthquakes, and tsunamis. The major landforms of these coasts are also discussed: rocky coasts, sand and gravel beaches, littoral cells, crenulate bays, tidal inlets, coastal dune fields, and so on. These coasts have an abundance of sand beaches, thus there is a detailed discussion of the beach cycle and a review of the erosional impacts of major coastal storms on beaches.

Section II is a discussion of the animals and habitats of the coastal zone, emphasizing birds, marine mammals, fish, and invertebrates that you will see during your visit. We discuss specific issues for threatened and endangered species and efforts to protect them.

In Section III, we discuss and identify opportunities for you to learn about and enjoy the diverse ecosystems that exist within the four geomorphological Compartments (as defined by us; see Figure 1) of the coastal zone. Each Compartment is further subdivided into individual geomorphological components called Zones. The description of each Compartment includes notes on its geological framework, beach morphology and sediments, and general ecology. A discussion of some key places to visit concludes the coverage of each of the Zones within the four Compartments, that are labeled on Figure 1 and listed below.

Compartment 1: SOUTHWEST COAST
(California border to Florence Jetties)

Compartment 2: CENTRAL UPLAND
(Florence Jetties to Tillamook Head)

Compartment 3: CENTRAL ESTUARIES
(Tillamook Head to Point Grenville)

Compartment 4: OLYMPIC PENINSULA
(Point Grenville to East End of Dungeness Spit)

SECTION I
Coastal Processes and Landforms

2 GENERAL MORPHOLOGY OF COASTS

INTRODUCTION

"The rivers and the rocks, the seas and the continents, have changed in all their parts; but the laws which direct those changes and the rules to which they are subject, have remained invariably the same." With these impeccable words, Playfair (1802) conveyed Hutton's doctrine of uniformitarianism (normally expressed as "the present is the key to the past"; Hutton, 1785), and it is upon the strength of that concept that all geologists, who study modern coasts with the intent of understanding their history, have built their science. With this in mind, we now begin our daunting task of relating the story of how the coasts of Oregon and Washington have evolved to their present forms.

All coasts have a rather complicated evolution, but the West Coast of the U.S. has been one of the most difficult ones for geologists to interpret, because of its relatively unstable position along the Pacific Ocean's "ring of fire," as well as the impact that major fault systems have had in its development. The ultimate form of the coast is dependent on a concept called tectonism, that is defined as the forces involved in or producing deformation of the Earth's crust, such as folding and/or faulting of the rocks, with the motion of features called tectonic (lithospheric) plates usually being the driving mechanism for that deformation. Key to understanding much of the tectonic evolution of the Earth hinges on the principles of a concept called **plate tectonics** that states that individual blocks of the rigid lithosphere, called plates, compose the outer portion of the Earth. The individual plates, illustrated in Figure 2, either

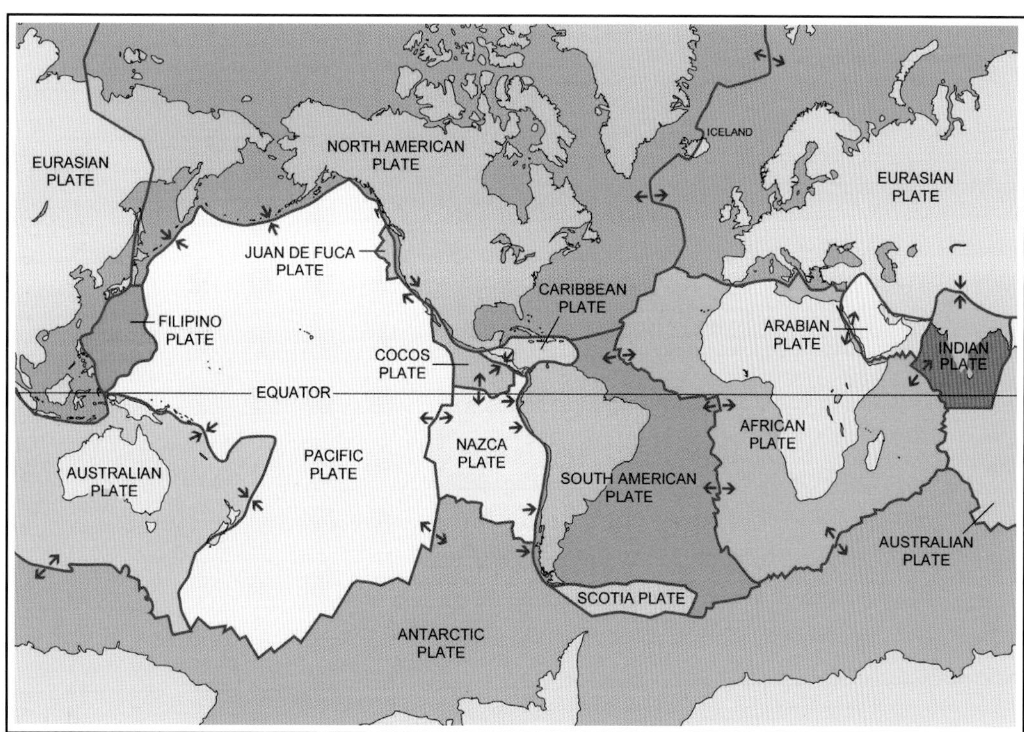

FIGURE 2. Tectonism. The tectonic plates that occur on the Earth and their relative movement.

move apart (as the seafloor of the oceanic plates spread away from mid-ocean ridges, a process usually called continental drift), into each other (as in **subduction zones** where oceanic and continental plates collide together with the oceanic plate moving beneath the continental plate), or parallel to each other in places along features called transform faults. The general illustration in Figure 3 demonstrates the process that takes place in a subduction zone. The plates move slowly above (riding on) the portion of the Earth's mantle called the asthenosphere, which behaves in a plastic manner.

When co-author Hayes was a graduate student at the University of Texas in the 1960s, very few of the professors in the elite Ivy League universities "believed in" **continental drift**, although the idea had been seriously proposed and discussed by scientists since almost the beginning of the 20th century. The problem for the "non-believers" was that, up to that time, no credible mechanism had been proposed to explain why the continents had drifted apart.

Now, of course, the concept of plate tectonics provides the basis for almost all current thinking relative to most topics in geology, including the mechanism for continental drift. Geophysical evidence, such as increasing age of parallel magnetic bands of volcanic rocks on the seafloor away from the mid-ocean ridges (Figure 4), provided indisputable evidence that the continents are indeed drifting apart. The exact mechanism causing this movement is more in doubt. The favored theory at this time, the convection cell theory (Figure 3), whereby huge individual slabs of rock, the plates, are slowly moved apart by convection cells of hot, softened mantle material below the rigid plates. There are other more complex theories. One fine day all

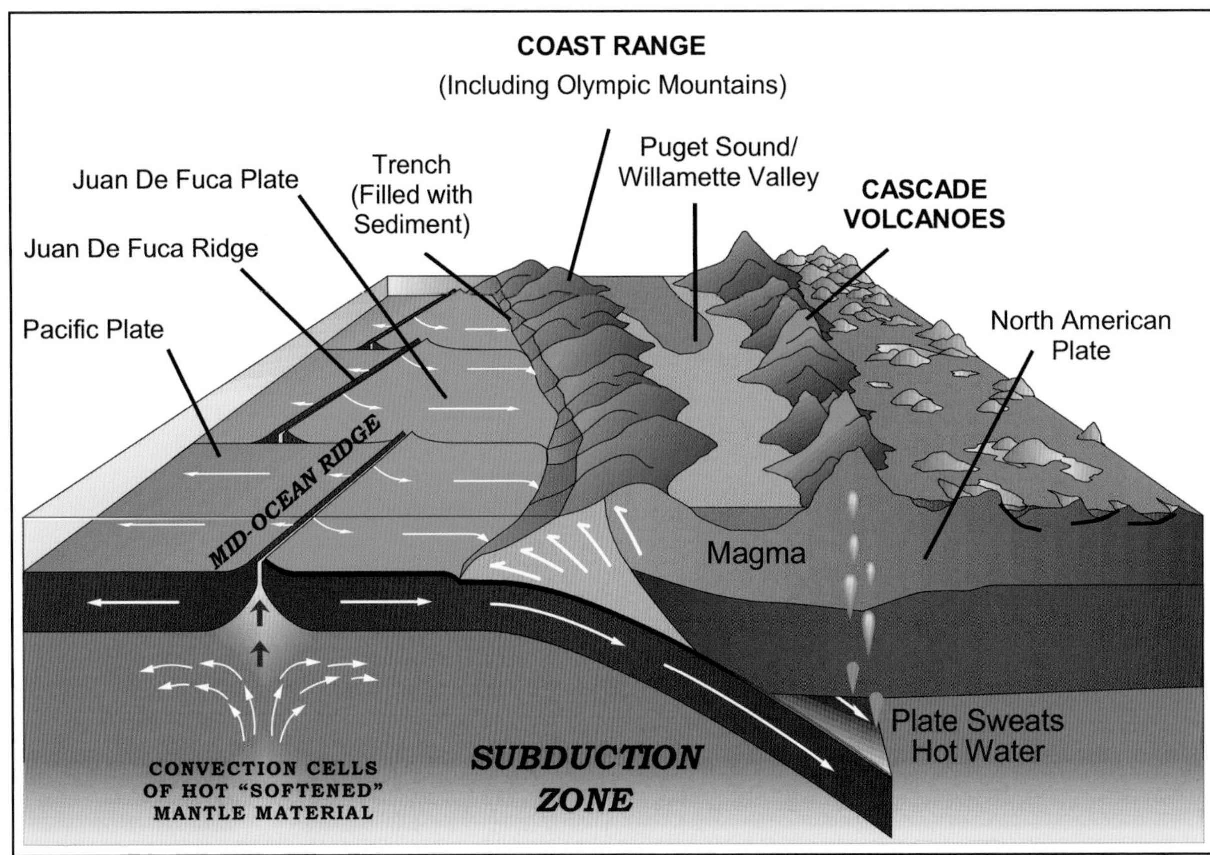

FIGURE 3. Illustration of many of the elements involved in the process of the migration (and collision) of tectonic plates and seafloor spreading. Two key elements in the ocean basin are: 1) The oceanic spreading ridge, where new material is added to the sea floor as the two adjacent oceanic plates move away from each other according to convection cell theory, explaining why the continents are drifting apart; and 2) The area where the continental and oceanic plates collide, resulting in the uplift of the Coast Range and the formation of the Cascade Volcanoes. Modified after Lillie (2015).

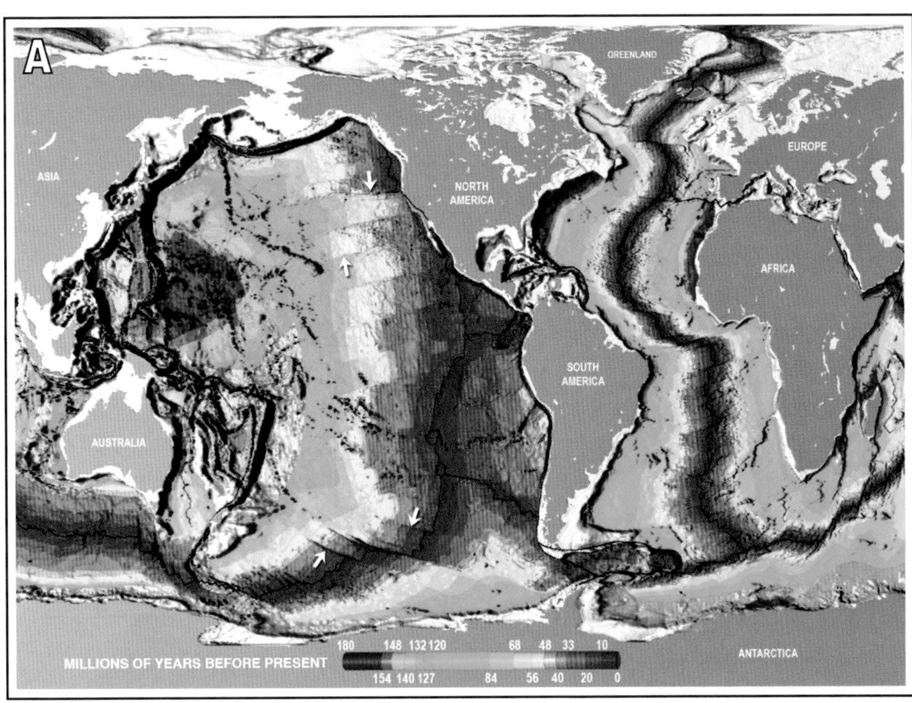

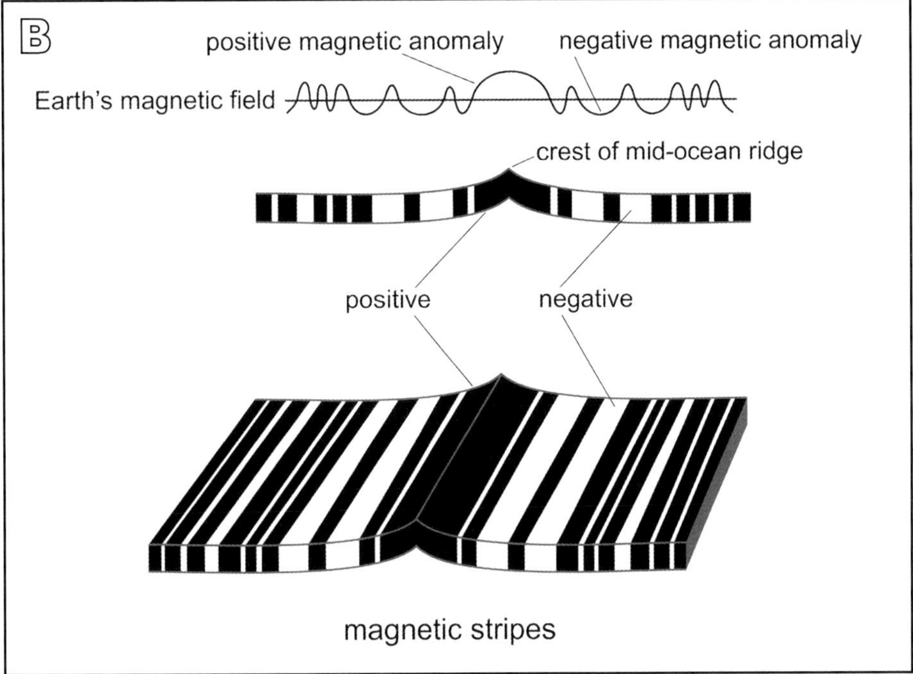

FIGURE 4. Mid-ocean ridges. (A) Sea floor topography, primarily of the Atlantic and Pacific Ocean Basins. The youngest lavas and rocks on the sea floors (shown in red) occur adjacent to the mid-ocean ridges away from which the ocean floors continue to spread. Note that the mid-ocean ridge is not centrally located in the Pacific Ocean; in fact, it has been partially overridden by the westward moving North American Plate. Arrows point to a few of the multitudes of transform faults that have occurred to accommodate the ocean floor's expansion on the spherical surface of the Earth. From Peter W. Sloss, National Geophysical Data Center (NGDC, 2009). (B) Occurrence of parallel magnetic bands of volcanic rocks on the sea floor away from the mid-ocean ridges, with the individual bands systematically increasing in age away from the crest of the mid-ocean ridges as is indicated by the colors in A.

of this will be figured out. Meanwhile, let us get back to the indisputable fact that the continents are moving.

Based on a study of the World's inner continental shelves that was sponsored by the U.S. Navy through the Defense Research Laboratory at the University of Texas, co-author Hayes (1964; 1965) proposed the following preliminary coastal classification based to a large extent on the concept of plate tectonics.

Class	Tectonism
A. Tectonic coasts	Rapid uplift
1. Young mountain range coasts	
2. Glacial rebound coasts	
B. Plateau-shield coasts	Stable
C. Depositional coasts	Rapid downwarp

In this case, the term rapid is used in a geological sense (measured in fractions of an inch per year). A similar concept that related coastal types to the plate tectonic postulate was presented in considerably more detail by Inman and Nordstrom in 1971. They used the term collision coasts, a category into which most young mountain coasts would fit, the term trailing-edge coasts, into which many depositional coasts would fit, and so on.

The cross-section given in Figure 3 shows one of the basic concepts of plate tectonics, in which a plate made up of a continental crust (e.g., the North American Plate, which is relatively light because of the abundance of aluminum, silica, sodium, etc. in the rocks) is riding over a sinking oceanic plate (e.g., the Pacific Plate, which is relatively heavy because of the abundance of iron, magnesium, etc. in the rocks). If we look at two of the simpler examples, the North and South American Plates, a striking contrast can be seen between the western shores of those continents and their eastern shores. Along much of the western shores, where the continental and oceanic plates collide, or the **leading edge** of the continental plates, young mountain ranges flank the coast. However, on the eastern shore, or the **trailing edge** of the continental plate, which is moving westward at a rate of around 4 inches/year (Monroe and Wicander, 1998), broad deltaic and coastal plains are present in places. To use this simple example, the western coasts of a significant amount of the North American continent and most of the South American continent are dominated by young mountain range coasts characterized by: 1) High mountains caused by relatively recent orogenic (mountain-building) activity; 2) Bedrock of variable age but dominated by relatively young sedimentary and volcanic rocks; 3) Active tectonic uplift (e.g., with old beach lines lifted up the sides of the mountains thousands of feet in some places); and 4) Mostly short, steep rivers emptying into the sea. This is a presentation of the leading-edge concept in its simplest terms. The eastern side of the North American Plate is dominated by depositional coasts characterized by: 1) Coastal zone made up chiefly of broad coastal and deltaic plains; 2) Bedrock composed generally of Paleogene/Neogene and Quaternary sediments (young geologically speaking; see geological time scale in Table 1); 3) Tectonically subsiding areas (e.g., thousands of feet of sediments have accumulated in some areas); and 4) Many large and long rivers emptying into the sea. A more elaborate discussion of coasts of that type is presented in the books - *A Coast For All Seasons: A Naturalist's Guide to the Coast of South Carolina (Hayes and Michel, 2008); and A Tide-Swept Coast of Sand and Marsh: Coastal Geology and Ecology of Georgia* (Hayes and Michel, 2013).

CHARACTERISTICS OF DEPOSITIONAL COASTS

Clearly, the coasts of Oregon and Washington occur along the leading edge of the North American Plate. Before beginning a more detailed treatment of leading-edge coasts in general, however, a brief consideration of the relevant aspects of depositional coasts is in order to highlight the contrasts between these two opposing types of coasts.

As already noted, global tectonic crustal movements are the fundamental cause of the major differences between depositional coasts and young mountain range coasts. Hayes (1965; 1976) concluded that the following hierarchy of factors determine the geomorphological makeup of depositional coasts – in decreasing order of importance: 1) Hydrodynamic regime; 2) Climate; 3) Sediment supply and sources; and 4) Local geological history and sea-level change. There are some significant differences in the tectonic impacts along trailing-edge and other types of depositional coasts; for example, some areas are sinking more rapidly than others because of heavy sediment loads (e.g., the Mississippi River delta). However, those differences do not appear

TABLE 1. Geologic time. Time intervals are listed in millions of years before the present [from a chart published by the International Commission on Stratigraphy (2008)].

Era/Period/Epoch			Time (Million yr ago)
Cenozoic era "Recent Life"	Quaternary period	Holocene epoch	0.012-0
		Pleistocene epoch	2.6-0.012
	Neogene period	Pliocene epoch	5.3-2.6
		Miocene epoch	23.0-5.3
	Paleogene period	Oligocene epoch	33.9-23.0
		Eocene epoch	55.8-33.9
		Paleocene epoch	65.5-55.8
Mesozoic era	Cretaceous period		145.5-65.5
	Jurassic period		199.6-145.5
	Triassic period		251.0-199.6
Paleozoic era	Permian period		299.0-251.0
	Carboniferous (Mississippian/Pennsylvanian) period		359.2-299.0
	Devonian period		416.0-359.2
	Silurian period		443.7-416.0
	Ordovician period		488.3-443.7
	Cambrian period		542.0-488.3
Precambrian	Proterozoic era		2,500-542
	Archaeozoic (Archean) era		4,600-2,500

to have a significant enough imprint on the regional geomorphology of depositional coasts to warrant further discussion here. Accordingly, we conclude that the factor that exerts the most control on the geomorphological nature of a depositional coast is the interaction of water in motion with the coastal sediments – typically called the **hydrodynamic regime** (*hydro* = water; *dynamics* = kinetic energy of the water). For purposes of interpreting the morphology of coastal features, the hydrodynamic regime is commonly expressed as the ratio of wave energy to tidal energy (i.e., how big the waves are versus how large the tidal range is). Generally speaking, on trailing-edge or epicontinental sea coasts, the tidal range is of utmost importance. On leading-edge coasts, wave action commonly plays a dominant role.

During Hayes' early study of the World's coasts for the U.S. Navy, he observed a striking correlation between tidal range (vertical distance between high and low tide) and the characteristics of depositional coasts. In 1976, he classified depositional coastal types based on this observation, differentiating among **microtidal** (tidal range less than 6 feet), **mesotidal** (tidal range = 6-12 feet), and **macrotidal** (tidal range greater than 12 feet) coasts.

[NOTE: These terms and their limits were also suggested by Davies (1973), but he did not make the kind of geomorphological correlations that Hayes did. In some cases, major inlets on microtidal, depositional coasts may have some of the characteristics more commonly found on mesotidal coasts (e.g., large shoals offshore of the inlets called ebb-tidal deltas). In that case, tidal prism (volume of water entering and leaving the inlets) becomes a factor (Davis and Hayes, 1984).]

Curiously, Hayes (1965) also noted a relatively equal distribution of the three defined classes of tidal range around the World's coasts, which was a surprise to him, because at that time, most of his field experience had been limited to the microtidal coast of Texas. Consequently, upon leaving Texas, much of his early research was focused on the mesotidal coasts of New England and South Carolina. Eventually, macrotidal coasts with tidal ranges greater than 30 feet were also studied in Alaska, France, and Chile. In addition, based partly on publications by the original guru of coastal geomorphology, W. Armstrong Price, he publicized two additional coastal types (on depositional coasts), wave-dominated coasts and tide-dominated coasts (Hayes, 1965; 1976).

As a generalization, most microtidal depositional coasts are **wave-dominated** and are characterized by such features as deltas with smooth outer margins and abundant long barrier islands (see Figure 5A). In North America, wave-dominated, depositional coasts occur on much of the coast of Texas, Outer Banks of North Carolina, and Kotzebue Sound, Alaska.

On the other hand, most macrotidal depositional coasts are **tide-dominated** and are characterized by such features as open mouthed, multi-lobate river deltas, abundant large estuarine complexes, and extensive tidal flats and salt marshes (Figure 5C). In North America, tide-dominated depositional coasts occur in Bristol Bay and Cook Inlet, Alaska and in the northern reaches of the Gulf of California.

To complicate matters somewhat, most mesotidal depositional coasts have intermediate-sized tides and waves; thus, we have recognized another class, **mixed-energy coasts**. Most mixed-energy depositional coasts are characterized by short, drumstick-shaped barrier islands with complex tidal flats and coastal wetlands on their landward flanks (see Figure 5B). In North America, mixed-energy, depositional coasts occur in Georgia and South Carolina, as well as along the Copper River Delta of Alaska.

As illustrated in Figures 5A and 5B, sand shoals created by ebb- and flood-tidal currents form on either end of the tidal inlets located between the barrier islands. However, as these Figures also show, on wave-dominated, typically microtidal coasts, the sand shoals on the landward side of the inlet, called flood-tidal deltas, are usually larger than the shoals on the seaward side of the inlet, called ebb-tidal deltas. On mixed-energy coasts, as a general rule, the ebb-tidal deltas are considerably larger than the flood-tidal deltas. The reasons for these differences are complex and discussed in detail in the Section on "Tidal Inlets."

A second major factor that determines certain key characteristics of depositional coasts is the regional **climate**. Wind patterns along a coast and offshore storms determine the average size of the waves, which have a major impact on the nature of depositional coasts. Storms, such as the hurricanes that commonly strike the Texas, Florida, and North Carolina coasts, typically generate the largest waves that occur in those areas. The type of sediments on tidal flats and beaches is strongly

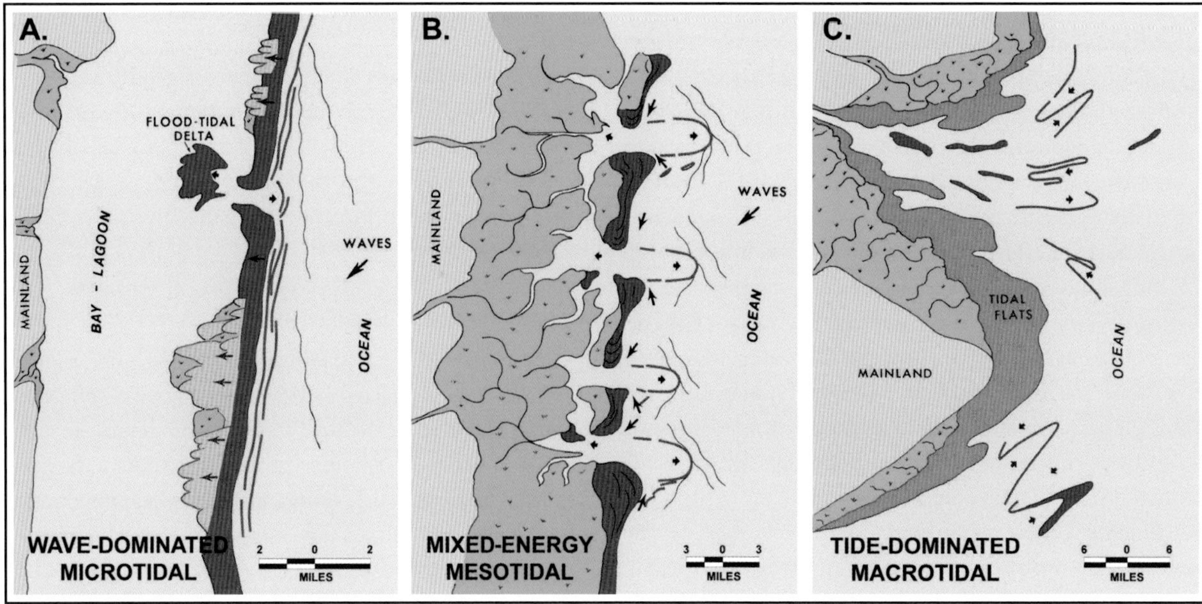

FIGURE 5. Generalized models of the three types of depositional coasts (excluding major deltas): (A) Wave-dominated (usually microtidal; tidal range <6 feet); (B) Mixed-energy (usually mesotidal; tidal range = 6-12 feet); and (C) Tide-dominated (usually macrotidal; tidal range >12 feet). The red lines in A represent nearshore, subtidal wave-formed sand bars, and those in B and C indicate intertidal and shallow subtidal sand deposits.

related to the regional climate. In tropical areas beach sediments are commonly composed of coral and algal fragments, carbonate precipitates, and shell. In temperate regions, quartz sand is usually the dominant sediment type, with rock fragments and feldspar being abundant in sand near river mouths and along coasts with eroding bedrock. Other obvious climatic influences include:

1) The presence or absence of major rivers;

2) Effects of glaciers and freeze/thaw processes. The glaciers in polar regions (past and present) bring an abundance of gravel-sized sediments to the coast;

3) Types of weathering (chemical or mechanical) that produce the sediments that make up the coastal systems. For example, chemical weathering produces an abundance of mud that accumulates along the coasts in the humid tropics;

4) Presence of beachrock in the tropics and permafrost in polar regions; and

5) Type of vegetation in coastal habitats. Intertidal zones sheltered from waves are commonly populated by mangroves in the humid tropics, salt marshes in temperate and subpolar regions, and algal mats and sediments called evaporates in the arid tropics.

Another control on the nature of depositional coasts is the **source and supply of sediments**. Much of the sand and gravel on the beaches of Oregon and Washington are delivered by rivers and streams, but wave erosion of coastal cliffs is an important contributor. The budget of beach sediments within defined littoral cells, a key concern to understanding causes of beach erosion along the Pacific Coast, is dicussed in detail later.

There are three major classes of sediments that occur along the shore (as illustrated in Figure 6): **gravel** (particles >2 mm in diameter, which consists of four

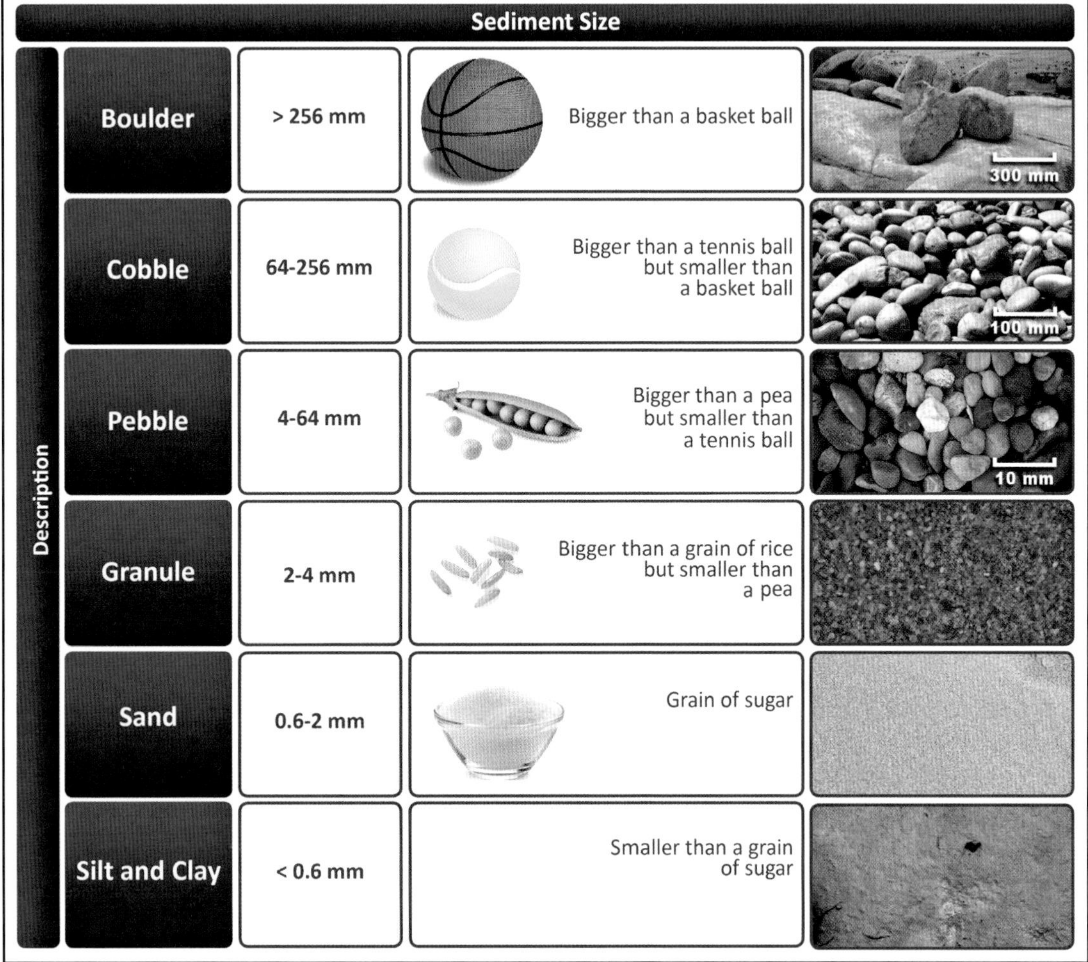

FIGURE 6. Definition of sediment types based on size in millimeters (mm).

separate classes, culminating with the largest - boulders), **sand** (particles between 0.0625 and 2 mm in diameter, ranging from very fine- to very coarse-grained), and **mud** (particles <0.0625 mm in diameter and consisting primarily of silt and clay). On the coasts of Oregon and Washington, there is a wide range in sediment grain size, from large boulders at the base of some of the eroding cliffs to mud on the tidal flats and salt marshes.

The last two factors affecting the nature of depositional coasts are **local geological controls** and **sea-level rise**. Examples of geological controls include glacial effects, faulting, earthquakes (that can abruptly raise or lower the coastal zone), and variations in coastal bedrock type. With regard to sea-level rise, sea level on a global scale has been relatively constant within the past 5,000 years. A key question at the moment is will sea level continue to rise globally as it has been in the past few decades (at the rate of about one foot/century), and, if so, how fast? This issue will be discussed in more detail later.

CHARACTERISTICS OF LEADING-EDGE COASTS

The coasts of Oregon and Washington are situated on the leading edge of the North American Plate. Classic features of collision coasts, such as young mountain ranges rising adjacent to a major tectonic trench under the influence of a subducting oceanic plate, have been present on the Pacific Coast throughout the past 200+ million years or so; some of the more recent phases are illustrated in Figure 7. However, as shown in Figure 7C, this is not the case at the present time along the Central California Coast, where the last collision coast configuration ended 25-30 million years ago, in that the present Central California sliding plate boundary is not a collision boundary in the classic sense, because of the presence of the San Andreas Fault. As also shown in Figure 7C, however, the coasts of Oregon and Washington are still along a subducting (collisional) margin because the Gorda, Explorer, and Juan de Fuca Plates continue to slide under the North American Plate.

To elaborate upon the characteristics of leading-edge coasts, we carried out a study of the entire length of the leading-edge coasts of North and South America (a straight-line distance of almost 16,000 miles), based on previous field work and imagery from Google Earth. Generally speaking, four attributes of these leading-edge coasts readily distinguish them from trailing-edge, depositional coasts:

1) The abundance of rocky shores: For example, the coast of western South America south from Chincha Alta, Peru to San Antonio, Chile, a straight-line distance of over 1,600 miles, has up to 50% rocky shores (mostly wave-cut rock cliffs and/or wave-cut rock platforms);

2) Scarcity of major river deltas: Only two significant major river deltas occur along the outer coasts of the leading edges of the North and South American Plates – the Copper River Delta in Alaska and the Colorado River Delta in Mexico. The Copper River Delta is there because of an overwhelming amount of sediment bought to the coast from glacial sources (annual sediment load of 100 x 10^6 metric tons – Reimnitz, 1966; Hayes and Ruby, 1994). The Colorado River has a rather unique position related to the opening of the Gulf of California as a result of an amazing lateral (south to north) motion along the trace of the San Andres Fault system.

3) Less extensive development of coastal wetlands; and

4) Relative lack of barrier islands.

Barrier islands are elongate, shore-parallel accumulations of unconsolidated sediment (usually sand), some parts of which occur above the high-tide line, except during major storms. They are separated from the mainland by bays, lagoons, estuaries, or wetland complexes and are typically intersected by deep tidal channels called tidal inlets (see Figures 5A and 5B). In order for barrier islands to exist, there must be a supply of sand in the longshore sediment transport system from which the islands can be formed, as well as waves large enough to build them. Also, the tidal range must be less than about 12 feet because the strong tidal currents in areas of large tides transport the available sand offshore, and waves need to be focused within a relatively limited vertical segment of the intertidal zone in order for them to construct the islands. In addition, barrier islands can only form on a relatively flat coastal zone that allows the sediments space to accumulate. Therefore, barrier islands never occur on steep rocky shores.

[NOTE: There are three very relevant terms regarding segments of coasts influenced by tides. *Supratidal* areas are those only very rarely covered with water, such as during significant storm surges. *Intertidal* areas are confined to within the zone of the highest and lowest astronomical tides. *Subtidal* areas are almost always covered with water. These distinctions are particularly important to the zonation of biological communities on rocky shores.]

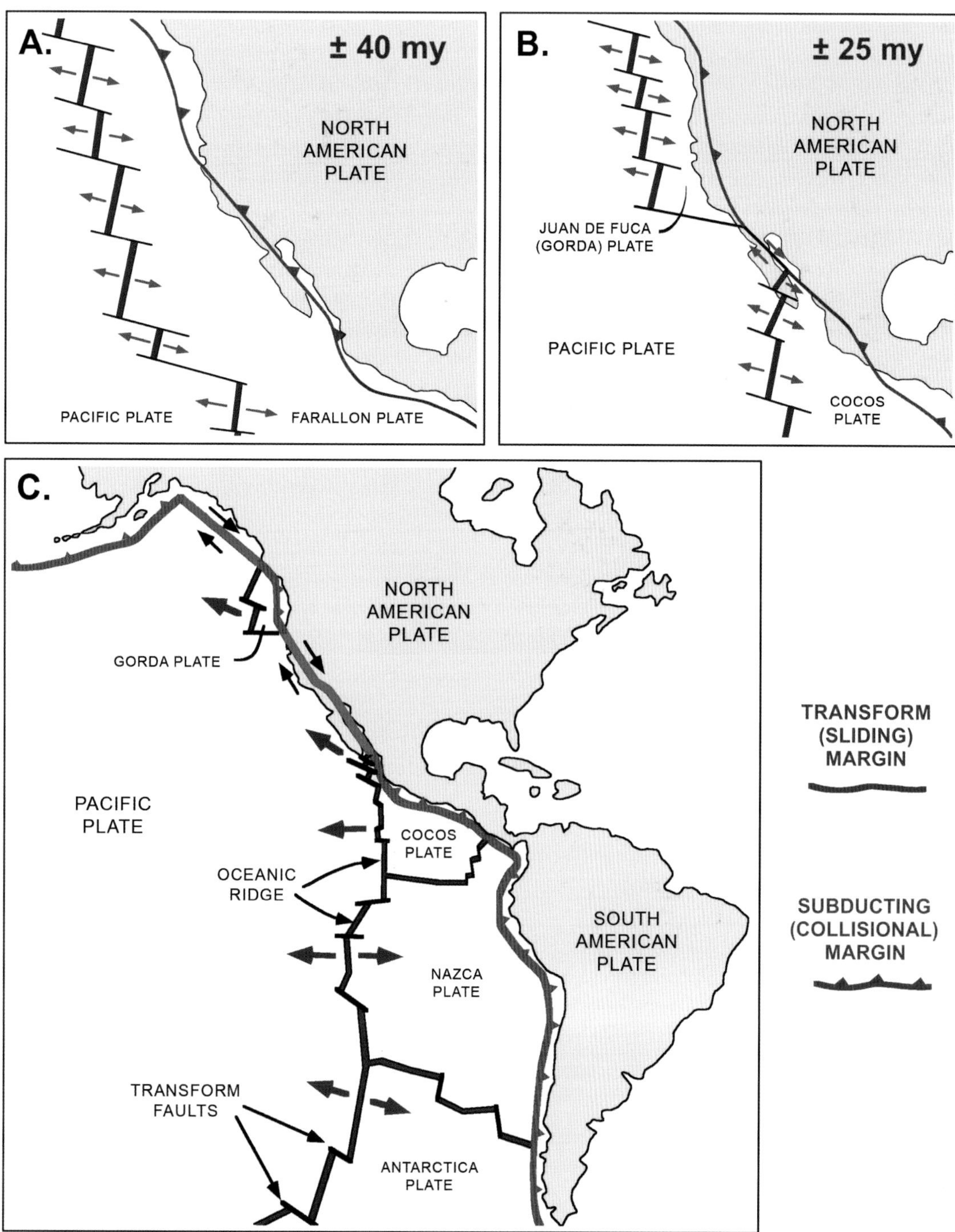

FIGURE 7. Evolution of the oceanic plates of the eastern Pacific Ocean. (A) At about 40 million years ago, the former mid-ocean (Pacific) ridge had not yet reached the leading edge of the North American Plate. (B) At about 25 million years ago, the mid-ocean ridge had met the shore of the Baja Peninsula and moved into what is now the Gulf of California area. The San Andreas Fault had already formed and the transform motion between the Pacific and North American Plates had begun. (C) The present configuration of the plates. Much of the Farallon Plate had disappeared under the North American Plate. The fragment of that plate still offshore to the north of Central California is now named the Gorda Plate. The active San Andreas Fault extends inland to the present Northern California along the transform margin (shown in blue). Note the subduction zones to the north and south, that are colored red. Diagrams courtesy of Dr. Ed Clifton.

There are six areas along the leading edges of the North and South American Plates that contain significant barrier islands, with one of the most interesting areas being along the Pacific coast of Colombia, where a continuous chain of 62 barrier islands occurs along 415 miles of the shore. The sand that composes those barrier islands is delivered to the coast by the abundant runoff of eight rivers generated by 200 inches of rainfall per year (Pilkey and Fraser, 2003).

There are altogether approximately 1,180 miles of barrier islands along the leading-edge coasts of North and South America, a somewhat surprising number considering the general models of the two contrasting coastal types that have been published to date. However, this number pales in comparison to the fact that barrier islands make up at least 80% of the trailing-edge coast of the North American Plate between Campeche, Mexico and Long Island, New York, a straight-line distance of almost 4,000 miles. A recent study of satellite imagery of barrier islands along the World's coasts by Matthew Stutz and Orrin Pilkey of Duke University revealed that 63% of the barrier islands occur on trailing-edge coasts, 16% on leading-edge coasts, and 21% on marginal seas. Going one step further, barrier islands compose only 1/16 of the total leading-edge coasts of the two continents that we studied (North and South America). However, the unexpected occurrence of over 1,000 miles of barrier islands on those two leading-edge coasts is an interesting puzzle that begs for explanation.

There are a number of sandy, spit-like features on the outer shores of Oregon and Washington that some may refer to as barrier islands. However, as pointed out by Dingler and Clifton (1994), *West Coast barriers have formed under a different geologic setting than the more familiar barriers of the Gulf and East Coasts, and, correspondingly, have a somewhat different appearance. All the West Coast barriers are spits, generally short because of the ruggedness of the coast, and have only one inlet because of the relation between their outflow (tidal prism and/or river discharge) and the local wave climate.* The terms generally used for those sandy features on the West Coast are **barrier spits** and **bay barriers**.

There are 36 barrier spits and bay barriers on the coast of Oregon according to Dingler and Clifton (1994). Those spits average 1.9 miles in length, with the longest one, at 7.8 miles, occurring at the mouth of Coos Bay. Fifteen of the spits are less than 1 mile long.

On the other hand, there are only 16 such spits and bay barriers on the outer coast of Washington (Dingler and Clifton, 1994). Those features average 2.8 miles in length, with the longest one, at 19.2 miles at the North Beach Peninsula across the entrance to Willapa Bay, just north of the Columbia River, a major source of sediment transport to the coast. Ten of those spits are less than 1 mile long.

By way of comparison, the barrier islands at the head of the Georgia Bight on the East Coast (Outer Banks of North Carolina and northeast coast of Florida) average around 20 miles in length (Hayes and Michel, 2013).

In the earlier discussion of the determining characteristics of depositional coasts, the effects of climate and sediment supply were considered to be less important than hydrodynamic regime as determined by the ratio of wave energy to tidal energy. On leading-edge coasts, however, the tidal range *per se* is not as important as a general rule. Of course, for leading-edge coasts on open oceans with an almost limitless fetch (on the windward side of the continents), the waves are usually quite large.

The general geological cross-section illustrating the collision of continental and oceanic plates is shown in Figure 3. At the collision point, the continental plate rides over the oceanic plate, creating a high, young mountain range. Material being transported along the ocean floor, which includes the oceanic crust of the Pacific Ocean in the case under discussion, is subducted downward to depths that allow melting of that material.

Some of this subducted solid crust that has been melted (called magma), which is a result of its reaching depths where temperatures exceed 2,192°F, then rises to the surface in the form of volcanoes (Figure 3). This chain of volcanoes typically forms above the sinking plate in a line parallel to the trench into which its parent material sank at a distance of 50 to 200 miles from the shore (Alt and Hyndman, 2000). According to this general model, the seaward edge of the continental plate is rising and the coast contains an abundance of eroding rocky shores.

However, close inspection of the two leading-edge coasts of North and South America reveals some significant variations. For example, the edges of the continental plates are not rising everywhere at the present time. The outer shore of the Kenai Peninsula, Alaska contains some irregular islands composed of sinking mountaintops. That area sunk as much as 3 feet during the Good Friday earthquake of 1964 (Plafker,

1969). In the long term, those mountains have risen in response to the major subducting system, but variations in the general trend makes for some unexpected coastal types.

PHYSICAL SETTING OF THE PACIFIC NORTH-WEST COAST

The western portion of Oregon consists of five geomorphic regions that are illustrated on Figure 8 (based on divisions drawn by Alt and Hyndman, 1978):

1. Klamath Mountains
2. Coast Range
3. Willamette Valley
4. Western Cascades
5. High Cascades

The western portion of Washington consists of five geomorphic regions that are illustrated on Figure 9 (based on divisions drawn by the Washington State Department of Natural Resources, 2014; and Lasmanis, 1991):

1. Willapa Hills
2. Olympic Mountains
3. Puget Lowlands
4. Southern Cascades
5. Northern Cascades

The origin and critical physical characteristics of those individual geomorphic units will be discussed in some detail later.

FIGURE 8. Shaded relief map of Oregon, courtesy of USGS, National Map.

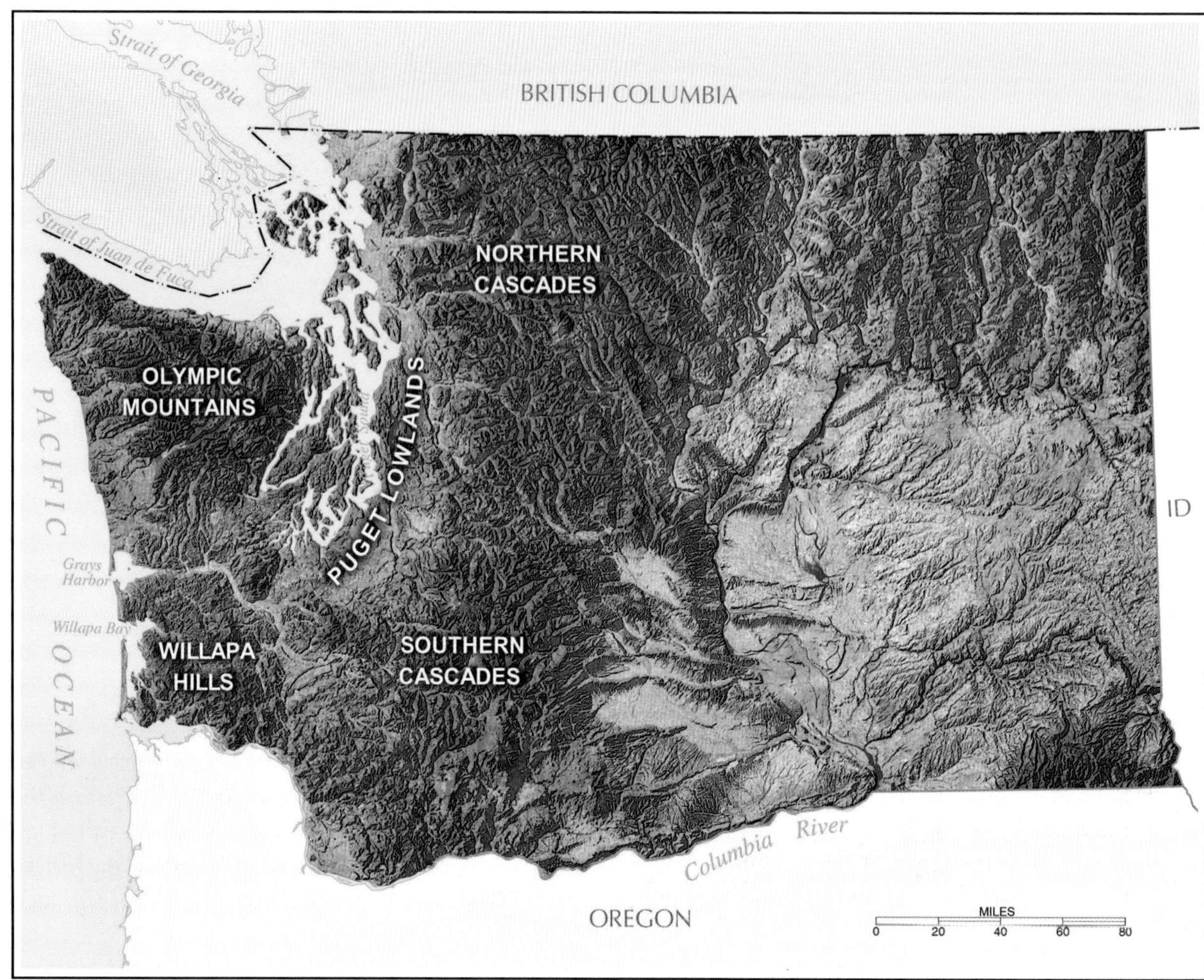

FIGURE 9. Shaded relief map of Washington, courtesy of USGS, National Map.

3 GEOLOGY OF THE PACIFIC NORTHWEST COAST

This history is designed now and ever to keep the sneers from the lips of sour scholars.

John Steinbeck - TORTILLA FLAT

INTRODUCTION

Unraveling the long history of how the present Pacific Northwest Coast of the U.S. originated has fascinated geologists for many decades. Before we get into the details, we want to define some terms.

The three major classes of rocks - igneous, sedimentary, and metamorphic - were first defined by the pioneering Scottish geologist James Hutton in the mid 1700s. He recognized a class of rocks called **igneous rocks** that crystallized from an extremely hot molten mass of material (magma) that was formed as if by fire (*ignis* = fire). One class of igneous rocks, that includes granite and diorite, develops by slow cooling of the molten mass at great depths within the Earth's crust. As a result of the slow cooling, large crystals of individual minerals form that can be as much as an inch or so in diameter. Those minerals, typically including feldspar and quartz, eventually coalesce to form a rock mass that may later be pushed up to the Earth's surface by mountain-building processes. A second class of igneous rocks, called volcanic rocks (e.g., basalt and rhyolite), crystallize rapidly as a result of molten lava flowing suddenly onto or near the Earth's surface. The crystals of these minerals form so rapidly that they are, for the most part, too small to be seen by the naked eye.

On the other hand, **sedimentary rocks** are most commonly formed by weathering and erosion of pre-existing rocks on the Earth's surface. This process creates sediment that can be transported by water, or wind in some cases, to be deposited in large masses on river deltas, beaches, offshore on the continental shelf, in deep ocean depths, and so on. Once deposited, those sediments may become buried where they are consolidated by chemical cementation and other processes. The sedimentary rocks formed by this process that are made up mostly of gravel-sized particles are called conglomerates, those composed of sand-sized particles are called sandstones, and those composed of silt and/or clay are called shales, siltstones, or mudstones. Limestones are composed of calcium carbonate ($CaCO_3$) and usually develop in marine waters by a combination of chemical precipitation and accumulation of the hard parts of some marine organisms, such as seashells. Another group of sedimentary rocks, called evaporates, usually form as the result of evaporation of seawater under unusually high surface temperatures. However, some sedimentary rocks in that class form at great depths within the ocean, such as the salt deposits now forming at the bottom of the Red Sea. Evaporite deposits are most commonly composed of halite (NaCl) and gypsum [$CaSO_4.(2H_2O)$].

The third major rock class, **metamorphic rocks**, results from dramatic changes in igneous and sedimentary rocks affected by heat, pressure, and water that usually result in a more compact and more crystalline condition. Those changes usually take place at significant depths below the Earth's surface. Examples of metamorphic rocks include slate, derived from a sedimentary rock composed of silt and/or clay; marble, derived from limestones; and gneiss, a banded, micaceous rock typically derived from granite.

The story of the origin of the Pacific Northwest Coast is rather complicated, and there are still many unanswered questions regarding details, such as the exact age and timing of the different geological episodes. Those events are related in a general way to the time line given in Figure 10. The following discussion is only a brief summary. For greater detail, the following books should be sufficient to satisfy your desire to learn more

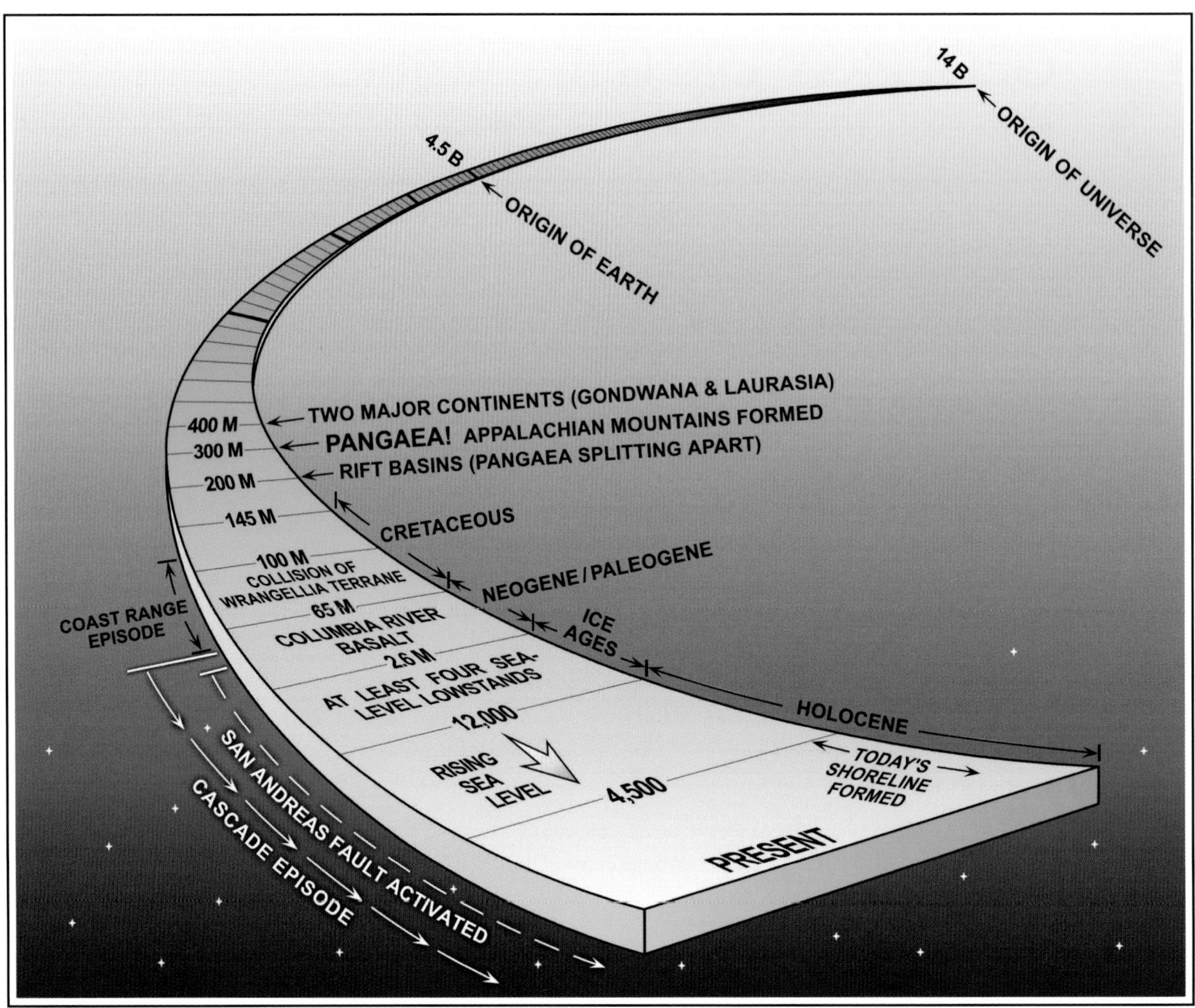

FIGURE 10. Time line for the origin of the coasts of Oregon and Washington. See Table 1 for nomenclature and timing of the geological events.

about this fascinating subject:

1) *Roadside Geology of Oregon* by David Alt and Donald Hyndman (1978)

2) *Roadside Geology of Washington* by David Alt and Donald Hyndman (1984)

3) *The Pacific Northwest Coast: Living with the Shores of Oregon and Washington* by Paul Komar (1998)

4) *Geology of Oregon* by Elizabeth Orr and William Orr (2000)

5) *Geology of the Pacific Northwest* by William Orr and Elizabeth Orr (2002)

A discussion of the geological history of the coasts of Oregon and Washington requires a slightly more detailed account of plate tectonics than the introduction given earlier. In order to do so, a generalized cross-section of the Earth with its three major components, crust, mantle, and core, is presented in Figure 11 and discussed below:

1) The surface material is called the Earth's crust. The crust is composed of igneous, metamorphic, and sedimentary rocks. The crust under the oceans, averaging 6 miles thick, is composed mostly of rocks that contain an abundance of iron and magnesium-bearing minerals (e.g., basalt), whereas the crust on the continents, that averages 20-30 miles thick, but can be up to 62 miles thick, is composed of rocks made up of lighter minerals (e.g., granite). Because the crust is relatively cold, it is made up primarily of brittle rocks (plus unconsolidated sediments in some areas). Those rocks and sediments are composed predominantly of a group of minerals called

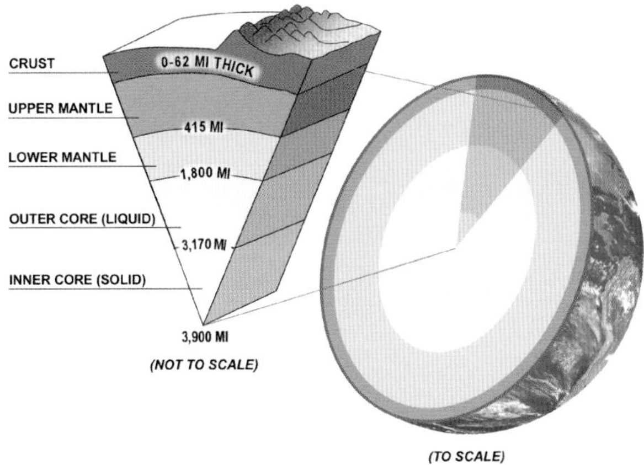

FIGURE 11. Cross-section of the Earth showing its internal structure [modified from a diagram by Jeremy Kemp, based on an illustration by the USGS (2005)].

silicates, the fundamental building block of the Earth's crust, with the atoms of silicon and oxygen combined making up about 75% of it. More information on this primary mineral group is given in the glossary. Because of their relative compactness, the rocks that make up the crust are susceptible to shearing and breaking, which, of course, causes the earthquakes that occur with some degree of regularity on the Pacific Coast.

2) The mantle contains more of the heavier elements than the crust, such as iron (Fe) and magnesium (Mg), and is 1,800 miles thick and makes up 70% of the Earth's total volume. The major component of the mantle is a dark colored, notably dense rock called peridotite. The most common minerals in that rock are olivine (magnesium iron silicate), pyroxene, as well as garnet.

Another important rock type initially derived from the mantle is a dark, greenish-colored rock called serpentinite composed of a number of complex, usually hydrated, iron- and magnesium-bearing minerals. As noted by Alt and Hyndman (2000), *serpentinite forms at the crests of oceanic ridges, where seawater sinks into fractures that open as the two plates separate. The tremendous pressure of the depths forces the water down into the extremely hot mantle rocks below the ridge crest. The water reacts with hot peridotite to make serpentinite, which constitutes a large proportion of the uppermost mantle.* Because of their location at the top of the mantle, those rocks are commonly squeezed or faulted (typically in a plastic state) into the overlying rocks and sediments in the course of a mountain-building event. Therefore, they now occur at the Earth's surface in several places in the mountains and along the outer coasts of Oregon and Washington, where they are easily recognized by their green color and slippery, soapy feel.

The uppermost part of the mantle, which is relatively rigid, is combined with the crust to form the lithosphere. The lithosphere is underlain by a part of the mantle known as the asthenosphere, which is more plastic than the overlying layer as a result of heating by radioactive decay of elements at depth. It is within this more fluid layer that the convection currents mentioned earlier (see Figure 3) transfer heat to the surface and initiate the process of the "drifting apart" of many of the individual tectonic plates, assuming that the convection cell theory is the correct one.

3) The mantle is underlain by the Earth's core, that is composed mostly of iron and makes up 30% of the Earth's volume. The core consists of two layers, an outer extremely hot layer, that is molten, and an inner layer, also very hot, that is solid because of the intense pressure at those depths within the Earth.

The lithosphere, which averages about 60 miles thick, is the material that makes up the individual tectonic plates on the Earth's surface. The plates that presently exist are illustrated in Figure 2. Innumerable individual plates have come and gone in the past, as the process of plate movement has apparently persisted far back into geological time. The general model of the evolution of an ocean basin (Figure 3) shows two oceanic plates moving away from a spreading center in the middle of the ocean called the **mid-ocean ridge**. As noted by Alt and Hyndman (2000), those *ridges mark the lines where plates pull away from each other. Every ocean has one, a long ridge with a deep rift along its crest that runs the length of the ocean*. As new lava flows spill out onto the ocean floor along fissures in the ridges (usually every few hundred years), new oceanic crust is created.

The plates on the opposite sides of the ridges move away from each other at an average rate of about 4 inches/year according to Monroe and Wicander (1998) and 2 inches/year according to Alt and Hyndman (2000). What's an inch or two among friends? Doris Sloan (pers. comm.) observed that the spreading rate depends on the spreading center. Moores and Twiss (1995) divide them into slow (0.4-2 inches/year), intermediate (2-3.5 inches/year), and fast (3.5-4 inches/year). The East Pacific Rise is fast spreading, at >3.5 inches/year.

The new crust thus created, that becomes an extension of the oceanic plates, moves away from the ridges until it encounters an approaching continental plate where it dives under the oncoming barrier (a rather slow dive at 2-4 inches per year!). The area where the two plates collide is called a subduction zone, and it is commonly marked by a deep, oceanic trench that, in places, has an arc-shaped zone of volcanic islands (called an island arc) on its concave side. In some areas, the volcanoes are located on an adjoining continental landmass (Figure 3), as is the case for the volcanoes of the Cascade Mountains in Oregon and Washington. The volcanoes are usually located between 50-200 miles away from the center of the trench, depending on the angle at which the oceanic plate plunges down into the trench.

As can be seen in Figure 4A, the Atlantic Ocean perfectly fits the model of a central mid-ocean ridge away from which tectonic plates diverge as the main ocean body grows (see Figure 3), but the Pacific Ocean does not. In the Pacific, the ridge known as the East Pacific Rise has been overtaken and partially covered by the westward-moving North American Plate. This is but one of the many factors that complicates the geology of the Pacific Northwest Coast.

The boundaries between two lithosphere plates may be of three possible types (Figure 12):

1) Divergent boundaries where the two plates are moving away from each other, for example along mid-ocean ridges;

2) Convergent boundaries where the two plates collide, for example, along subduction zones; and

3) Transform boundaries, which are referred to as "conservative plate boundaries," where no material is created or lost. The two plates simply slide past each other along a boundary called a transform fault (without colliding or pulling apart).

As pointed out by Alt and Hyndman (2000), *all transform boundaries connect oceanic trenches and ridges in some combination.... Transform boundaries define the edges of plates moving away from oceanic ridges and oceanic trenches.* Many of these are visible adjacent to the mid-ocean ridges illustrated in Figure 4A.

A fault was defined by Lahee (1952) in his classic book, *Field Geology, as a fracture along which there has been slipping of the contiguous masses against one another. Points formerly together have been dislocated or displaced along the fracture.* Perhaps the most common and most easily understood faults are those that show vertical movement, such as the two classes illustrated

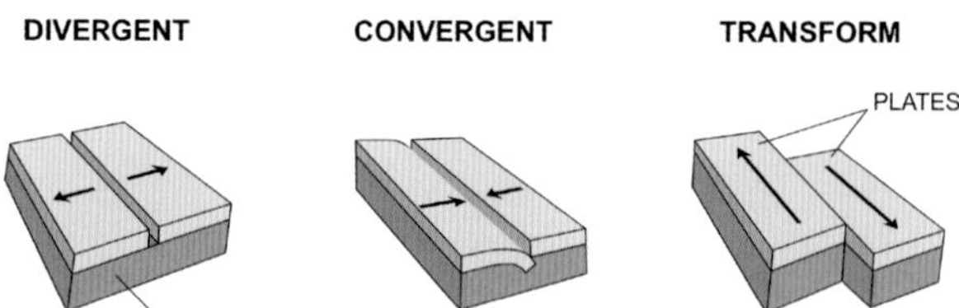

FIGURE 12. The three types of motion of plate boundaries: 1) Divergent – pulling away from each other, as along mid-ocean ridges; 2) Convergent – colliding with each other, as along subduction zones; and 3) Transform – where the two plates slide past each other with little vertical movement. Modified from José F. Vigil, USGS.

in Figure 13. In the type of fault shown in Figure 13-Top, the side of the fault on the left, called the hanging wall, slides down the fault surface under the influence of gravity (and/or tension) relative to the side on the right, called the foot wall. That type of motion creates a normal or gravity fault. On the other hand, where the hanging wall moves up the fault line over the foot wall under the influence of compression, the resultant motion creates a reverse or thrust fault (Figure 13-Middle). When you take your walks along the rocky shores in Oregon and Washington, you will have ample opportunity to observe both of these types of faults in the wave-cut rock cliffs.

The importance of **transform faults** was not recognized until a geologist named J. Tuzo Wilson used the concept to explain away some of the major problems facing the proponents of the theory of drifting continents. He did this in a classic paper published in 1965 titled *A New Class of Faults and Their Bearing on Continental Drift*. These newly discovered faults allowed the fresh material created along the mid-ocean ridges to conform to the spherical surface of the Earth by sliding and adjusting in such a way as to fit it. Even before Wilson, a geologist named Harry Hess, making observations of the seafloor while captaining a ship for the U.S. Navy during World War II, had concluded that somehow the floor of the Pacific Ocean was in motion and "expanding." After the war, he proposed a theory that was eventually called seafloor spreading (published in 1962). And by 1965, the pioneering oceanographer Bruce Heezen extended Hess's work by making detailed bathymetric maps of the oceans. Those maps showed not only the amazing lines of mid-ocean ridges with their central rift valleys, but also that the mid-ocean ridges are commonly offset into horizontal segments bounded by Wilson's transform faults (see Figure 4A). Wilson's

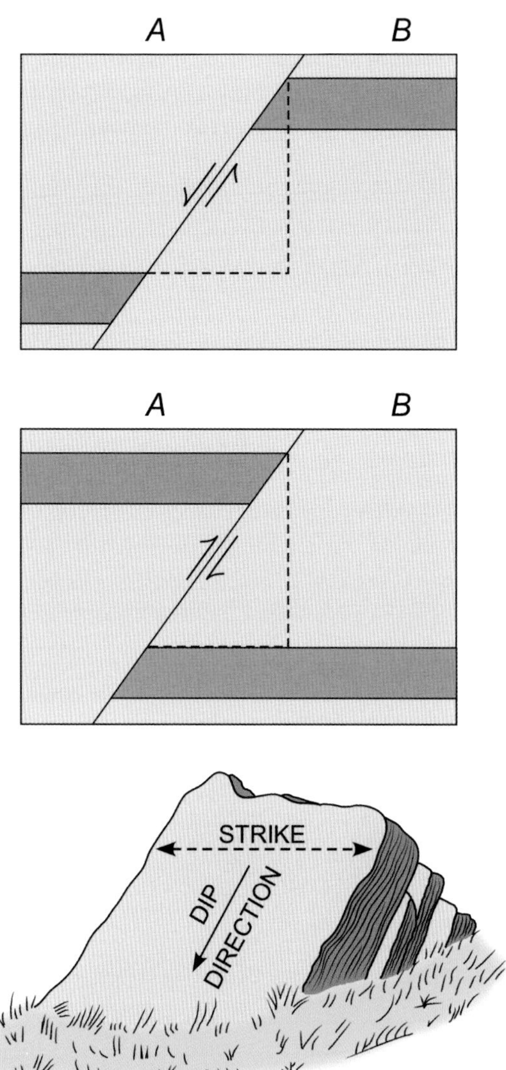

FIGURE 13. Faulted and dipping strata. (Top) A gravity, or normal fault, in which the hanging wall (A) slides down the fault plane under the influence of gravity. Thus, the orange layer is shifted lower along the fault. (Middle) A thrust, or reverse fault, in which the hanging wall (A) slides up the fault plane, over beds younger than itself, under the influence of compression (after Lahee, 1952). (Bottom) Terms used in the description of dipping rock layers, strike and dip. A strike-slip fault is one that moves laterally along the strike line without marked vertical motion.

transform faults explained how those offsets are possible. He was also the first to use the term plate tectonics and drew the first plate boundaries. As a geology professor at the University of Massachusetts in the late 1960s, co-author Hayes was able to attend some of Wilson's ground-breaking lecture tours (and those by Heezen as well), and even though a mere coastal geomorphologist, it was clear to him that a new day had dawned for the geological sciences.

So what? Well, it turns out that the infamous San Andreas Fault system that has worked its way right through the middle of the Central California coastal area is a transform fault where the boundaries of the Pacific and North American Plates meet (Figure 7C). Wilson also recognized that tectonic plates don't just glide smoothly past each other; instead, over time stress builds up along the boundary between the plates to the point that accumulated potential energy is released suddenly and the rocks on the opposite sides of the fault move abruptly, commonly causing earthquakes.

The San Andreas transform fault is a **right-lateral fault**. That is, if you are standing on one side of the fault line looking across it, the *terra firma* on the other side of the fault line has moved relative to the right during the earthquake (as shown in Figure 7B). The same thing is true if you step across the fault line and look back at the side you were standing on (i.e., that side also has moved to the right in a relative sense). A more general term is strike-slip fault. To understand that term you have to be familiar with the geometry of dipping rock layers. To measure the angle of dip of the layer, one lays an instrument called a Brunton compass on the rock surface, pointing it in the exact direction in which the rock dips and turning a pendulum-like arm within the instrument until its upper surface is level, which allows you to read the angle of dip. If you drew a line along the rock layer exactly perpendicular to the direction of dip, that would be the strike of the rock layer (see sketch in Figure 13-Bottom). Making such readings, one of the first things a beginning geology major learns to do, is a fundamental step in constructing the geological history of the rocks on the Earth's surface. A strike-slip fault, then, is one that moves exactly parallel to the imaginary line you "drew" on the rock layer, with relatively little up or down motion in the process.

HOW OLD IS IT?

The most elemental question in the interpretation of the origin of rock sequences is: How old are the rocks? Such information is necessary before the evolution of any significant portion of the Earth's rocks can be properly interpreted. Three methods are commonly used to answer that question: 1) Applying the law of superposition; 2) Paleontology correlations; and 3) Radiometric dating.

In the 11th century, a Persian Muslim philosopher named Avicenna came up with an idea for determining the age of rock sequences that was later formulated more clearly in the 17th century by the Danish scientist Nicolaus Steno. That concept later became known as Steno's Law or The Law of Superposition, and states *Sedimentary layers are deposited in a time sequence, with the oldest at the bottom and the youngest at the top.*

In other words, layers of sedimentary rocks are laid down in a sequence from bottom to top like you would do if baking a cake composed of three distinct layers. The first layer you put in the pan is "oldest" and the third one, which is placed on top of the second layer, is "youngest." The major difference is that, while it takes only a few seconds to create the three layers in the cake, geological processes may take millions of years to create stacked layers of rocks. The problem becomes much more complicated if the rock layers have been tilted, or even completely overturned, as is true for many rock sequences, especially those in tectonically active areas like the coasts of Oregon and Washington.

The rock layers exposed in the sides of the Grand Canyon in Arizona are a classic example of relatively simple "layer-cake geology." Rocks at the bottom of the canyon consist of granites and schists (metamorphosed rocks) overtopped by some tilted rock layers in places, all of which are Precambrian in age (see Table 1). Those rocks were scoured off into a relatively flat surface during a major erosional event. The resultant eroded surface is called an unconformity and, because many of the rocks under the eroded surface in the Grand Canyon are tilted, that unconformity is more specifically referred to as an angular unconformity.

The major angular unconformity near the bottom of the Grand Canyon, that represents a significant gap in geological time, was later covered by over 5,000 feet of horizontal layers of sedimentary rocks, beginning with Cambrian sandstones more than 500 million years old at the base. Coming up the sides of the canyon, younger and younger relatively horizontal layers of sedimentary rocks are encountered until one reaches the top of the canyon, where rocks of late Triassic age (around 225 million years old) are found (Bues and Morales, 1990). Those Triassic rocks, named the Chinle Formation, consist of abundant siltstones and some sandstones that were originally deposited mostly in river floodplain and lake environments. Notable examples of petrified forests and dinosaur remains are also contained within the Chinle Formation.

There are some time intervals missing from the sedimentary layers in the Grand Canyon, but this is a remarkably complete record of 300 million years of relatively steady deposition of horizontal rock layers, oldest at the bottom and youngest at the top. That technique of examining layered rocks has been used to interpret the history of the Earth's rocks since the earliest geological maps were constructed, even as far back as the 11th century!

[NOTE: Many important economic products, such as oil, gas, and coal, have accumulated in sedimentary rock layers (or strata) similar in some respects to those exposed in the sides of the Grand Canyon. Consequently, the branch of geology known as stratigraphy, the study of sedimentary rock layers or strata, particularly their distribution, mode of deposition, and age, has provided careers for more geologists than any other branch of the geological sciences.]

Once the sequence of sedimentary rock layers had been worked out sufficiently enough to build a rudimentary time scale, it became apparent that over time thousands of organisms had come and gone through evolutionary trends and extinctions. The somewhat limited life spans of some of species were related to rock sequences around the World early in the pursuit of the science of geology. Therefore, it became possible to determine the relative ages of the sedimentary rock sequences based upon their fossil content, or their paleontology.

Neither the law of superposition nor paleontological correlation works for determining the ages of intrusive masses of rocks like granites and highly metamorphosed rocks (because they are not layered, nor do they contain significant fossil remains), or markedly tilted and overturned strata, such as many of the rock sequences in Oregon and Washington. Therefore, radiometric dating is commonly used to date igneous rocks, as well as organic remains associated with geologically young materials (e.g., by archaeologists). Some of the basic elements and concepts of that process include:

1) Naturally occurring radioactive materials break down into other radioactive materials at known rates (a

process known as radioactive decay);

2) Radioactive parent elements eventually decay to stable daughter elements. Each radioactive isotope has its own **half-life**, the time it takes for half of the radioactive parent element to decay to the daughter product;

3) Each radioactive element decays at its own nearly constant rate. Examples include Potassium 40 (radioactive parent) to Argon 40 (stable daughter) with a half-life of 1.25 billion years, and Uranium 235 (radioactive parent) to Lead 207 (stable daughter) with a half-life of 700 million years;

4) Therefore, measuring (in a rock sample) the amount of radioactive parent element and the amount of stable daughter elements allows the geologist to estimate the length of time over which the decay has been occurring; and

5) A method called Carbon-14 dating is used for preserved formerly living plant or animal remains less than 70,000 years old. All living things contain a constant ratio of Carbon 14 to Carbon 12. After death, any Carbon 14 in the tissues decays to Nitrogen 14 with a half-life of 5,730 years. The specimen is dated on the basis of the change in the Carbon 14 to Carbon 12 ratio.

Radiometric age dating, using one of the most fundamental precepts of nuclear physics, radioactive decay, is quite accurate, considering the long time spans involved in some of the measurements. The other two methods, superposition and fossil correlation, are relatively precise also, if the rocks under study are exposed well enough for detailed work to be carried out. Therefore, you can expect the age dates cited throughout this book to be fairly accurate in situations where the geological evidence is sufficient and the radiometric measurements on the samples have been accurately carried out. In some cases, those dates may be adjusted somewhat at a later time based on new data.

EARLY EVOLUTION OF THE PACIFIC NORTHWEST COAST

There is still some question about the exact timing of all this, but conventional wisdom is that, at about 400 to 300 million years ago, there were two main land masses on Earth separated by the Rheic Ocean. In the south sat **Gondwana**, a supercontinent consisting of what is now South America, Africa, India, Australia, and Antarctica, and in the north sat Laurasia, made up of what is now North America, Greenland, Europe, and part of Asia.

Some time around 300 million years ago, during the Carboniferous Period of the Paleozoic Era (see geological time scale in Table 1), these two landmasses (Gondwana and Laurasia) collided to form a single large continent called **Pangaea** (Figure 14A). In the process, the Appalachian Mountains were formed (Dietz and Holden, 1970). Those mountains are said by some to have been as high as the Himalayas and by others at least as high as the modern Rocky Mountains.

[NOTE: According to The Natural History Museum (2014), *the largest extinction event (known to science) and the one that affected the Earth's ecology most profoundly took place 252 million years ago. As much as 97% of species that leave a fossil record disappeared forever*. We know of two hypotheses to explain this: 1) When the continents came together to form Pangaea, a large number of habitats around the margins of the original two major continents were compressed together, eliminating many habitats that originally existed and contained a variety of life forms; and 2) A large asteroid or comet impacted Earth creating that extinction, similar to the one that occurred around 65 million years ago causing the extinction at that time of the dinosaurs and a multitude of other life forms.]

Between 225-190 million years ago, during the Triassic Period, the major landmass of Pangaea started to split apart along the old collision zones. That pulling apart created some rift basins in parts of the area now known as the eastern U.S. similar to the ones presently occurring in south-central Africa. Those rift valleys filled with sediments eroded from the uplifted blocks along the rift margins. From that point onward, the continents continued to drift apart, eventually reaching their present locations (see Figure 14B).

[NOTE: In the winter of 1965/66, co-author Hayes led a geological expedition in Antarctica to examine a sequence of sedimentary rocks called the Beacon Sandstone of Permian age (around 300 million years old). Those rocks are very similar to other Permian units in South Africa and India, containing huge petrified trees that grew in tropical forests, among other artifacts. The correlation of those very similar sandstones around the southern hemisphere was supporting evidence for the makeup of the original "southern" supercontinent Gondwana, that collided with the "northern" supercontinent Laurasia to form Pangaea, that was starting to break up into separate components during early phases of the Triassic period, after which the newly created fragments of Pangaea, including the ones that originally made up Gondwana,

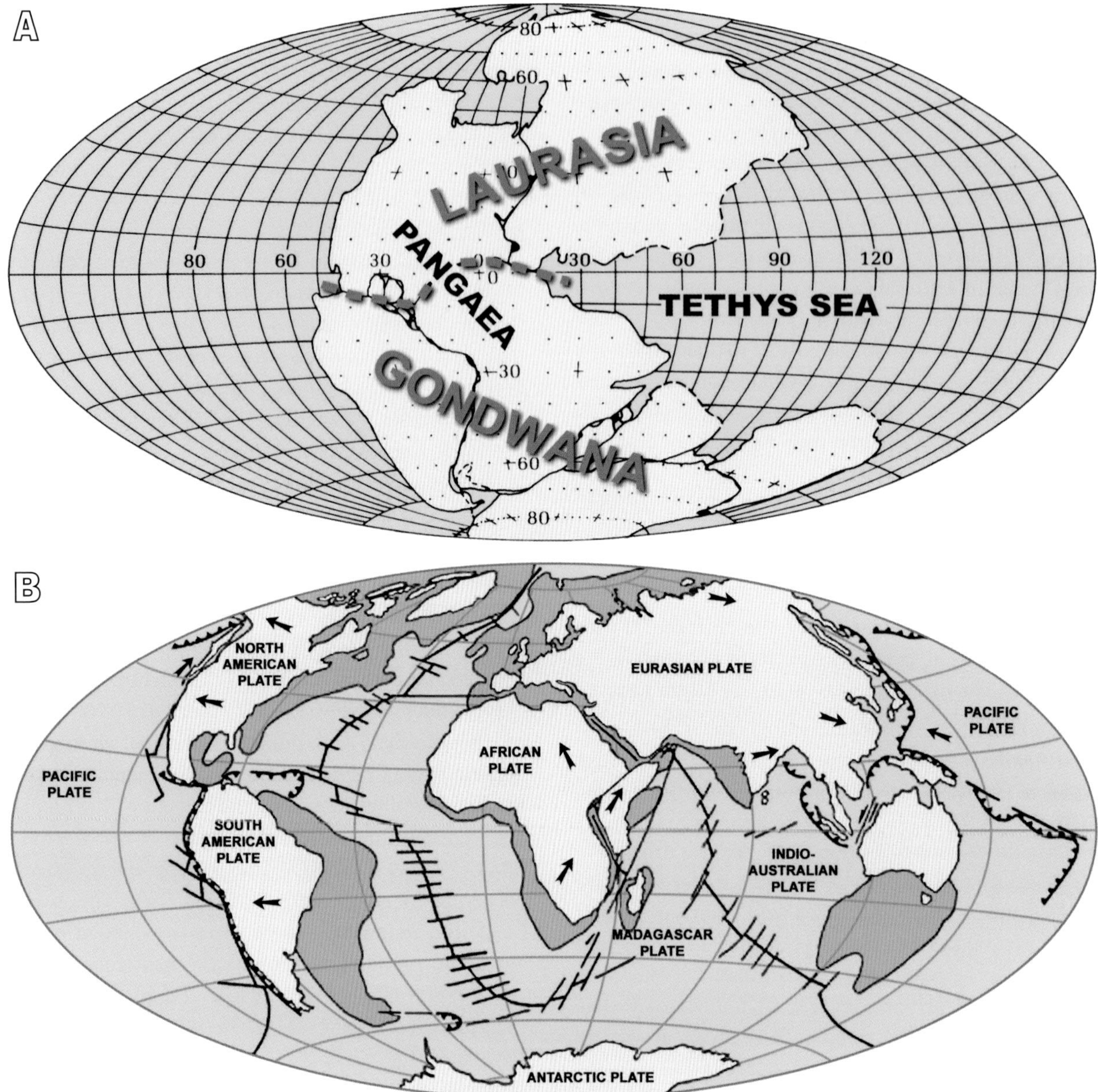

FIGURE 14. Two of the basic elements in the geological history of Oregon and Washington. (A) The continents that assembled to form a universal landmass at the end of the Permian Period (Table 1) around 250 million years ago (as defined by Dietz and Holden, 1970; modified after Glen, 1975). (B) The major continents in their present position. Arrows indicate the directions the different landmasses have moved since they first started "drifting apart" around 200 million years ago. The blue "shadows" represent previous locations of the continents as they moved along.

eventually migrated to their present positions. That is, the southern parts of South America and Africa, as well as the peninsula we now call India, began to move far away from each other as well as from the present continent of Antarctica. Many studies by others showed that the rocks of those four areas were so similar they had to have been joined at one time. Therefore, that study by Hayes and his students on the Beacon Sandstone in Antarctica was one of many that provided small bits of supporting evidence for the theory of continental drift.]

EVOLUTION OF THE PACIFIC NORTHWEST COAST 200-2.6 MILLION YEARS AGO

Tectonic Processes

After the breakup of Pangaea around 225-190 million years ago, the North American and South American continents moved away from Europe and Africa, being forced to do so by the newly formed lavas extruded along what eventually became the Mid-Atlantic Ridge as the seafloor "spread" (Figures 3 and 4). As noted by Connor and O'Haire (1988), *as North and South America moved west, they pushed against the oceanic crust of the Pacific. The oceanic plates moving east or north from oceanic ridges in the Pacific Basin began to sink through trenches along the margins of the American continents.*

As the North American Continent moved west, the area now known as the Pacific Northwest Coast grew wider as a succession of volcanic island chains and assorted ocean-floor rocks was added along the continental margin expanding the edge of the continent by 400 miles to the west. At least four major blocks of oceanic rock, including the *remains of two volcanic island chains and two extensive belts of ocean-floor rocks were added to the Pacific Northwest over the last 200 million years* (Burke Museum, 2014a).

The second phenomena that shaped the geological history of the state of Washington was the formation of a series of continental volcanic arcs. The continental arcs are regions where molten rock (magma) intrudes and moves toward the surface, often to form volcanoes (as shown in Figure 3), volcanic rocks at the surface, or large plutons of granite at depth. Four distinct continental arcs have formed in the Pacific Northwest in the last 200 million years. From the oldest to the youngest, these include: 1) Omineca Arc; 2) Coast Range Arc; 3) Challis Arc; and 4) Cascade Arc. The Cascade Arc (located where the Cascades mountain ranges are shown in Figures 8 and 9) extended north-south from Mt. Garibaldi in British Columbia to Mt. Lassen in northern California about 37 million years ago, and is still active today (Burke Museum, 2014a). They note:

It is important to realize that the accumulation of these Cascade Arc volcanic and plutonic rocks is not the origin of the modern Cascade Range. The modern range is a much younger feature dating from only the last 5-7 million years. The volcanoes of the Cascade Arc before that time occupied positions along a divide of only modest relief, one that did not constitute a major weather barrier.

Columbia River Basalt

Around 17 million years ago, a huge area of flood basalts developed in eastern Oregon. Figure 15 shows their surficial distribution. According to Alt and Hyndman (2000), these basalt flows, that ended around 15.5 million years ago, were *so overwhelming that they add an entirely new dimension to our ordinary notions of volcanic catastrophe.*

State of Oregon (2014) listed the following characteristics of these flows:

1) The first eruptions were a series of gigantic lava floods erupted from great fissures near the Oregon-Idaho-Washington border;

2) These lava flows are among the largest to have occurred anywhere on Earth;

3) Some of these rapidly moving sheets of lava traveled as far as 400 miles from their fissure vents; and

4) Individual flows cover as much as 10,000 square miles, equivalent to one tenth of the state of Oregon, and many are over 100 feet thick. The combined flows reached a total thickness of 5,900 feet.

Those lava flows have generated a great deal of interest and enthusiasm among geologists, who have made a variety of interpretations of why they occurred, particularly at that specific time and place. Four of the many theories that have been proposed are:

1) Alt and Hyndman (1978) proposed that a descending slab of seafloor broke off and sunk rapidly into the Earth's interior. *If so, the soft rocks beneath the crust would have flowed rapidly, filling in behind the sinking slab... The resulting westward movement of material beneath the crust would have pulled Oregon westward, perhaps opening the cracks that produced the basalt floods*;

2) Alt and Hyndman (2000) postulated that this cataclysm was triggered by an asteroid that struck the Earth and created a crater when it exploded;

3) The most popular idea at the present time is that the Hot Spot that now underlies Yellowstone National Park affected the area in question 17-13 million years ago as the North American Plate migrated its way westward. This was a mantle-plume much like the one that has

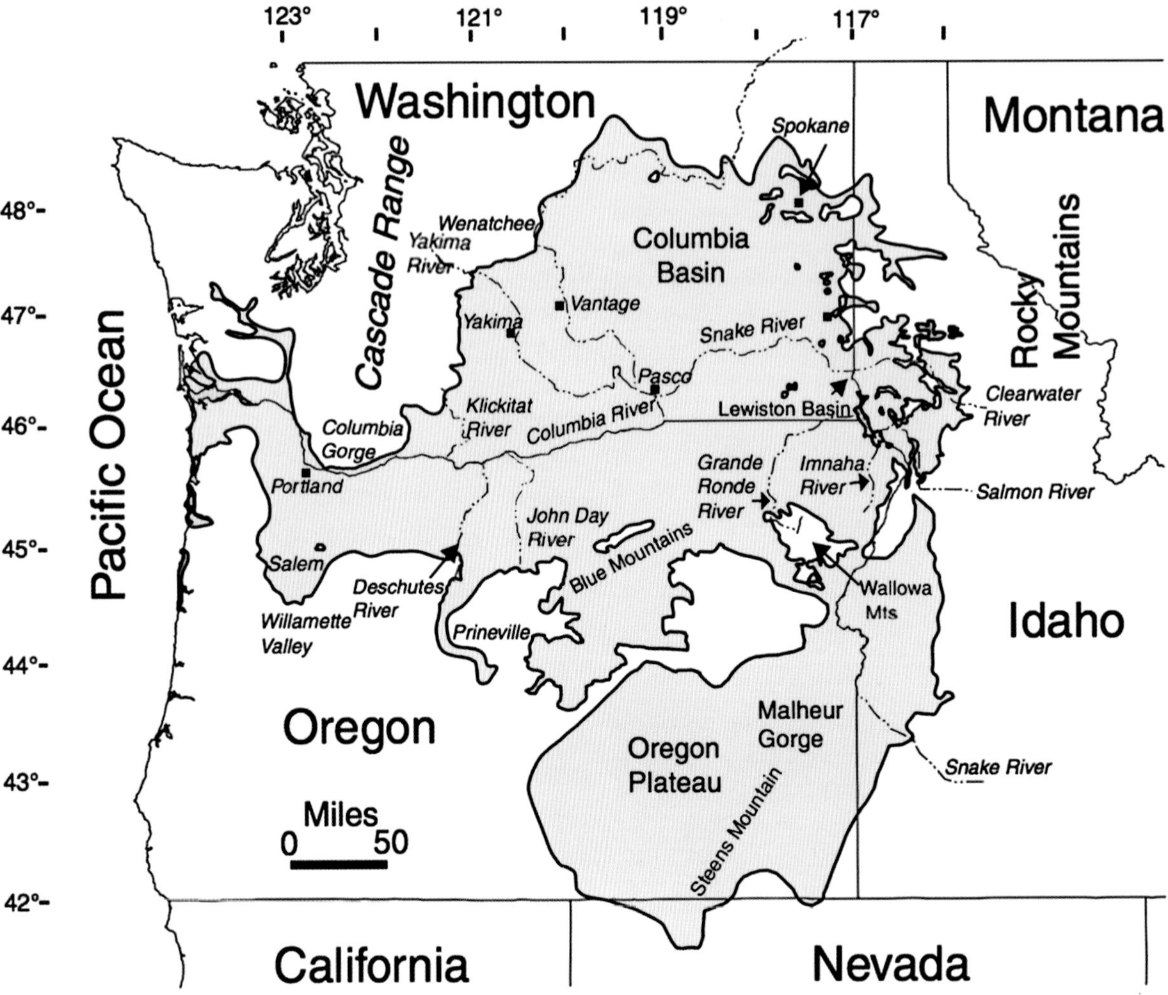

FIGURE 15. The maximum extent of the 17-13 million year old Columbia River basalt flows that covered parts of four states. Map courtesy of Cascades Volcano Observatory, USGS.

formed the Hawaiian Islands, except that this one now lies under the continent near Yellowstone National Park; and

4) Burke Museum (2014b) added another wrinkle that we had not encountered before. They brought up the fact that the East Pacific Rise, with it's presumed rising basalt is now buried a ways to the east under the westward migrating North American Plate (as is indicated in Figures 3 and 4A). Therefore, they concluded that *we believe that the basalt flows erupted from the coincidence of the Yellowstone Hot Spot and this spreading center.* Sounds pretty good to us.

[NOTE: Geological hot spots are defined as the surface manifestations of plumes, that is, columns of hot material, that rise from deep in the Earth's mantle. Hot spots are believed to be fixed in the deep mantle so that the age and orientation of these hot spots provide information on the absolute motions of the plates (e.g., the explanation number 3 given above for the location of the Columbia River Basalts). Geologists have identified some 40-50 such hot spots around the globe, with Hawaii, Reunion (in the Indian Ocean east of Madagascar), Yellowstone, Galapagos, and Iceland overlying the most currently active ones.]

Modern Cascade Range

To understand the evolution of the modern Cascade Range, one should first consult the present tectonic plate situation of the Pacific Northwest (Figure 16). The last

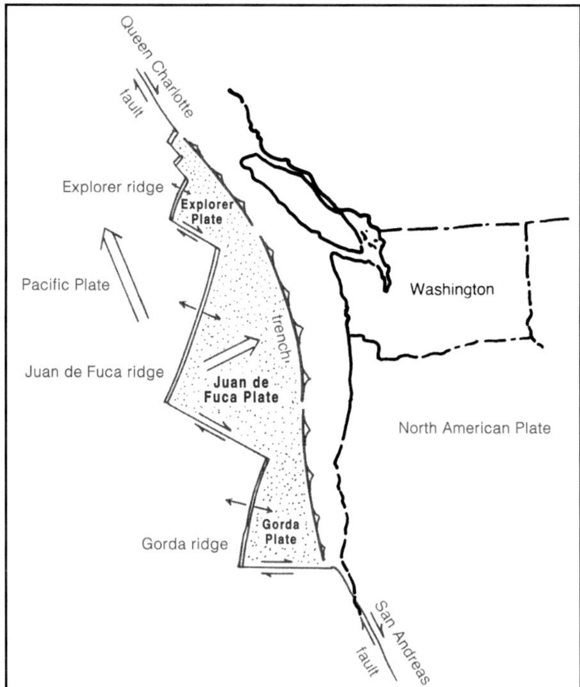

FIGURE 16. The present plate tectonic situation in the Pacific Northwest. The stippled area covers the three micro-plates in the area, which are the remnants of the old Farallon Plate. The Explorer Plate broke off from the Juan de Fuca Plate between 5 and 7 million years ago, and, as it did, the Cascade Arc resumed and the modern Cascade and Olympic Mountains began to rise. Modified after Burke Museum (2014e) and Alt and Hyndman (1984).

of the Farallon Plate is now made up of three small fragments, the small **Gorda Plate** to the south, equally small **Explorer Plate** to the north, and the **Juan de Fuca Plate** in the middle. According to Burke Museum (2014b), the *Explorer Plate broke off from the Juan de Fuca Plate between 5 and 7 million years ago. As it did, the Cascade Arc* (defined above) *resumed and the modern Cascade and Olympic Mountains began to rise.* These two mountain ranges are the crests of two parallel folds, with the Puget Sound Basin positioned in the trough between them.

Further analysis of Figure 16 shows that, as these three micro-plates move eastward, the Pacific Plate continues to slide to the north along two transform, right-lateral, strike-slip faults, the San Andreas Fault to the south and the Queen Charlotte Fault to the north (Alt and Hyndman, 1984). The three oceanic ridges generate new oceanic crust (shown in stippled pattern in Figure 16) destined to sink into the adjacent trench.

The 18 volcanoes of the modern Cascade Range, from the Bridge River cones in British Columbia in the north to Lassen Peak in California in the south, are shown in Figure 17. Orr and Orr (2002), after pointing out that over 20 centers within the Cascade volcanic chain have erupted for hundreds of thousand of years, observed that *hazards from future Cascade eruptions must be assessed with respect to the volcanic history of the range as well as the growth of cities in the Pacific Northwest.* Figure 17 also shows estimates of the hazard zones (from mud and lava flows). Orr and Orr (2002) noted that *damage and destruction from a major volcanic event today would depend on the proximity of towns, population density, direction of the blast, and type of volcanism, but it is*

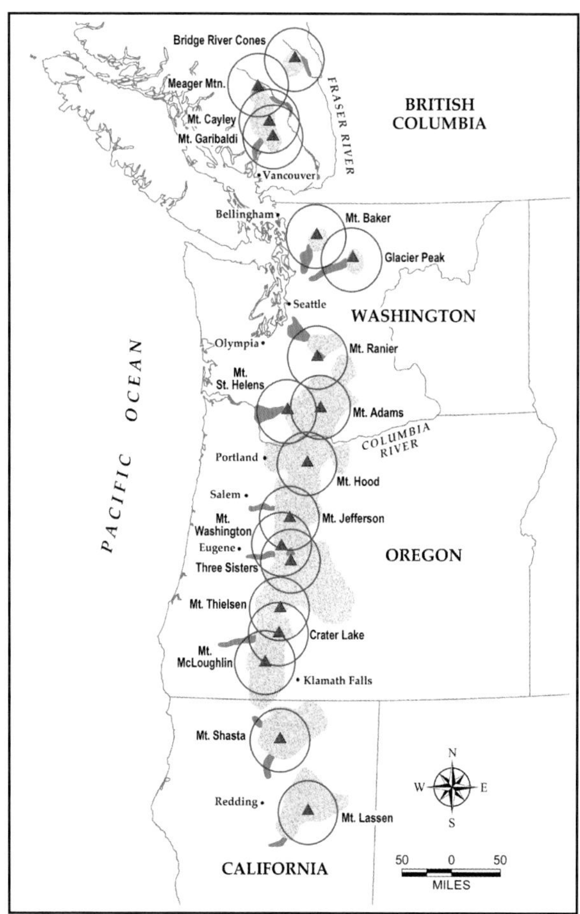

FIGURE 17. The volcanoes of the modern Cascade Range, including estimates of volcanic hazard zones for each. The circles are 60 miles in diameter. The light blue areas are potential lava flows, and dark blue areas lie in the pathway of potential mud flows (after Miller, 1990). Modified after Orr and Orr (2002).

possible that eruptions similar to those of the past would seriously affect thousands of people miles from the main volcanic center. They observed further that:

1) Serious potential hazards that typically accompany eruptions along the Cascade Range include lava, mud, and ash;

2) Lava tends to be slow moving but burns whatever is encountered;

3) Mud flows move quickly down stream valleys with tremendous force;

4) Mixtures of mud, volcanic debris, and melting ice would carry away bridges, buildings, and roads;

5) In fact, during the 1980 eruption of Mount St. Helens, *incandescent gas and rock fragments moved at incredible speed killing 57 people, destroying almost 300 square miles of forest, and causing over $950 million in losses to the state of Washington*; and

6) *The debris avalanche from Mt. Shasta 300,000 years ago reached more than 40 miles northward and covered over 250 square miles.*

One of the most notable volcanic eruptions in the modern Cascade Range occurred 5,670 years ago when Mount Mazama erupted catastrophically to form Crater Lake. Wikipedia (2014b) and Orr and Orr (2002) listed the following significant relatively recent eruptions, among others: Mount Hood in 1781-82, 1865, and 1859; Mount Shasta in 1786; Mount Ranier in 1854, and Mount Lassen in 1914-1915. Some of these eruptions were *spectacular displays.*

THE QUATERNARY PERIOD

Introduction

A major change in climate initiated the beginning of the Ice Ages (the Pleistocene epoch). Earlier workers suggested that this occurred around 1.8 million years ago, but the latest thinking is that that date will probably be reset to 2.6 million years ago (see Table 1). Within the Pleistocene epoch, four major glacial events took place, during which ice covered large areas of the North American continent (as far south as the Ohio River), as well as many of the high mountain areas in the Pacific Northwest. The approximate median peak times for these four major glacial episodes, expressed in years before the present, have been generally estimated to be: 650,000 years ago for the oldest one, called the Nebraskan Glaciation; 450,000 years ago for the Kansan Glaciation; 150,000 years ago for the Illinoian Glaciation; and 21,000 years ago for the last one, the Wisconsin Glaciation (some estimates imply that there were three episodes of significant glacial advance during the Wisconsin glacial sequence, one being at around 60,000 years ago).

During each of these major glaciations, sea level dropped to levels much lower than it is today; during the Wisconsin Glaciation, it was 410 feet (+/- 16 feet) lower, according to Fleming, et al. (1998)! Clark and Mix (2002) suggested that the drop of worldwide sea level during the last glacial maximum (Wisconsin) was 410-440 feet below the present level. Peltier (2002) proposed a low of 394 feet. All three of these estimates are fairly close.

As shown in Figure 18, during the peak of the last major glaciation approximately the northern half of the state of Washington was covered with ice, with one large lobe protruding down into the Puget Lowland. As recently at 14,000 years ago, the ice was 3,000 feet thick over the area where Seattle is now located (Burke Museum, 2014b). That ice only receded out of the state around 10-12,000 years ago.

The explanation of why these four distinct major glaciations occurred during the Pleistocene epoch is linked to the Milankovitch (1941), or astronomical (orbital forcing), theory of climate change. *This theory postulates that as the Earth travels through space around the Sun, cyclical variations in three elements of Earth-Sun geometry combine to produce variations in the amount of solar energy that reaches Earth* (Kaufman, 2002). Those three elements that have cyclic variations are **eccentricity** (the shape of the Earth's orbit around the Sun), **obliquity** (the tilt of the Earth's axis away from the orbital plane), and **precession** (change in orientation of the Earth's rotational axis – like the wobble of a spinning top) (Howard, 2002). *Using these three orbital variations, Milankovitch was able to formulate a comprehensive mathematical model that calculated latitudinal differences in insolation* (intensity of incoming solar radiation) *and the corresponding surface temperature for the 600,000 years prior to the year 1800* (Kaufman, 2002). This theory, which its author thought explained the repetitive cycles of glaciation that occurred during the Pleistocene epoch, has been questioned and modified over the years, but some form of it probably answers the question of the cause of the Pleistocene glaciation cycles.

To complicate matters a little more by considering

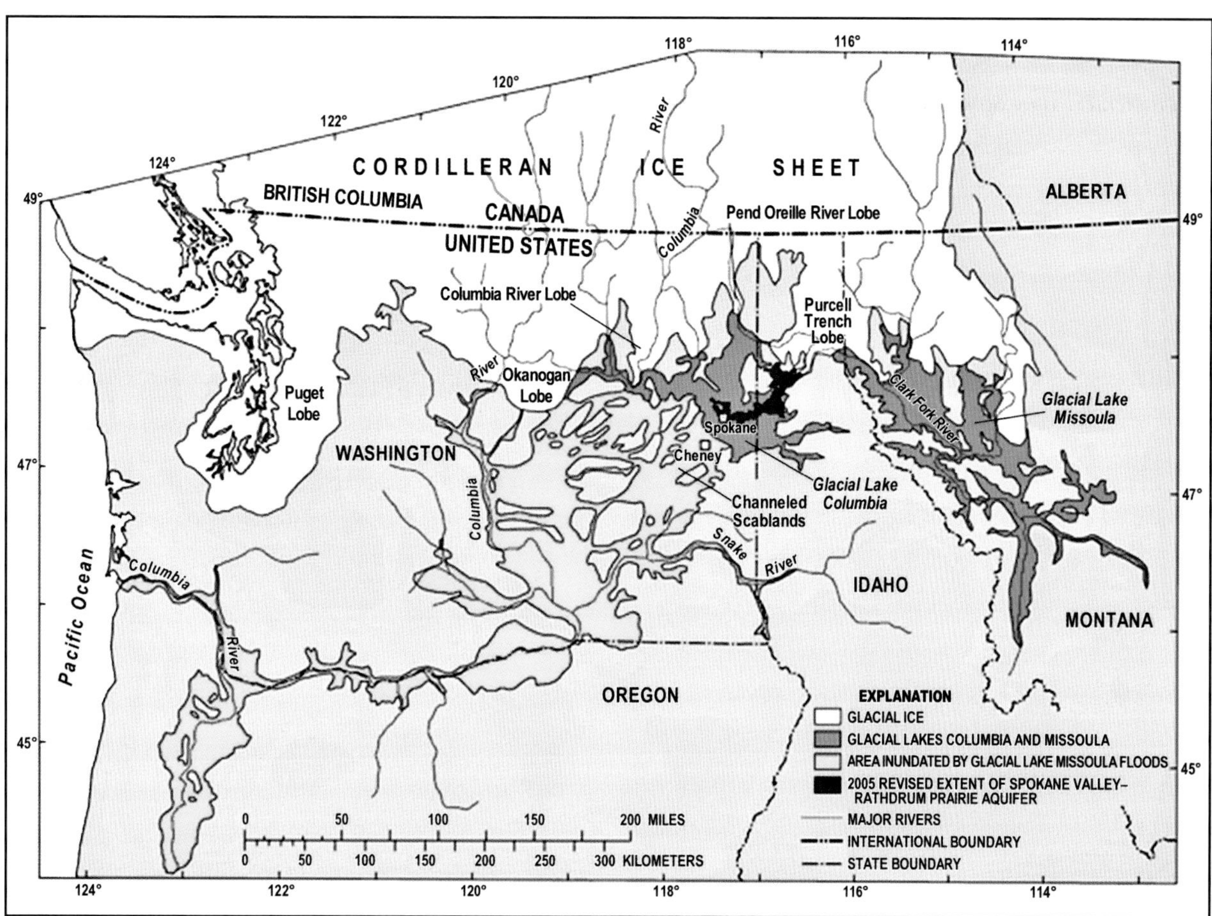

FIGURE 18. Maximum advance of ice into the Pacific Northwest during the Wisconsin Glaciation. One lobe of ice dammed Glacial Lake Missoula in Montana. This dam breached periodically, *sending the largest floods ever recorded through the channeled scabland of Washington and down through the Columbia Gorge* (Burke Museum, 2014b, source of the quote; Kahle et al., 2005, source of the map).

conditions further back in geological time than the 2.6 million years since the Pleistocene epoch started, some other major changes in the Earth's so-called boundary conditions took place, such as: 1) Continental geography and topography; 2) Oceanic gateway locations and bathymetry; and 3) The concentration of atmospheric greenhouse gases. *These boundary conditions are controlled largely by plate tectonics, and thus tend to change gradually* (Zachos et al., 2001). Therefore, it should be no surprise that other major glacial episodes have happened throughout geological time, a noteworthy one having occurred during the Permian period (251-299 million years ago).

The Glacial Lake Missoula Floods

As can be seen in Figure 18, during the Wisconsin glaciation, the intrusions of ice dammed up rivers in the area and created lakes that eventually resulted in floods, some of which were fairly spectacular. However, as noted by Figge (2009) and Alt and Hyndman (1984), those floods paled in comparison to what happened further to the northeast when the Purcell lobe of the continental ice sheet advanced south into what is now the Idaho Panhandle, where it dammed the Clark Fork River (Figure 18). A very large lake, known as Glacial Lake Missoula, developed behind that ice dam. The resulting lake was nearly 2,300 feet deep, containing over 500 cubic miles of water (Alt and Hyndman, 1984) – *a body about the size of Lake Erie*. That vast lake was held back by *a relatively small ice lobe, which eventually started to break up under the pressure of the water*. Once it started to break up, the dam rapidly disintegrated, *freeing the unimpounded waters of Glacial Lake Missoula to head for the Pacific Ocean*, some 428 miles to the west. *This unleashed what are considered the most cataclysmic floods ever known on the planet.* They are known widely as the Spokane Floods. Figure 19 shows the area thus affected by those major floods, of which there were quite a few. According to Alt and Hyndman (1984), *the valley walls above the Clark Fork River in western Montana bear numerous lake shoreline scars, perfectly horizontal benches daintily etched in the hillslopes. Those shoreline scars are too faint to count with confidence, but their number is between 25 and 40. Each one probably corresponds to a separate filling of the lake*, and hence to the source of another flood.

From the geological perspective, one of the most interesting features formed during those floods is the Channeled Scablands, shown in Figure 20. During the major floods, the huge flow of water *completely filled the normal stream valleys, and then washed across the divides in a headlong sweep across southeastern Washington to the Columbia River. In several places, the dry channels cross divides several hundred feet above the level of the modern drainage* (Alt and Hyndman, 1984). The Channeled Scablands give the appearance of a giant braided stream, consisting of multiple, relatively shallow channels that divide and recombine numerous times forming a pattern resembling the strands of a braid. Some channels contain stream-rounded boulders of basalt as much as several feet in diameter.

Co-author Hayes led a team studying the processes of braided streams on the outwash plains of Iceland in the early to mid-1970s. One of his graduate students, Jon Boothroyd and his post-doc, Dag Nummedal, wrote a classic paper on these streams (Boothroyd and Nummedal, 1978). This team expanded interest in braided streams to include comparisons between the Channeled Scablands and the preserved channels on the planet of Mars. Based on their study of the Mars channels, Nummedal, Gonsiewski, and Boothroyd (1976) concluded that the *shape of the channels of the Channeled Scablands corresponds well to that of numerous erosional remnants observed within the channel environments on Mars.*

GOLD PLACERS AND NATURAL GAS

No discussion of the geology of the Pacific Northwest would be complete without at least some mention of the economic benefit the rocks have brought to the area.

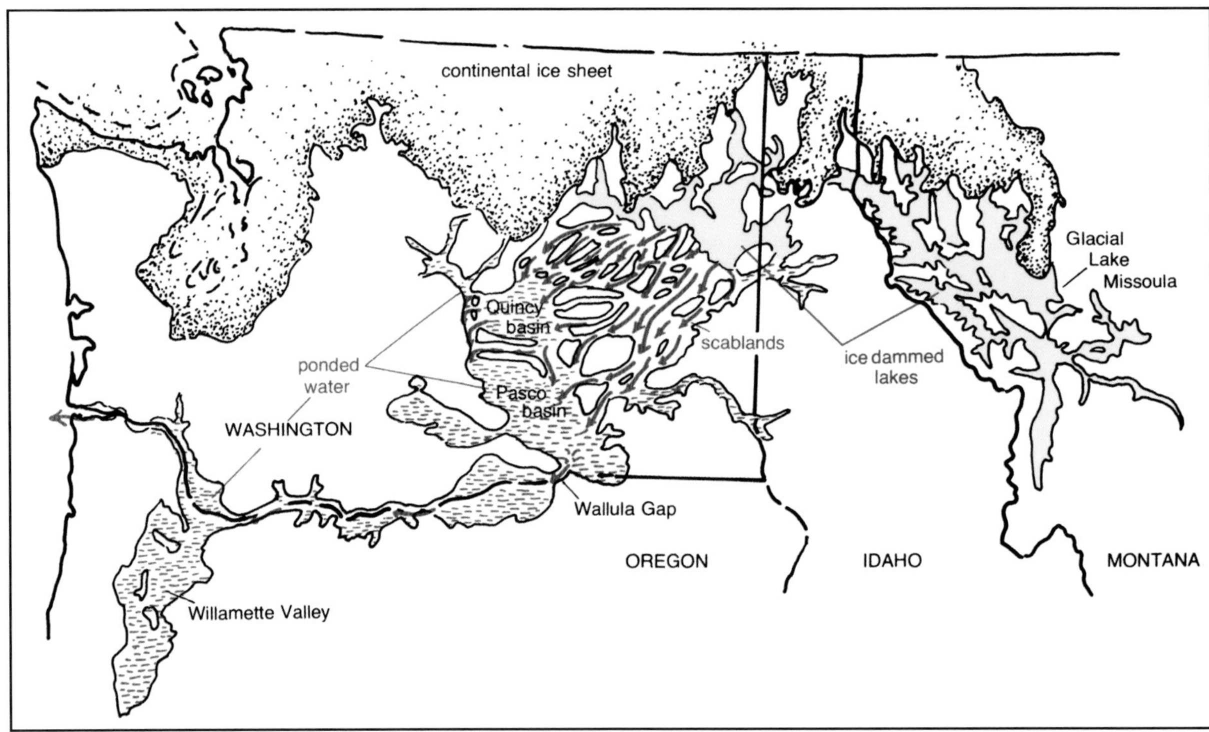

FIGURE 19. The area affected by the floods of Glacial Lake Missoula. Modified after Alt and Hyndman (1984).

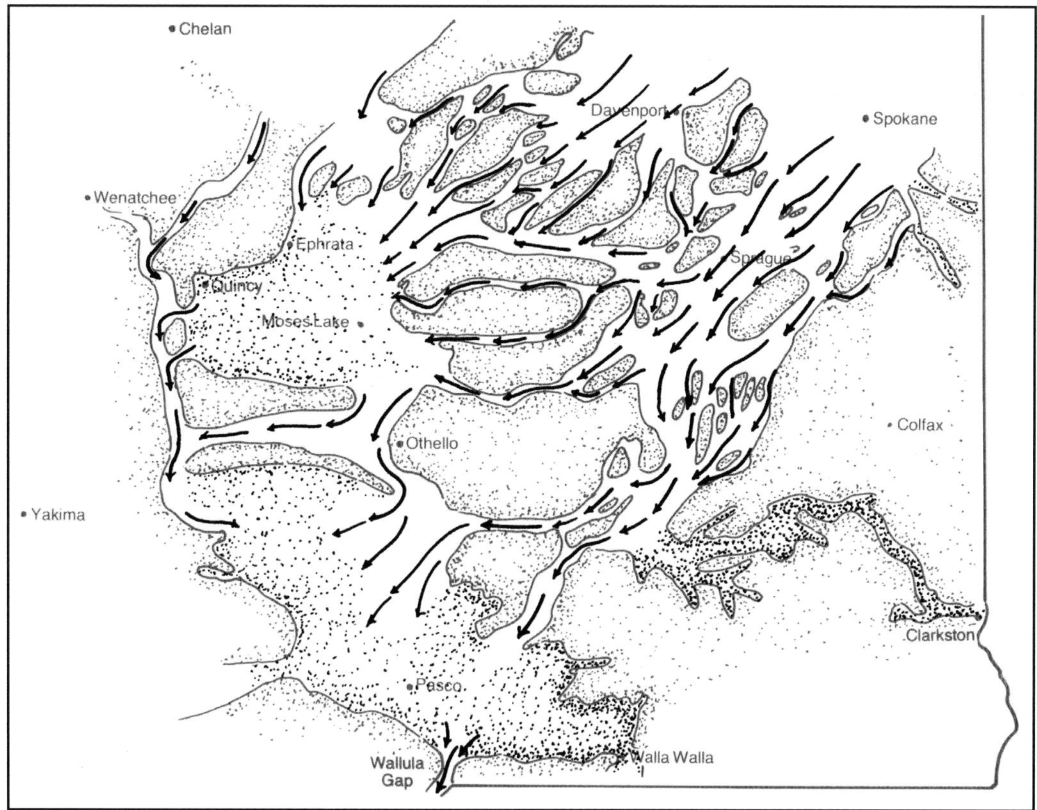

FIGURE 20. The Channeled Scablands of western Washington (located on Figure 19). The black arrows show flood-scoured channelways, the black dots show areas where floodwaters backed up to form slackwater lakes. The orange-stippled areas stood above the floodwaters. Modified after Alt and Hyndman (1984).

Compared to states like Alaska and California, gold mining has been a relatively minor activity in these two states. Wikipedia (2014d) gave the following information on gold deposits:

Oregon:

1) Placer gold was discovered in the Klamath Mountains in southwest Oregon in 1850, starting a gold rush to the area, where lode gold deposits were also discovered; and

2) Although gold mines are located in many areas in Oregon, almost all of the gold produced has come from either the Klamath Mountains or Blue Mountains in the northeast.

Washington:

1) Placer gold deposits were discovered in the Yakima Valley in 1853;

2) Production from the state never exceeded 50,000 troy ounces per year until the mid-1930s, when large hard-rock deposits were developed near the Chelan Lake and Wenatchee deposits in Chelan County, and the Republic deposit in Ferry County; and

3) Production through 1965 is estimated to be 2.3 million ounces.

The Bureau of Land Management (BLM, 2014) pointed out that the *U.S. Geological Survey has estimated as much as one trillion cubic feet of natural gas occurs in the Columbia Basin of south-central Washington and north-central Oregon. They have also estimated that there are numerous small gas fields throughout the Pacific Northwest, such as the Mist Gas Field in northwestern Oregon*. Not surprisingly, thick basalt flows overlay the potential reservoirs, *making exploration and development costly*.

The BLM also stated that *no new oil and gas leases are being explored or developed as of October 2013*. But looking ahead, in *September 2013, 145,476 acres of land in Eastern Oregon were nominated by industry for a future O&G lease sale*. Needless to say, this new nomination marks renewed interest in natural gas in the state.

4 COASTAL HAZARDS

FIRST, A WARNING

Anyone who lives along or visits this coast should be aware of the following coastal hazards: earthquakes, tsunamis, sneaker waves, and erosion and flooding that occur during extreme winter storms, particularly during strong *El Niños*. Each of these hazards has its own, unique risks in terms of frequency, magnitude, and life- and property-threatening consequences. Below we discuss earthquakes, tsunamis, and sneaker waves. Erosion and flooding during extreme storms are discussed under the section on Coastal Processes. Another major coastal hazard is sea-level rise, which is discussed as its own section.

EARTHQUAKES

As noted by Harden (2004), *most of the World's earthquakes occur at plate boundaries; in fact, the zones of concentrated seismicity* ...(another way of saying the rocks are faulting frequently within some limited area)... *are one of the main lines of evidence used to identify plate boundaries.* Did you notice that the contact between two plate boundaries (Gorda and North American Plates) runs offshore of the Pacific Northwest Coast (Figure 7B)? In fact, the present Cascadia subduction zone, an obvious location for earthquakes, is 750 miles long, running along the Pacific Coast from Northern California up to southern British Columbia (Figures 7C and 21).

It is certain that we do not have to remind anyone of the importance of the impact of earthquakes on the Pacific Northwest Coast. As discussed by Wikipedia (2014b), *the Juan de Fuca Plate is capable of producing megathrust earthquakes of moment magnitude 9: the last such earthquake was the 1700 Cascadia earthquake, which produced a tsunami in Japan and may have temporarily blocked the Columbia River with the Bonneville Slide. More recently, in 2001, the Nisqually earthquake (magnitude 6.8) struck 16 km (10 mi) northeast of Olympia, Washington, causing some structural damage and panic.*

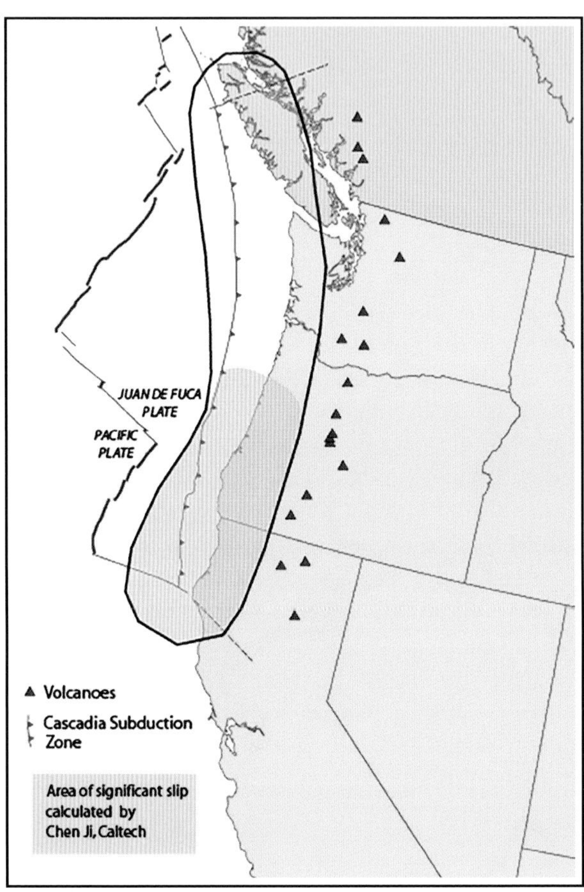

FIGURE 21. The Cascadia subduction zone. Modified after Wikipedia (2014c).

For further reading about earthquakes in the region, we recommend *Full-Rip 9.0: The Next Big Earthquake in the Pacific Northwest* by Sandi Doughton (2013), and *Earthquake Hazards of the Pacific Northwest Coastal and Marine Regions* by Goldfinger et al. (2012).

Some of the most fundamental concepts and issues related to the earthquakes that might take place along the Pacific Northwest Coast include:

1) Earthquakes occur when the strain that has accumulated in the rocks on either side of a fault is suddenly released as the rocks snap apart and slide past each other. The amount of movement along the fault is usually not more than a few feet, but an offset of up to 31 feet was measured along the San Andreas Fault during the Fort Tejon earthquake of 1857 (Collier, 1999).

2) This sudden release of energy into the surrounding Earth radiates away from the location of the fracture (place where the movement starts), called the hypocenter (or focus) of the quake, in the form of two seismic waves. The *P*, or primary seismic wave, moves through the Earth in a fashion *similar to sound waves due to compression of the rock particles, with each particle moving backward and forward in a straight line from the origin of disturbance. P* waves are the faster moving of the two (about 4 miles per hour in the Earth's crust and 5 miles per hour in the mantle). The *S*, or secondary seismic waves, *are due to shearing or transverse motion that moves rock particles not forward and backward, but side to side or up and down the way ripples radiate out when a rock breaks the surface of a calm pond* (Collier, 1999). *S* waves, that move at about half the speed of *P* waves, are only transmitted through solids. Because they are shear waves, the *S* waves are the ones that produce the rolling motion of the ground during an earthquake and are, therefore, more damaging to structures than *P* waves.

3) These seismic waves are recorded by an instrument called a seismograph, that runs continuously, recording all incoming waves. Because of their different speeds, the *P* waves from a specific earthquake will arrive ahead of the *S* waves. Although their rates of speed may vary significantly depending upon what they are moving through, the ratio of the difference between the speeds of the two waves are quite constant. *This fact enables seismologists to simply time the delay between the arrival of the P wave and the arrival of the S wave to get a quick and reasonably accurate estimate of the distance of the earthquake from the observation station. Just multiply the S-minus-P (S-P) time, in seconds, by the factor 8 km/s to get the approximate distance in kilometers* (University of Nevada-Reno, 2009). The smaller the interval of time between the arrival of the two waves, the shorter the distance between the epicenter and the recording seismograph. If data are recorded by several seismographs at different locations on the Earth's surface, the exact location of the epicenter of the quake can be determined by triangulation.

4) The Richter scale and the Moment-Magnitude scale are most commonly used to describe the magnitude of earthquakes. The Richter scale, invented by Charles F. Richter in 1935, is *based on the calculation of the maximum amplitudes of the shear waves recorded by seismographs. Seismographs record a zig-zag trace that shows the varying amplitude of ground oscillations beneath the instrument* (USGS, 2016). Both the Richter and the Moment-Magnitude scales are logarithmic (base 10). For example, a magnitude 5 earthquake on the Richter scale would result in ten times the level of ground shaking as a magnitude 4 earthquake (and 32 times as much energy would be released). Quakes with Richter scale values of 3.5 to 5.4 are often felt, but rarely cause damage. Those above 7.0 on the scale are classified as major earthquakes and can cause widespread damage. The Magnitude-Moment scale (expressed as M), defined as the amount of energy released on the fault, is based on several factors, including: a) How far the fault slipped (i.e., how far a fence line was offset; b) Length of the fault line that was affected; c) Depth at which movement was initiated; and d) "Stickiness" of the surfaces being faulted (rock rigidity). It is the one most commonly used in reporting earthquake magnitudes at this time (2018).

The greatest concern for the Pacific Northwest Coast is the future occurrence of a **major subduction earthquake**, such as the 1700 earthquake (discussed below). Using layers of turbidites in offshore sediments, Goldfinger et al. (2012) identified 41 earthquakes along the Cascadia subduction zone over the last 10,000 years (they did an amazing study of turbidite deposits, submarine landslides triggered by large earthquakes, determining their mass and age). Over the entire subduction zone, the recurrence interval is 530 ± 260 years; however, the recurrence intervals decrease towards the south to 240 ± 120 years along the southern half of Oregon. The last one was 1700; you can do the math. Some people say that the region is overdue for an earthquake. Goldfinger et al. (2012) say *It is also highly likely that the next event will be a southern-margin event* (e.g., southern Oregon/northern California).

Komar (pers. comm., 2018) said:

- *Note that the colliding plates are presently locked, storing energy that "spring-like" is causing the coast to slowly rise. This uplift is occurring along the full length of the Cascadia subduction zone, though not at a uniform rate along the coast, some areas rising faster than the rate of sea-level rise, while other areas are rising more slowly, being transgressed by the uniform rise in sea level.*

- *The occurrence of a subduction earthquake releases this stored energy, resulting in the abrupt drop-down of the coast. This amounted to 1 to 2 meters along the shore, according to the field studies by Atwater and a number of other investigators. This drop in the coast due to the earthquake in 1700 would have induced extreme erosion along the coast, and this shoreline retreat is evident in the coast's morphology (e.g., most apparent at Bandon on the southern Oregon coast).*

TSUNAMIS

On 26 December 2004, an earthquake off the coast of Sumatra, Indonesia produced the largest trans-oceanic tsunami in over 40 years and killed more people (225,000) than any tsunami in recorded history. We likely all remember the videos of people running from the advancing waves, hanging onto tree trunks, and so on. That particular tsunami was restricted mostly to the Indian Ocean; however, the coasts of Oregon and Washington are not exempt from the effect of tsunamis.

A tsunami, a Japanese word translated as *harbor wave*, is a large ocean wave usually generated by a submarine earthquake, although they may also be generated by volcanic eruptions, landslides (both those initiated on shore and submarine), and even by a comet or asteroid impact (not by meteorites; too small). Historically, such waves have been called tidal waves, but astronomical tides are not involved in their formation. *Seismic seawaves* is another term that scientists commonly use.

These are typically very long waves (up to several hundred miles) with relatively small wave heights (commonly less than 3 feet), and long periods (over an hour from crest to crest in some cases) that have sufficient energy to travel across entire oceans, as was noted during the Sumatran earthquake in 2004 when the waves crossed the total width of the Indian Ocean. They move very fast in the open water at speeds measured in hundreds of miles per hour; however, as they approach shore, they usually slow down to 40-50 miles per hour. Like their smaller wind-generated cousins, when they enter shallow water, they abruptly peak and break, reaching heights of well over 100 feet.

Inasmuch as a tsunami has a much smaller wave height offshore, and a very long wave length, they generally pass unnoticed at sea, forming only a slight swell in the normal sea surface. Obviously, a tsunami can occur at any stage of the tide, but even at low tide, it will still inundate coastal areas if the incoming wave's surge is high enough.

Tsunamis are most commonly generated by earthquakes where converging plate boundaries generate faults in which at least one side of the fault line undergoes abrupt vertical movements that displace the overlying water. It is very unlikely that major ones can form at divergent plate boundaries or along transform faults because they do not typically have significant vertical movement and do not usually displace the water column vertically. Thus, subduction zone related earthquakes generate the majority of all tsunamis (i.e., where major plates converge as along major oceanic trenches around the Pacific Ocean "ring of fire"). Such a subduction zone is located off the coasts of Oregon and Washington (the Cascadia Subduction Zone; Figure 21).

Satake et al. (1996) were the first to determine that a major tsunami that affected Japan in 1700 was triggered by a large earthquake in the Cascadia Subduction Zone. Atwater et al. (2005), in the report *"The Orphan Tsunami of 1700,"* summarized the evidence for a magnitude 9 earthquake in the Northwest on 26 January 1700, known as the 1700 Cascadia Earthquake. The earthquake produced a tsunami so large that there are records in Japan mentioning it. These records were used to assign a precise date and approximate magnitude to the earthquake. There was dramatic drop in the elevation of the coastal land, as evidenced by: 1) Buried marsh and forest soils that have since been buried by modern sediments; 2) A layer of tsunami sand in some areas; 3) The death or injury of coastal forests; and 4) Descriptions of the earthquake and tsunami in regional Amerindian legends.

We have been conducting research along the earthquake/tsunami-prone coast of Alaska for many years, thus we have observed the effects of these phenomena up close in many localities along that coast. As noted by Dr. George Pararas-Carayannis (2009a) in his detailed account the "Good Friday" Alaska earthquake (on 27 March 1964), that earthquake generated catastrophic tsunami waves that devastated many towns in the Prince William Sound area, along the Gulf of Alaska, along the West Coast of Canada and the United States, and in the Hawaiian Islands. In Alaska, the tsunami horizontal run-up measurements were 20 feet at Kodiak Island, 30 feet at Valdez, 80 feet at Blackstone Bay, and 90 feet at the native village of Chenega in Prince William Sound. That 1964 earthquake caused 119 deaths in Alaska alone, with 106 of them due to tsunamis that were generated by tectonic uplift of the seafloor, and by localized upland-generated and submarine landslides. According to Dr. Pararas-Carayannis, there were two different types of "tsunami generation mechanisms" associated with that earthquake:

1) Tectonic movements of the seafloor caused the tsunamis along the open shore of the Gulf of Alaska; and

2) Tectonic movements that uplifted some of the shores over 30 feet and the resulting landslides were responsible for most of the local tsunami waves that caused the destruction and deaths in Prince William Sound. Those tsunamis destroyed local towns and fishing villages, killing 82 people. The maximum tsunami wave height recorded in Prince William Sound was over 200 feet.

According to the USC Tsunami Research Group (1964), *the 1964 Alaskan tsunami is the largest and most destructive recorded tsunami to ever strike the United States Pacific Coast. Along the Washington Coast, tsunami waves destroyed houses, cars, boats, and fishing gear, causing an estimated 80,000 dollars in damage to roads and bridges alone. At Ocean City, 5- to 6-foot tsunami waves collapsed the bridge over Copalis River.*
[NOTE: The authors of that report had obviously not yet had the opportunity to read *The Orphan Tsunami of 1700*.]

Oregon was also hit hard by the tsunami (on Good Friday 1964), *which killed four people and caused an estimated 750,000 to one million dollars in damage to bridges, houses, car, boats, and sea walls. The greatest tsunami damage in Oregon did not occur along the ocean front as one might expect, but in the estuary channels located further inland, where those with the correct bathymetry actually amplified the wave tsunami wave heights. Of the communities affected, Seaside, struck by a 10-foot wave, was the hardest hit. Tsunami wave heights reached 10 to 11.5 feet in the Nehalem River, 10 to 11.5 feet at Depoe Bay, 11.5 feet at Newport, 10 to 11 feet at Florence, 11 feet at Reedsport, 11 feet at Brookings, and 14 feet at Coos Bay.*

More recently, the March 2011 magnitude 9 earthquake off the coast of Japan generated a tsunami that impacted the Pacific Northwest Coast, as described by Allan et al. (2012):

- Tsunami waves were recorded from San Diego, CA to Neah Bay, WA.
- The maximum wave heights in Oregon were at Port Orford (11.2 feet; the eighth wave to approach) and waves greater than 5.2 feet occurred at an average of every 16.3 minutes for 13 hours.
- The maximum wave heights in Washington were at La Push (4.7 feet).
- Damage from the tsunami waves mostly impacted vessels in harbors, because the largest waves arrived during a relatively low tide. The most damage occurred at the harbor at Brookings, Oregon, where much of the commercial part of the harbor was destroyed.
- Damage would have been considerably greater had the largest waves arrived during high tide.

Oregon officials have been identifying and mapping the tsunami inundation hazard along the coast since 1994, using the best tsunami science available today. According to the Oregon Department of Geology and Mineral Industries (2018):

The evacuation zones on these maps were developed by the Oregon Department of Geology and Mineral Industries in consultation with local officials. Evacuation routes were developed by local officials and reviewed by the Oregon Department of Emergency Management. Note that the tsunami inundation limits shown on these maps represent the worst-case scenario for the two types of tsunami shown (local and distant). *The inundation limits are intended only to guide tsunami evacuation for these two extreme events.*

These maps have been used to post signs along coastal roads when you enter and leave one of the tsunami inundation zones (see Figure 22).

SNEAKER WAVES

Another sign that you will see on the shoreline is "Watch for Sneaker Waves." A sneaker wave, also known as sleeper wave, is a sudden, unusually large single wave. When the wave strikes the shoreline, the previously dry shoreline can quickly become inundated, and people on the shoreline can be swept to sea. García-Medina et al. (2018) analyzed thirteen sneaker wave events along the Oregon Coast between 2005 and 2017, which resulted

FIGURE 22. State of Oregon road sign to indicate that you are entering a tsunami hazard zone. There is also one for when you are leaving the hazard zone.

in fifteen fatalities. All of these events occurred between October and April and were associated with long-period swell waves. So, please be very careful when you are recreating on the shoreline and watch for sneaker waves.

5 SEA-LEVEL RISE: PAST AND FUTURE

He said, "I'm surprised they don't lock you up – a reasonable man. It's one of the symptoms of our time to find danger in men like you who don't worry and rush about. Particularly dangerous are men who don't think the World's coming to an end."

John Steinbeck - SWEET THURSDAY

A key issue in the understanding of the makeup of the modern coasts of the World is the history of changes of sea level that took place during the Pleistocene and Holocene epochs of the Quaternary period (2.6 million years ago to the present). When sea level was lower during the major glaciations, during what is referred to as **lowstands**, the rivers along many coastal areas of the World carved deep valleys across the coastal zone and out onto the present continental shelf. During some of the lowstands, aeolian processes built large sand dunes on the surface of what is now the continental shelf.

The glaciations were separated by warming periods, called interglacials, during which time sea level rose. During those warming periods, the highest sea levels were higher than it is today, with each succeeding interglacial having lower and lower **highstands**. Sea level during the last interglacial before the present one (peak about 120,000 years ago) stood around 10-12 feet higher than it is today. Of course, it appears to have been higher than that along parts of the Pacific Coast of North America because of the constant rising of the Coast Ranges in those areas.

Although understanding the history of the Quaternary period is still in its relative infancy and research continues, progress has been made with regard to comprehending the changes of sea level during that interval as the result of new developments in Quaternary dating methods. The two primary new methods are: a) Measurements of uranium-series disequilibrium, particularly of preserved coral deposits; and b) The fact that certain configurations of amino acids within the proteins of living organisms (such as marine mollusks) convert to different configurations upon the death of the organism.

As noted by Muhs et al. (2004), *the direct dating of emergent marine deposits is possible because U* (uranium) *is dissolved in ocean water but Th* (thorium) *and Pa* (protactinium) *are not. Certain marine organisms, particularly corals (but not mollusks), co-precipitate U directly from seawater during growth.* Because the other two elements are not co-precipitated, if they are present in the coral, they are recognized as a decay product. Inasmuch as three of the naturally occurring isotopes of U are incorporated into living corals, the potential for radioactive decay sequences of the uranium to occur is possible (i.e., radioactive parent elements decaying to stable daughter elements). In fact, three different radioactivity ratios (e.g., $^{231}Pa/^{235}U$) occur in corals, that provide *three independent clocks for dating the same fossil coral* (Gallup et al., 2002). That method can date materials as old as 500,000 years. It is an excellent method to use when fossil corals are available.

If not, there are two other methods to rely on. Carbon-14 dating is commonly used for preserved formerly living plant or animal remains less than 70,000 years old. For older materials, amino-acid measurements have provided a complementary method. Muhs et al. (2004) pointed out that *the basis of amino-acid geochronology is that proteins of living organisms (such as marine mollusks) contain only amino acids of the L configuration. Upon the death of an organism, amino acids of the L configuration convert to amino acids of the D configuration, a process called racemization.* The racemization process continues at a rate that can be measured, hence the organism's age can be determined. The use of that technique is best applied to fossil mollusks, which are widespread.

With regard to Pleistocene and Holocene sea-level

histories, the record for both glacial and interglacial episodes and, hence, sea-level changes, is reflected in oxygen-isotope composition (i.e., relative amounts of ^{16}O and ^{18}O) of the carbonate shells of bottom-feeding foraminifera preserved in deep-sea sediments (Imbrie et al., 1984). The oxygen-isotope compositions of foraminifera *are a function of both water temperature and the oxygen isotope composition of ocean water at the time of shell formation. Foraminifera precipitate shells with more ^{18}O in colder water. The oxygen isotope composition of ocean water is a function of the mass of glacier ice on land, because glacier ice is enriched in ^{16}O. Thus, relatively heavy (^{18}O-enriched) oxygen isotope compositions in foraminifera reflect glacial periods, where light compositions (^{16}O-enriched) reflect interglacial periods.* That is, with the ^{16}O taken up in the ice, the relative amount of ^{18}O in the ocean increases during major glacial episodes.

[NOTE: Foraminifera are small (usually less than 0.04 inches in size), single-celled marine organisms that float freely in the ocean in large numbers. They secrete a calcareous shell, or test. About 275,000 species are recognized, both living and extinct. The first ones appeared about 500 million years ago. Many extinct species existed for geologically short periods of time and both extinct and living species can be limited to a very narrow environmental range (UCal Berkeley, 2004). Consequently, geologists use them for both dating relative ages of rocks and sediments and for interpreting past environments.]

To put this all in perspective, Figure 23A shows the changes in deep-sea oxygen isotope records based on measurements on foraminifera tests that are present in deep-sea sediment cores. The peaks in the curve represent times of warming, and the lows represent periods of cold that are usually accompanied by widespread glaciation. That record, which goes back 500,000 years, shows five major periods of warming (interglacials?). The last major interglacial showing a low ice volume similar to that of today *started 130,000 years ago, peaked at 125,000 years ago, and terminated 116,000 years ago* (Kukla, 2000; indicated by the P on Figure 23A; Pamlico highstand).

Another way to determine these major climatic cycles is to study cores of ice taken in areas such as the Antarctic (Petit et al., 1999) and Greenland icefields (Kukla, 2000). In January 1998, a collaborative cooperative ice-drilling project among Russian, U.S., and French scientists carried out at the Russian Vostok station in East Antarctica resulted in the taking the deepest ice core ever recovered, at nearly 12,000 feet. That core allowed recovery of ice up to 450,000 years old. The peaks and valleys in the blue curve in Figure 23B show Earth temperatures based on measures of either irradiance and/or hydrogen and oxygen isotopes of the ice cores.

Also shown in Figure 23B is the green curve, which is a record of the historic concentrations of CO_2 in the atmosphere determined by measuring the CO_2 content of bubbles of air trapped in the ice. Clearly, the graph shows that increases in temperature and CO_2 are correlated. Glacial cycles that are thought to be caused by changes in the Earth's orbit, or the Milankovitch cycles, must also influence the carbon cycle. The red curve in Figure 23B shows the volume of dust in the air for the same time interval as the other two curves based on the Vostok core. The periods of highest dust appear to correlate to some extent with the glacial episodes (cold dry periods?).

The Pleistocene epoch ended around 12,000 years ago, after the last of the four major glaciations ceased and sea level was already rising at a relatively rapid rate. This major ice-melting episode that got under way about 12,000 years ago marked the beginning of the Holocene (modern) epoch, at which time sea level was approximately 400 feet below its present position. As the melting proceeded, sea level rose rapidly (see Figure 23A), as much as 2 feet/century, reaching near its present level 5,000 years ago. Because a near stillstand that occurred at that time, the bulk of the major Holocene landforms on the relatively stable trailing-edge coasts of the southeastern U.S. and Gulf of Mexico (e.g., deltas, including the modern Mississippi delta, and barrier islands) have formed. During that same time interval, a considerable amount of Holocene landforms also developed on the Pacific Coast, including Coos Bay, Tillamook Bay, Nehalem Bay, Willapa Bay, and Grays Harbor. Therefore, much of the later discussion in this book deals in large part with events taking place during the Holocene epoch (12,000 years ago until now).

When coastal geomorphologists like ourselves use the term sea-level rise, we normally insert the word **relative** as a modifier, because there are two possible factors that create major changes in sea level: 1) Changes in the volume of water in the World Ocean; and 2) The moving up and down of the Earth's surface (both the land and the seafloor). Changes in the volume of water

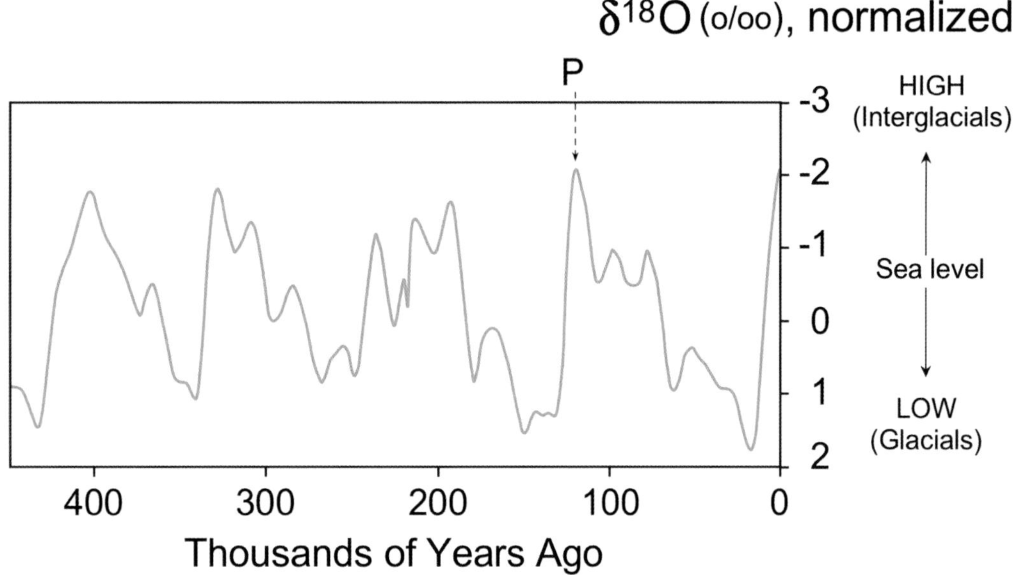

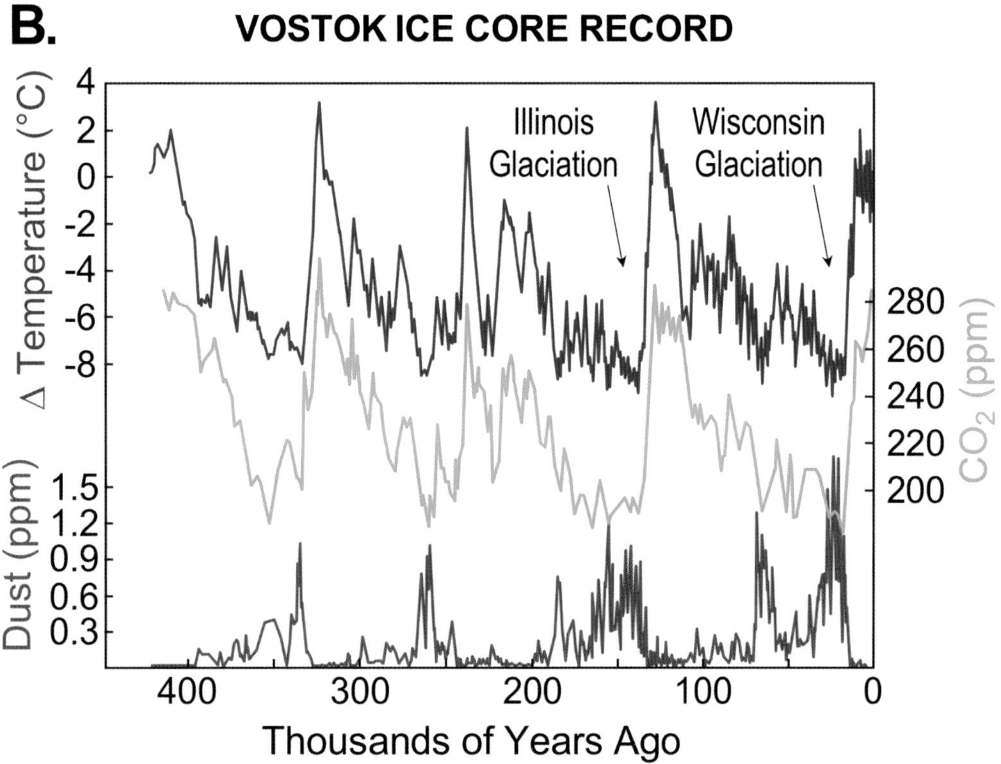

FIGURE 23. Data indicating the occurrence of glacial and interglacial periods extending from the present to 450,000 years ago. (A) Relative sea level based on oxygen-isotope data collected from foraminifera in deep-sea sediments. The exact levels in feet are not given. "P" indicates the Pamlico highstand that peaked at 125,000 years ago. Modified after Muhs et al. (2004). (B) Graph of temperature (blue), CO_2 (green), and dust concentration (red) measured from the Vostok, Antarctica ice core as reported by Petit et al. (1999).

in the World Ocean are mostly related to major changes in the World's climate. For example, during the peak of the last Ice Age, around 20,000 years ago, the surface of the World Ocean was about 400 feet below its present level, because so much of the Earth's hydrosphere was taken up in the huge ice sheets that covered large land areas of some of the continents. Also, with a warming Earth, the water in the Ocean increases in volume as it warms (called thermal expansion).

A number of reasons can cause a land mass to rise or fall, including (to name but a few):

1) Tectonic activity such as mountain building which causes the land to rise;

2) Elevation of the land as a result of the melting of the large ice sheets present during the Pleistocene epoch, because the Earth's surface was depressed under the heavy load of the ice; and

3) Sinking of land where large rivers have dumped huge sediment loads on the continental shelf. This kind of sinking, aided by compaction of the sediments, has caused sea level to rise up to 4 feet per century in certain areas of the Louisiana Coast, where former lobes of the Mississippi River Delta were abandoned as the River shifted positions (Yuill, Lavoi, and Reed, 2009).

Available data suggest the following historical rates of sea-level rise along tectonically stable shores, such as those that occur on many trailing-edge coasts (partly from Pilkey and Pilkey-Jarvis, 2006):

1) Between 12,000 and 6,000 years ago – Overall rate of **3 feet per century**, with possible "blips on the curve" of as much as 10 feet per century. Also, there were some fairly well documented stillstands during that rise, at which time river deltas, barrier island systems, and so on were deposited, the remnants of which are still present in places on the continental shelf;

2) Between 6,000 and 4,500 years ago – probably around **1.5 feet per century**; and

3) Between 4,500 and the present – This is the time interval we refer to as the present **"stillstand"** of sea level. It is during that period of time that most of the modern deltas, barrier islands, and estuaries on the trailing-edge coast of the North American continent have been formed. Others have suggested 5,000 to 6,000 years ago as the start of the stillstand on the West Coast. Numerous sea-level curves show that sea level has bumped up and down several times during this "stillstand" period, sometimes more than 3 feet.

Since the industrial revolution starting around 1900, the burning of fossil fuels had caused a dramatic increase in CO_2 in the atmosphere that has continued to increase since that time (Figure 24), a scary prospect indeed. In May 2013 scientific instruments at Mauna Loa Volcano on the island of Hawaii showed that carbon dioxide had reached an average daily level above 400 parts per million (Gillis, 2013); in April 2018, it was 411.9 ppm (NOAA, 2018). Carbon dioxide gas in the air has not been that high for at least three million years. As seen in Figure 23B, the highest level reached during the interglacial periods of the Pleistocene epoch was only 280 parts per million.

Based on tide gauge measurements from around the World, the Fifth Assessment Report of the IPCC (Intergovernmental Panel on Climate Change, 2013) estimated that global sea level rose an average of about 1.7 mm (0.07 inches) per year between 1901 and 2010, for a total of 0.19 m (0.62 ft). Between 1993 and 2010, using a combination of precise satellite sea surface measurements and tide-gauge records, the rate of sea-level rise nearly doubled, to 3.2 mm (0.13 inches) per year.

We don't want to be the prophets of doom, but here is what the 2017 National Climate Assessment, the most authoritative assessment of the science of climate change with a focus on the U.S., concluded (US Global Change Research Program, 2017):

- Global annually averaged surface air temperature has increased by about 1.8°F over the last 115 years (1901-2016). **This period is now the warmest in the history of modern civilization**.
- The magnitude of climate change beyond the next few decades will depend primarily on the amount of greenhouse gases (especially carbon dioxide) emitted globally. **Without major reductions in emissions, the increase in annual average global temperature relative to preindustrial times could reach 9°F or more by the end of this century**.
- Global average sea level has risen by about 7-8 inches since 1900, with almost half (about 3 inches) of that rise occurring since 1993.
- Global average sea levels are expected to continue to rise—by at least several inches in the next 15 years and by 1-4 feet by 2100. **A rise of as much as 8 feet by 2100 cannot be ruled out**.
- There is broad consensus that the further and the faster the Earth system is pushed towards warming,

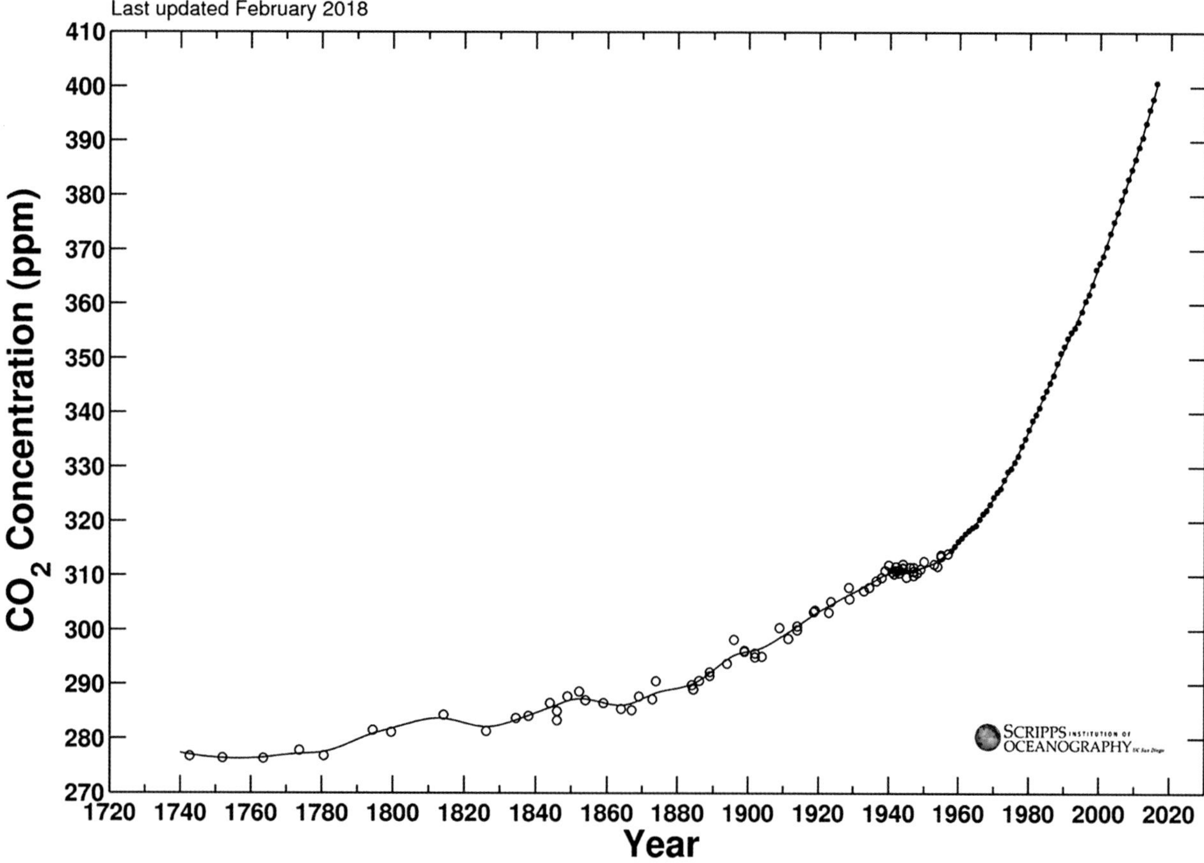

FIGURE 24. CO_2 concentration (in parts per million, ppm) in the Earth's atmosphere, from air extracted from an Antarctic ice core (open circles) merged with direct atmospheric measurements at Mauna Loa and the South Pole (black circles). From Scripps CO_2 Program (Keeling et al., 2001).

the greater the risk of unanticipated changes and impacts, some of which are potentially large and *irreversible*.

The biggest uncertainty is not if, but when the ice sheets in Greenland and Antarctica will melt at rates that will significantly increase their contribution to sea-level rise. Over the period of 1993-2010, contributions to sea-level rise were: 1) Thermal expansion (39%); 2) Melting of glaciers and ice caps (27%); 3) Greenland ice sheet (11.8%); 4) Antarctica ice sheet (9.6%); and land water storage (13.6%).

Future sea-level predictions vary widely based on which models are used. The problem, as with a lot of models used to predict change in nature, is the assumptions, or data input, used to create the model. In the earlier models, the data input related to rising sea level was the melting of glaciers, which is well documented, and the physical expansion of water as it warms. They did not take into account large-scale melting of the big ice fields in Greenland and Antarctica, which is pretty scary when you consider that in 2002, Antarctica's 1,255 square-mile Larsen B ice shelf broke off and disappeared in just 35 days. An iceberg the size of Delaware spilt off from Antarctica's Larsen C floating ice shelf in mid-July 2017. Supposedly it was not large enough to affect sea-level rise when it melts. However, if the ice shelf continues to disintegrate, it could cause the grounded glaciers that are behind it to release more ice into the Ocean, with amounts that could increase sea levels.

However, the Greenland ice sheet is what we really have to worry about. The 2013 IPCC report states *The available evidence indicates that sustained global warming greater than a certain threshold above pre-industrial would lead to the near-complete loss of the Greenland ice sheet over a millennium or more, causing a global mean sea level rise of about 7 m* (23 feet). That would definitely be scary. And it has happened before. The maximum global mean sea level during the last interglacial period (129,000 to 116,000 years ago) was at least 16 feet higher than present, mostly from melting of the Greenland ice sheet and some contribution from the Antarctica ice sheet. This rise occurred when air temperatures were at least 2°C warmer than today. Recent studies of the high-end projections considering rapid ice loss from Antarctica predict additional sea-level rise of 3.6-8 feet (Le Bars, Drijfhout, and de Vries, 2017).

All of that being said, there is no doubt that an abrupt rise in sea level of only 2 or 3 feet (not to mention 8 feet) would affect an unimaginable amount of structures that are now built close to the beach in many areas populated by millions of people.

Along the Pacific Northwest Coast, relative sea levels vary due to the rising of the land due to tectonics, as described by Komar, Allan, and Ruggiero (2011):

The trends in RSLs (relative sea levels) *measured by the (tide) gauges are strongly affected by the tectonics of this region, the collision and subduction of the oceanic Juan de Fuca and Gorda Plates beneath the continental North American Plate. The active tectonics have resulted in significant alongcoast variations in land-elevation changes and trends of RSLs that range from -1.89 mm/y on the emergent shore of Neah Bay on the north coast of Washington, to the submergent shore along the north-central Oregon coast, with an RSL rise of 1.33 mm/y determined from the Yaquina Bay tide gauge.*

As shown in Figure 25, on the southern half of the Oregon Coast, the land had been rising faster than sea-level rise (land rising at the rate of <1 mm/year faster, as of the early 1990s), whereas on the northern half of the Oregon Coast, the land is being submerged by sea-level rise (land rising at the rate of 1-2 mm/year slower than the rise of the sea). As discussed above, future sea-level rise is expected to increase at a faster rate; thus, eventually all of the Oregon coast could experience higher sea levels.

The open ocean waters presumably follow the general worldwide pattern of a sea-level rise of around 1 foot per century. Except for the general rise of up to 400 feet that has occurred Worldwide in post-glacial times, attempts to relate detailed sea-level curves for this coast are difficult because of the fairly complex tectonic influences just mentioned.

In a recent detailed analysis for the state of Washington, Miller et al. (2018) determined the vertical land movements for 171 locations that covered the entire coastal region of the state (detailed spreadsheets can be downloaded for each location). These land movements varied widely, as shown in Figure 26, with some examples as follows for land movement changes by 2100:

1) – 0.5±0.2 feet at Tacoma, in Puget Sound;

2) + 1.1±0.3 feet at Neah Bay, on the Olympic Peninsula, which is rising the fastest in the state; and

3) + 0.3±0.5 feet at Taholah, along the central

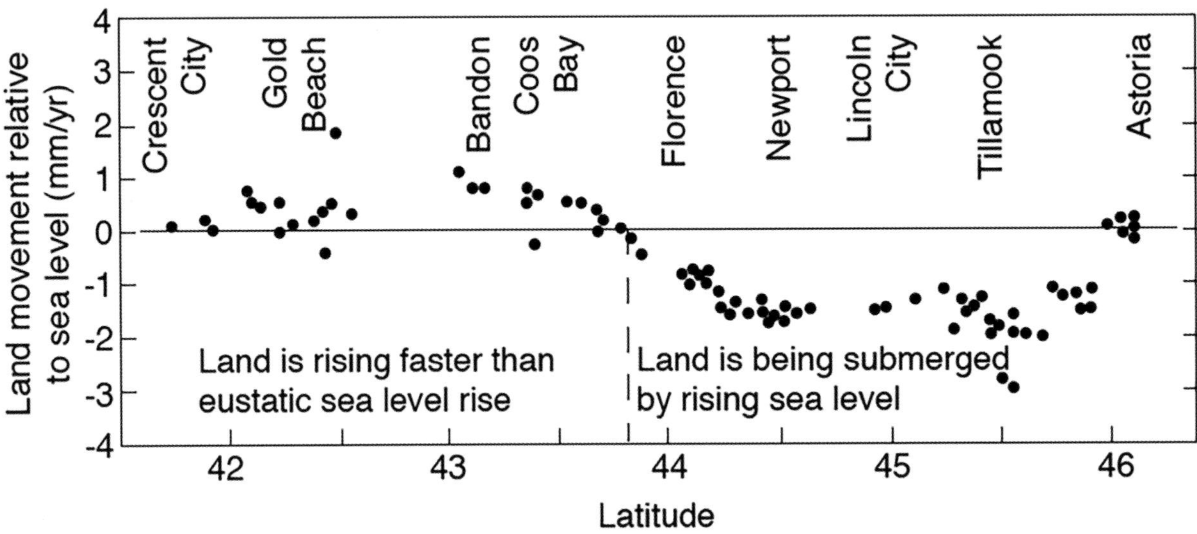

FIGURE 25. Vertical land movements along the Oregon Coast relative to mean sea level (Komar, 1992).

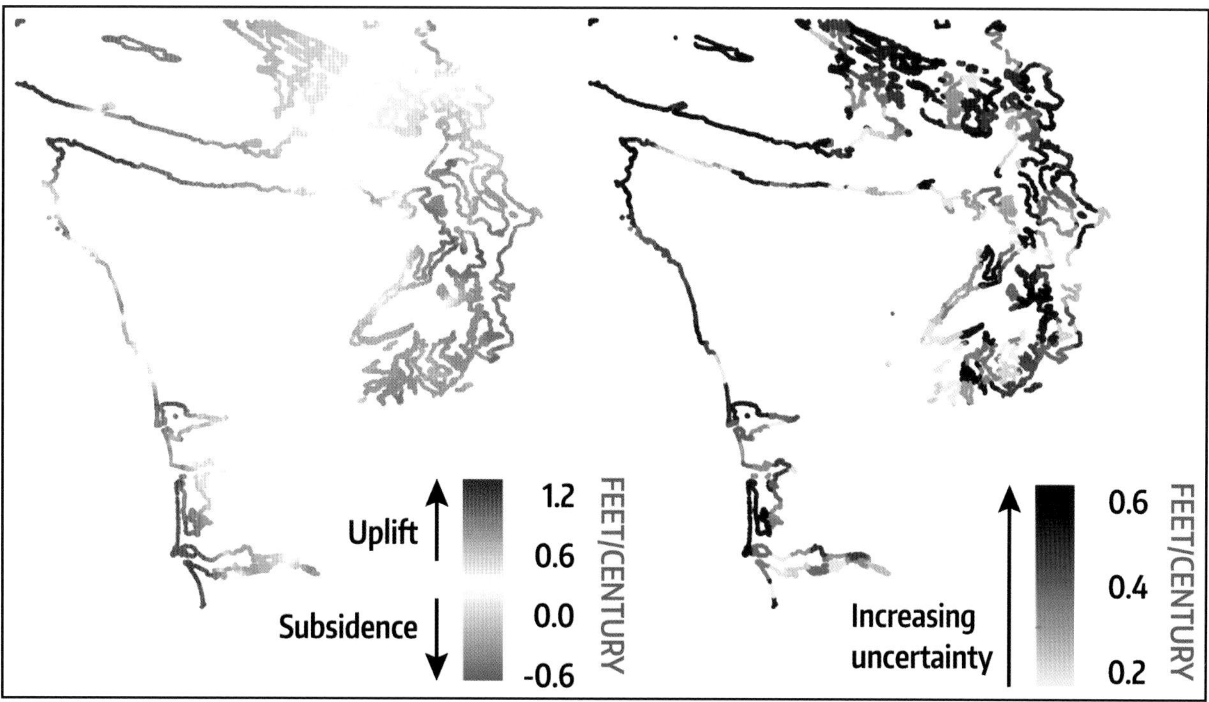

FIGURE 26. Vertical land movements best-estimate rates in feet/century (left) and their uncertainties (right). From Miller et al. (2018).

coast.

The National Research Council (2012) published a report titled *Sea-Level Rise for the Coasts of California, Oregon, and Washington: Past, Present, and Future.* In their analysis, they considered the regional geology that is causing land uplift as well as global sea-level rise. For the Washington and Oregon Coasts, they project sea level to be between -0.13 feet (sea-level fall) and +0.75 feet by 2030, between -0.1 feet and +1.6 feet by 2050, and between +0.3 and +4.7 feet by 2100. Note that the ranges get larger over time, because of the uncertainty discussed above. Depending on which global sea-level rise rates you want to use, based on the Miller et al. (2018) analysis, most of the Washington Coast is being submerged by rising sea level. That is, even where the land is rising the most due to tectonic processes, global sea level is increasing faster. Furthermore, as discussed earlier, large earthquakes along the subduction zone can instantaneously lower the land surface. Miller et al. (2018) included in their analysis the land-level changes from a subduction zone earthquake with a 500-year recurrence interval for each of the 171 sites, with their model results indicating changes along the coast from -1.3 to -8.5 feet, which would be in addition to the relative sea-level rise numbers cited above.

"Do you know why I live out there on the cliff?" I've only told a few. I'll tell you, because you're coming to stay with me." He stood up, the better to deliver his secret. "I am the last man in the western world to see the sun. After it is gone to everyone else, I see it for a little while. I've seen it every night for twenty years. Except when the fog was in or the rain was falling, I've seen the sun set."

John Steinbeck – TO A GOD UNKNOWN

---

6 COASTAL PROCESSES

TIDES

He consulted the tide chart in Thursday's Monterey Herald. There was a fair tide at 2:18 P.M., enough for chitons and brittlestars if the wind wasn't blowing inshore by then.

John Steinbeck - SWEET THURSDAY

Changing water levels along the shore, generally referred to as tides, can be caused principally by three mechanisms: 1) Process called wind set-up during which wind stress causes the water levels to increase downwind and decrease upwind; 2) Combination of wind set-up and decreasing barometric pressure that accompanies major storms, such as hurricanes (usually referred to as storm surges); and 3) Gravitational attraction of the Moon and Sun on major water bodies, called **astronomical tides**. As shown in Figure 27, when the Sun and Moon are in line (full Moon and new Moon; syzygy), the gravitational attraction of the Moon and Sun on the water bodies of the World Ocean are combined, giving rise to the maximum tides of the month (called spring tides). Minimum tides (called neap tides) occur during the first and third quarter of the Moon (quadrative), when the two forces work in opposition. During some lunar cycles, the tidal range during spring tides can be almost twice that during neap tides. Because of its location so far from the Earth, the gravitational pull of the Sun on the World Ocean is only 0.455 that of the Moon. Therefore, the Moon has the greater influence on the tidal cycle, controlling the timing of high and low tides.

In the much-simplified diagram shown in Figure 27, during syzygy, two bulges of the high tide occur on opposite sides of the Earth, one facing toward the Moon and another one facing away from it. The bulge facing the Moon is the result of the gravitational pull of the Moon and Sun. The bulge on the opposite side is much more difficult to explain. In actuality, the Earth and the Moon revolve together around the center of mass (barycenter) of the combined Earth-Moon system (Figure 28A). The barycenter is displaced a distance from the center of the Earth toward the Moon, and it is always located on the side of the Earth turned momentarily toward the Moon, along a line connecting the individual centers of mass of the Earth and Moon (Wood, 1982). To put it simply, as this system revolves around the barycenter, centrifugal force comes into play. This force is defined as the apparent force, equal and opposite to the gravitational force, drawing a rotating body away from the center of rotation, caused by the inertia of the body. Thus, the combined Earth-Moon system acts as a lever as the two equal and opposing forces revolve about a fulcrum, the barycenter. Therefore, the gravitational attraction of the Moon (and Sun) creates the bulge facing the Moon and centrifugal force creates a bulge of relatively equal size on the opposite side of the Earth (Figure 28B).

The diagram in Figure 27 shows that, as the Earth rotates about its axis, a given shore location on the Earth's surface will pass by the peak of the two bulges every 24 hours and 50.4 minutes. The extra minutes are required because that is how long it takes that beach location to "catch up" with the Moon as it revolves all the way around the Earth in about 28 days.

Because of the intervening continents, a single bulge cannot pass all the way around the Earth during its daily rotation. Instead, individual **amphidromic systems** of different sizes are set up within the Earth's ocean basins. These systems are separate, more-or-less circular, tidal systems with a nodal point in the middle. They are gyrating standing waves, analogous to the wave you would create by sloshing a layer of water around the sides of a circular pan. The nodal point in the middle of the pan (the amphidromic point) would maintain a constant level, with the highest vertical motion of the water layer occurring around the edge of the pan. The amphidromic systems present in the Pacific Ocean are illustrated in Figure 29.

Ed Clifton gave the following succinct discussion

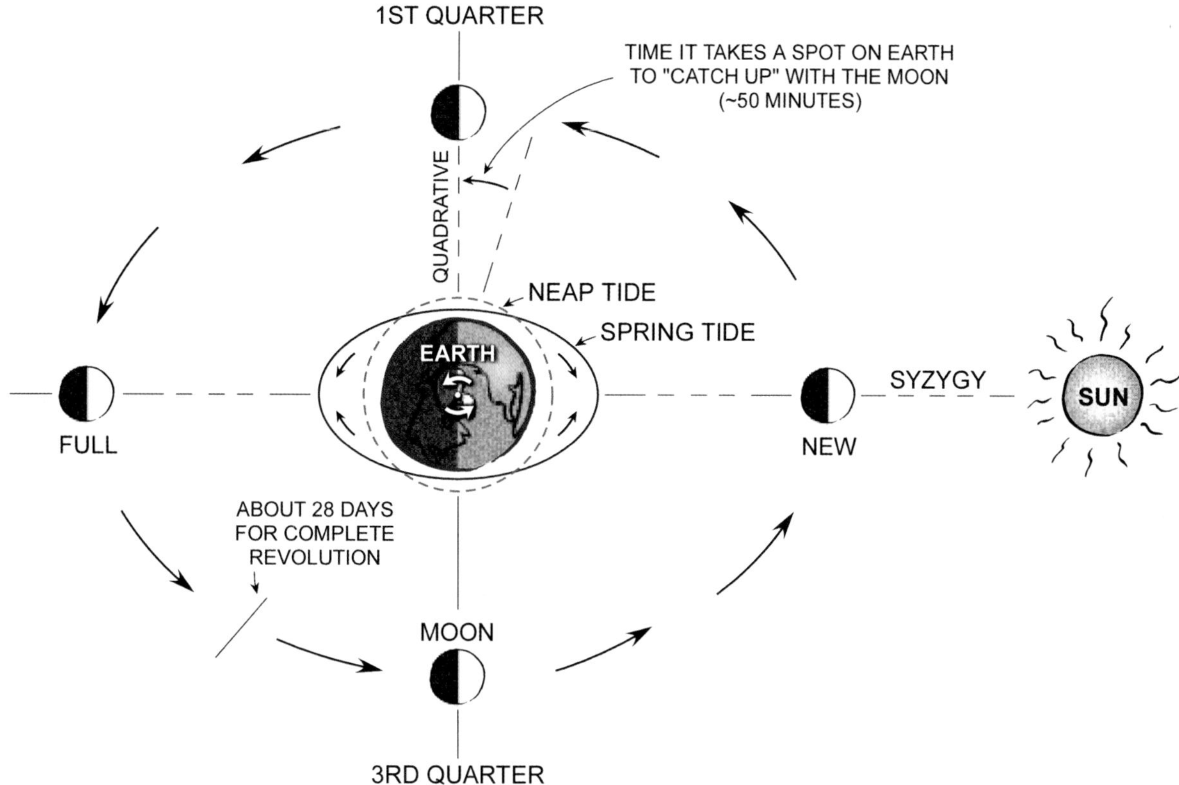

FIGURE 27. Forces involved in the generation of astronomical tides.

of amphidromic systems that occur within the Pacific Ocean:

Amphidromic systems are large rotary tidal systems created by the behavior of tides as giant low waves that are affected by the shape of the World's oceans and the rotation of the Earth. The height of the tide depends on its location within the system. Tides at the axis of the system (nodal point) are nil, but they increase outward with distance from the axis. In the Northern Hemisphere, the tides rotate counterclockwise about the axis; in the Southern Hemisphere, the rotation is clockwise. Both diurnal (one complete rotation of the tides daily) and semi-diurnal (two complete rotations of the tides daily) can coexist in the same ocean basin. The Pacific contains two diurnal systems and 8 partial or complete semidiurnal systems.

There are many other complications to the simple story presented in Figure 27. For example, because the Moon revolves in a slightly elliptical, rather than a circular, orbit around the Earth, *the distance of the Moon from the Earth may vary between extreme limits separated by some 31,278 miles* (Wood, 1982). When the Moon is closest to the Earth, this close encounter is called the **perigee**, which is 13% closer than when it is furthest away (called the **apogee**). Thus, perigean tides may be 20% larger than normal mean tides. Furthermore, if the perigean tide corresponds with the syzygy (full and new Moons), the so-called perigean spring tides (also called king tides), the tidal range can increase by 40% (Wood, 1982).

The actual final tidal curve is the result of the complex interaction of a number of **tidal constituents**, the primary one of which is the gravitational attraction of the Moon. Other constituents include the gravitational attraction of the Sun, the tilt of the Earth's rotational axis, the inclination of the lunar orbit, and the ellipticity of the orbits of the Moon about the Earth and the Earth about the Sun. Wind set-up is a key factor in generating changing water levels that is difficult to predict.

Along most of the east coast of the U.S., the tides are termed semi-diurnal, because two complete tidal cycles occur within 24 hours and 50.4 minutes. In other areas in the World, diurnal tides that have only one high and one low tide a day occur (e.g., the Texas Coast during some parts of the month). The tides on the coasts of Oregon

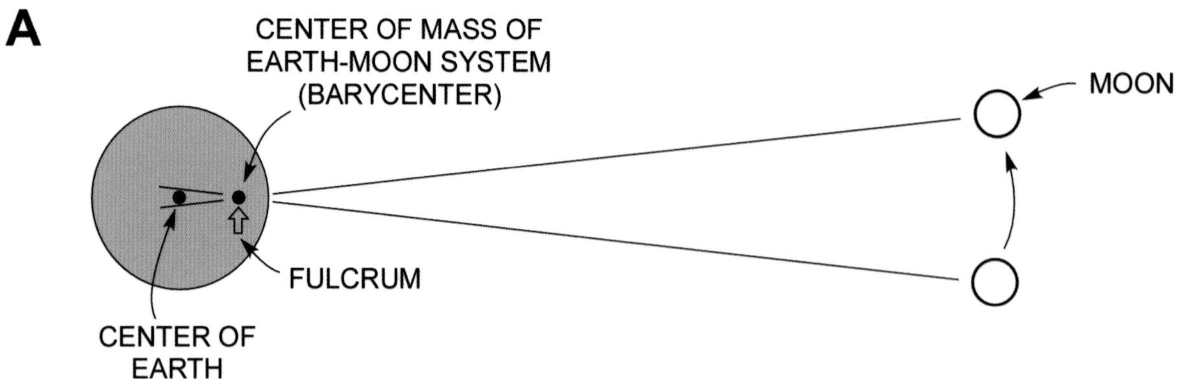

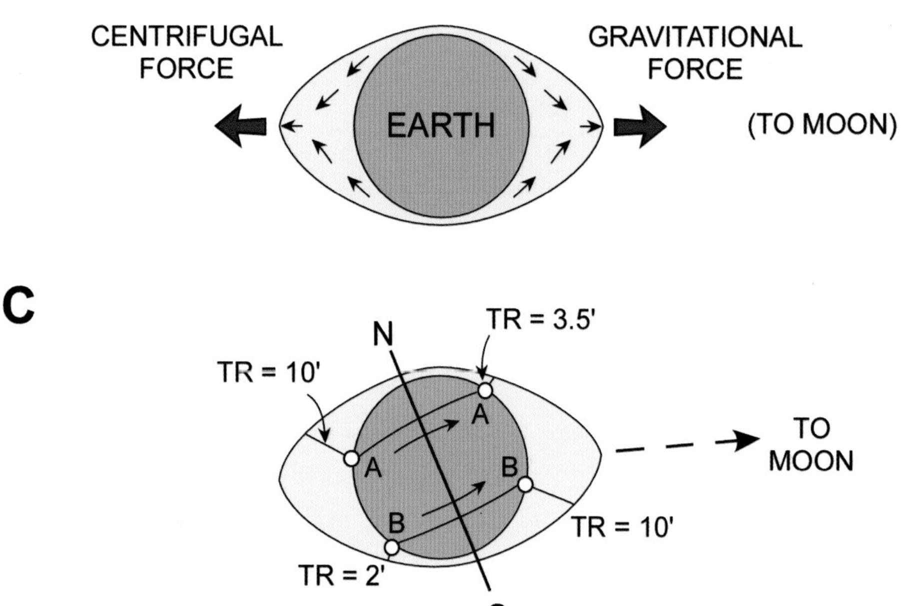

FIGURE 28. Complexities of the Earth's tides. (A) Location of the barycenter. (B) Creation of tidal bulges. (C) A factor involved in creating diurnal inequalities of the tides during a complete daily tidal cycle.

and Washington are called **mixed, predominantly semi-diurnal tides**, as is illustrated by the tidal curve for Astoria, Oregon in Figure 30. Although two complete tidal cycles occur within the lunar day on these coasts, and the tides are roughly semi-diurnal, there is a strong **diurnal inequality** of the different tidal levels, with four significantly different levels occurring during one tidal day (referred to as higher-high, lower-high, higher-low and lower-low tides on Figure 30).

A frequently cited reason for the diurnal inequality of tides is illustrated in Figure 28C. Because of the tilt of the Earth's axis, the depth of the water under the tidal bulge will be different on opposite sides of the Earth when the Moon is not directly over the equator. However, on the coasts of Oregon and Washington, these "mixed" tides are more accurately related to the fact that those two coasts are impacted by a diurnal (24-hour) amphidromic system with a nodal point located close to the middle of the Pacific Ocean (Figure 29B), and a smaller system that is semi-diurnal (12-hour) with a nodal point located a good distance straight off the coast of Central California (Figure 29A). Therefore, the resultant tidal

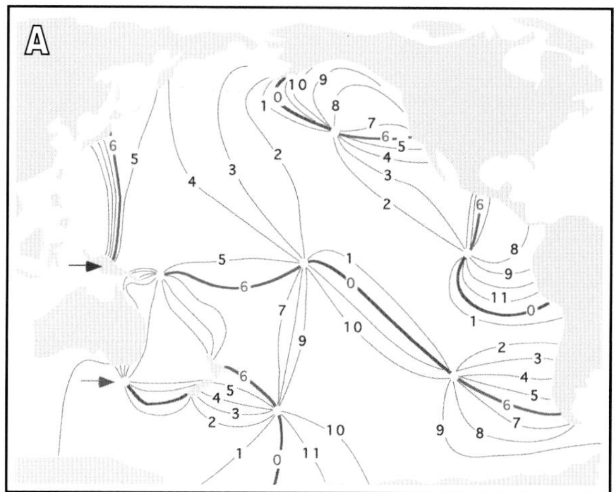

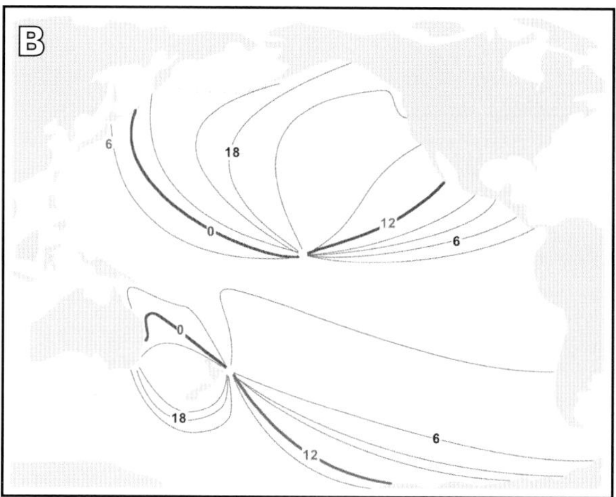

FIGURE 29. The amphidromic tidal systems of the Pacific Ocean. The tidal range is zero at the points in the middle of each system, called the amphidromic points. The tidal range increases outward from the amphidromic points. The numbers represent time in hours. The tidal phase is the same all along each line (from the amphidromic point to the end of the line). (A) The semi-diurnal systems, modified after Cartwright (1969). (B) The diurnal systems, modified after Dietrich (1963). Diagram courtesy of Dr. Ed Clifton.

curve is a product of the combination of the curve of the one diurnal tide with the curves produced by the two semidiurnal curves that occur during a 24-hour period. The result is the diurnal inequality illustrated by the tidal curve in Figure 30.

The fact that the Pacific Northwest Coast is located so far from the nodal points of the two amphidromic systems illustrated in Figure 29 causes the tidal range to be relatively large. Keep in mind that the further a point is located away from the nodal point of the amphidromic system, the larger the tidal range.

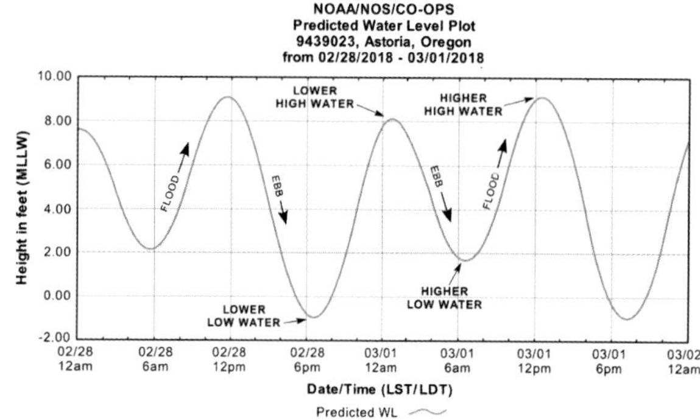

FIGURE 30. The complex tidal curve at Astoria, OR on 28 February and 1 March 2018. Note the diurnal inequality during that period. Based on NOAA Tide Tables.

On the coasts of Oregon and Washington, in November 2017, spring tidal range increased slightly from south to north, increasing from 8.1 feet at Brookings at the southern end of Oregon to 9.7 feet at Cape Flattery in Washington. That places this coastal area in the mesotidal class (tidal range of 6-12 feet) during certain spring tides. Also, the tides are progressively later from south to north.

[NOTE: The explanation of why the tides are progressively later from south to north along the these coasts can be seen by observing the semi-diurnal amphidromic system illustrated in Figure 29A (the system that is located closest to the western shore of North America). The stage of the tide (high, mid-tide, low, etc.) is the same all along each of the individual lines drawn on the diagram (cotidal lines), all the way from the amphidromic point to the end of the line. That particular system is rotating in the counterclockwise direction; therefore, the individual cotidal lines indicating similar tidal phases over time move progressively from south to north along the coasts of California, Oregon, and Washington. In Figure 29A, the line labeled 6 joins the shoreline at Baja California and the line labeled 0 (same as 12) joins the Alaska Peninsula. That means that when it is dead low tide on the Baja Coast it is dead high tide on the Alaska Peninsula. Note that low tide in Washington is three hours (more or less) later than low tide at Baja. Therefore, the low tide (or high tide for that matter) occurs a little later in Washington than it does in Oregon.]

As a general rule, notably along trailing-edge coasts, the average tidal range in an area is dependent upon several factors, with the largest tides occurring where: 1) Water body has on open connection with the World Ocean; 2) Continental shelf is wide; and 3) Coastal embayments occur. On the other hand, position within the ampidromic system is the key cause for the larger tides on the coasts of Oregon and Washington (Figure 29). Some of the amphidromic systems around the World are rather small. Komar (1976) cited an example in the North Sea, which has resulted in tides on the coast of Norway that are nearly nil and those on the open coast of Great Britain that can reach 13 feet.

With the exception of the Bay of Fundy, an area where resonance plays a role, most the larger tidal ranges in the World occur in coastal embayments, such as the Bay of Bengal (tidal range >17 feet at the upper reaches); the English Channel shore of France (tidal range approaching 50 feet); and the head of the Gulf of California (tidal range over 30 feet). Komar (1976) gave the following explanation for this – *as the tidal front approaches a narrowing indention of the coastline, … the enveloping shores constrict its movement and wedge the water together.*

Along the Pacific Northwest Coast, the actual water levels can be higher than the predicted tidal levels, which can increase risks from erosion and flooding along the coast. Komar et al. (2013) identified the following factors that can lead to higher-than-predicted water levels:

- On a seasonal basis, water levels are about 4 inches higher in winter, for two reasons. Water levels are lower in summer when coastal currents flowing parallel to the coast are directed to the south. These currents turn to the right (offshore) due to the Coriolis effect, resulting in upwelling of deep, colder water (which is more dense). Alternately, in winter, the coastal currents are mainly northerly and again turn to the right, resulting in higher water levels along the shore. This seasonal pattern is shown for the actual water levels as recorded by the Yaquina Bay tide gauge in Figure 31.
- During major *El Niño* events, which occur in winter, water levels can increase by 1-2 feet due to strong southeast winds that pile the water against the shoreline and low atmospheric pressures. Also, during an *El Niño*, the Coriolis effect is increased due to the stronger current flows, piling additional water against the shore. See Figure 31. In addition, water temperatures during *El Niños* are warmer than during normal years, which raises the water levels more. In contrast, water temperatures during *La Niñas* are colder, and denser than normal, which lowers the water level.
- Strong winter storms can create surges that also raise the water levels along the coast for hours to days, on top of the seasonal pattern. As many as fifteen significant storm surges can occur in winter, with magnitudes reaching 5 feet.

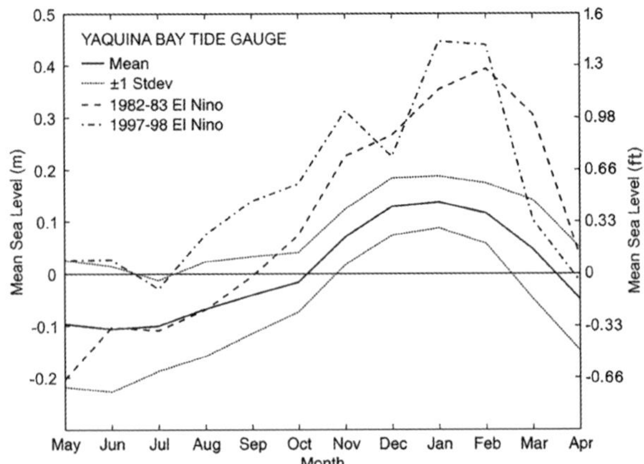

FIGURE 31. Water levels on the mid-Oregon coast at Yaquina Bay: Monthly means (+ 1 standard deviation) showing the long-term seasonal cycle of higher levels in winter and lower levels in summer; and During two of the major *El Niño* storms (1982-1983 and 1997-1998) when the water levels were much higher. From Komar et al. (2011).

The combination of all these factors (which occur in winter) can result in serious shoreline erosion. For example, the 19-20 November 1997 *El Niño* storm raised total water levels by 1.3 feet at Newport, Oregon and 2 feet at Toke, Washington, which caused beach and property damage, mostly along beaches backed by dunes (Allan and Komar, 2002).

WAVES

You've seen the sun flatten and take strange shapes just before it sinks in the ocean. Do you have to tell yourself every time that it's an illusion caused by atmospheric dust and light distorted by the sea, or do you simply enjoy the beauty of it? Don't you see visions?

"No," said Doc.

John Steinbeck – SWEET THURSDAY

General Introduction

Water waves are generated by the transfer of kinetic energy to the water surface as the wind blows over it. As shown in Figure 32A, a typical wave is defined by its wave length and wave height. The amount of time it takes two succeeding wave crests to pass a single point is called the wave period. The factors that determine the size of waves include:

1) The fetch of the wind (length of water surface over which the wind blows to form the wave);
2) Wind velocity;
3) Duration of the wind; and
4) Incident swell.

As indicated in Figure 32B, wind generates two different types of waves. Wave conditions in the area where the wind forms them are referred to as **seas**. Waves that travel out of that area, perhaps ultimately breaking on distant shores, are called **swell**. In seas, the waves can be over-steepened by the wind, causing them to break in the form of whitecaps. Swell typically does not break until it reaches shallow water, where it forms the familiar breakers that surfers love. A choppy sea surface with whitecaps indicates "sea conditions," whereas a smooth sea surface punctuated by breakers near the shore implies conditions of swell.

An observer standing on a beach may have difficulty distinguishing between the two types of waves, and it is very common for both types of waves to interact to form the resulting waves in the surf zone. To identify sea conditions, note the direction the wind is blowing on that day and look for the presence of whitecaps. Also, the period (time between the passage of two successive wave crests) of swell waves is typically considerably longer than for those waves generated by local onshore winds. Therefore, swell waves are usually formed by offshore storms that generate large waves. The waves generated by the storms move away from their zone of formation at a certain velocity, with longer waves moving faster than shorter ones. A classic example of this escape of long waves from their zone of formation is the forerunner waves that precede the landfall of hurricanes. Komar (pers. comm.) notes that *By definition, "forerunners" are the long-period waves, on the order of 20 seconds, well-developed regular swell having low wave heights (commonly a couple of feet or less) that arrive hours to days before the highest waves of the storm that typically have periods on the order of 10 to 15 seconds. As such, in past centuries these low long-period forerunners served as a warning of the high storm waves that were to arrive later.* Surfers flock to the shore to take advantage of these huge, long-period "forerunner" waves.

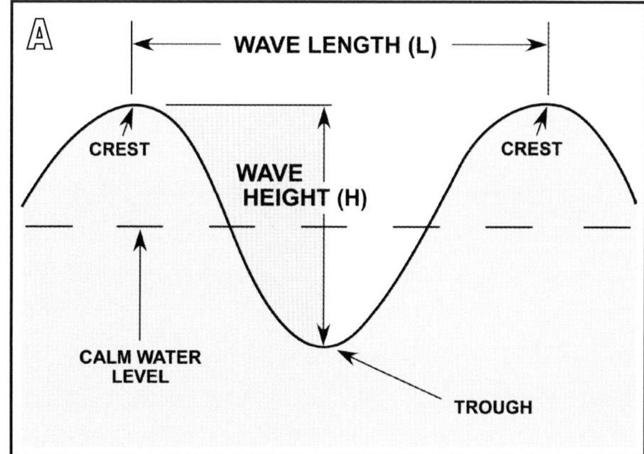

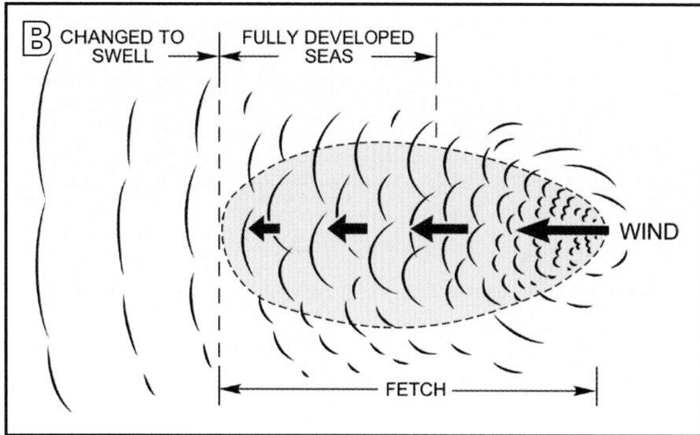

FIGURE 32. Waves. (A) Components of a typical water wave. (B) General pattern for the generation of waves by wind blowing across the water surface.

As waves approach the shore, they increase in steepness and break in different ways, to be discussed later. To understand why this happens, consider a cork on the water surface as a wave passes by. The cork ascribes a circular motion as the wave passes, in the process moving a small distance in the direction the wave is moving. As shown in Figure 33A, the water directly beneath the wave also undergoes a circular orbit, with the size of those orbits decreasing exponentially with depth until completely ceasing to orbit at a depth of about one half the wave length.

As the wave moves onshore, the wave begins to "feel the bottom," that is, the orbital motion of the water under the wave begins to bump up against the bottom where the water becomes shallower than one half the wave length (Figure 33B), moving landward until they eventually break. Ed Clifton (pers. comm.) pointed out

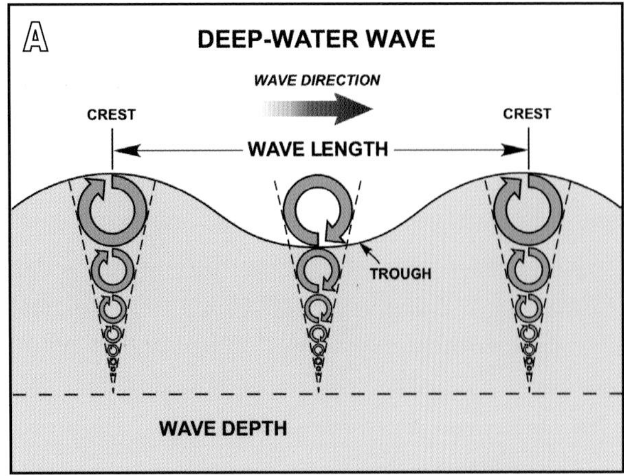

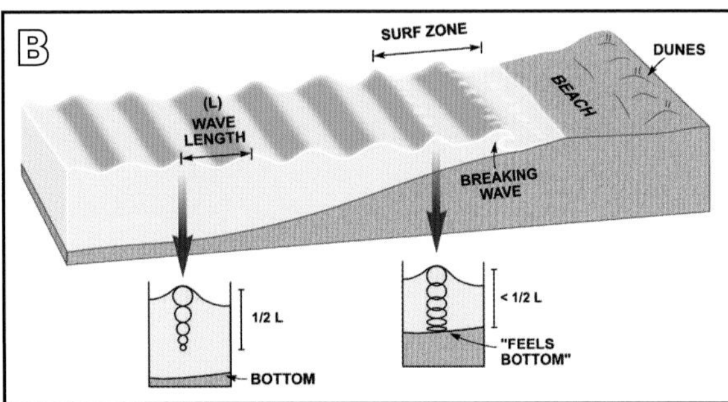

FIGURE 33. Waves approaching shore. (A) Orbital water motion at depth under a passing wave. (B) Characteristics of the surf zone under the influence of wave action. Note that in water shallower than approximately one half the wave length, the orbits of the wave-generated water motion under the waves (illustrated in A) flatten out as the wave begins to "feel the bottom." This causes the waves to slow down and become steeper and steeper until they eventually break.

several important aspects of both breaking waves and their interaction with the seafloor:

1) Sea waves break when they are over-steepened by the force of the wind driving the crest of the wave faster than the preceding trough is moving.

2) As a swell wave advances toward the shoreline, it becomes asymmetric as it approaches the break point. Breaking swell results where bottom friction on the orbital wave motion slows the speed of the wave in the shallow wave trough relative to the speed of the following crest.

3) On the swell-dominated Pacific Northwest Coast, where the average wave period is around 11.5 seconds, the average deep-water wave length would be around 650 feet, which means that the wave would "feel bottom" at about 330 feet. Even in the summertime, in the absence of storms, waves with the average period of 10 seconds will "feel bottom" at around 250 feet.

4) With regard to the common swell waves, not storm waves, he pointed out that the orbital motion of the water under shoaling (10-second period) waves shows a shoreward asymmetry. As the waves shoal, they deform from their sinusoidal shape and the crests become sharp, separated by broad flat troughs. As a result, the time available for the onshore flow (of the orbiting water cells, Figure 33A) under the crest lessens and, as a result, the water moves faster than it does under the broader trough. This velocity asymmetry does several things. It drives the coarser material (on the seafloor) landward (by overcoming the threshold of transport limits) and thus tends to drive most of the sand landward (since the rate of sand transport is generally considered to be a function of a power of the velocity of the moving water), thus restoring beaches that were eroded by storm waves, presumably because of enhanced rip currents during storms. However, there may be other factors as well, such as the presence of sea cliffs that cause the water to pile up along the shoreline. That is an important point with regard to erosional and depositional patterns on sand beaches, which will be discussed in some detail later.

The zone where the waves are breaking is referred to as the surf zone (Figure 33B). After breaking, the water released by the broken wave impinges on the portion of the beach known as the beachface (also referred to as the swash zone). The water moving up the relatively steep beachface, known as the wave uprush, eventually loses its momentum and that portion of the water that doesn't percolate into the beachface returns back down the beachface slope under the influence of gravity (known as the backwash).

Breaking waves are usually of three distinct types – plunging, spilling, and surging. As pictured in Figure 34A, plunging waves take on a cylindrical shape (Hawaii Five-O type) and fall abruptly down with considerable force, usually breaking in the near vicinity of the beachface. The surfaces of spilling waves, on the other hand (shown in Figure 34B), start disintegrating into foaming lines fairly far offshore and continue to foam their way to shore, gradually decreasing in height as they go. Surging waves approach shore and wash directly up the beachface without breaking. As a generalization, for waves of any given height, the slope of the nearshore zone

FIGURE 34. Two types of breaking waves. (A) Plunging wave on the Mexican coast near Acapulco. The idiot running into the surf to collect a sediment sample is co-author Hayes (summer 1961). Photograph by Robert L. Folk. (B) Spilling waves at Cowell Ranch Beach, CA on 16 August 2008.

determines the type of breaking wave present. Plunging waves occur more commonly where the nearshore slopes are quite steep and spilling waves where the nearshore slopes are flat, in a relative sense. Under some conditions, a single wave may have the characteristics of both plunging and spilling waves.

When offshore swell waves are viewed from the air, their crests are usually linear and parallel, forming what are known as wave fronts. In many situations, the wave fronts bend as they approach underwater features, this bending being referred to as refraction and diffraction. Wave **refraction** is defined by the U.S. Army Corps of Engineers (2002) as "any change in direction of a wave resulting from the bottom contours." As noted by Woodroffe (2002), *the speed with which the wave travels in shallow water is related to water depth. Those parts of the wave which enter shallow water move forward more slowly than those in deeper water, causing the wave crests to bend toward alignment with the bottom contours.* During the process of refraction, a segment of the wave front may bend around an island or bar, as is illustrated Figure 35. In the purest sense, **diffraction** is defined as the lateral spread of wave energy along the wave crest from a point (an example is the wave formed when a pebble is dropped into a pool of still standing water). The U.S. Army Corps of Engineers (2002) defined diffraction as

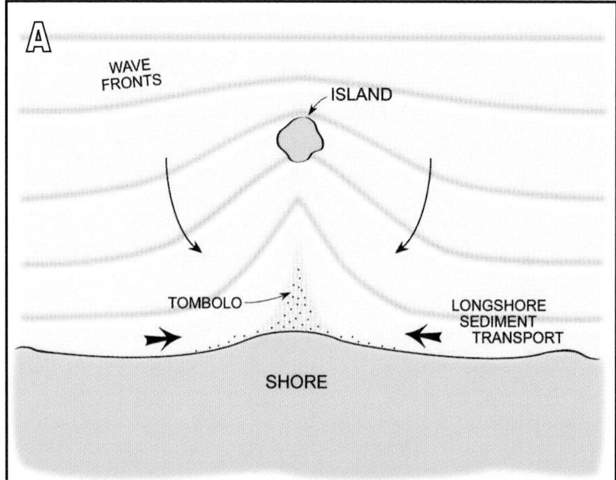

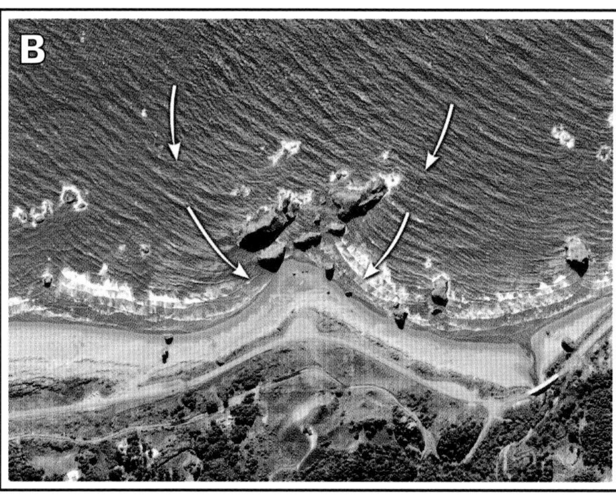

FIGURE 35. Illustrations of the creation of a tombolo. (A) Wave fronts bending around an offshore island, generating two opposing directions of sediment transport in the lee of the island. Where the two directions of transport meet, a triangular offshore projection of sand called a tombolo is formed. (B) A similar process at Pistol River State Scenic Viewpoint south of Gold Beach, OR. The arrows, which indicate the direction of motion of the waves, are oriented perpendicular to the wave crests, or fronts. NAIP 2011 imagery, courtesy of Aerial Photography Field Office, USDA.

the phenomenon by which energy is transmitted laterally along a wave crest. When part of a wave in interrupted by a barrier, the effect of diffraction is manifested by propagation of waves into the sheltered region within the barrier's geometric shadow. During incidents of wave diffraction near the shore, a segment of the wave front may bend into something like a navigational channel or into the lee of an offshore breakwater or reef, relative to the rest of the front. As a result, an arc, or bow, develops in the wave front as it moves into the channel or into the lee of the offshore structure.

When waves approach and break in the surf zone adjacent to the beach, they set up two very distinct, and in some cases very strong, nearshore currents – longshore currents and rip currents. As illustrated in Figure 36 (Upper Left and Lower Left), for a wave that approaches a beach at an angle, which many do, two types of water flow are generated that have the capability of moving sediment along shore. That type of longshore sediment transport is a key consideration with regard to the stability of beaches. An illustration of wave-generated longshore sediment transport is given in Figure 35, which shows waves bending into a sheltered zone in the lee of an offshore island. The result of this process

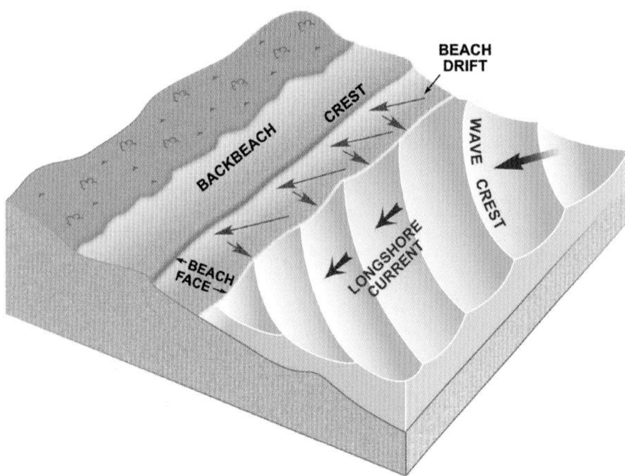

FIGURE 36. Longshore sediment transport. (Lower Left) General model for the generation of beach drift and longshore currents by waves approaching the beach at an angle. Under these conditions, the sand grains move in a sawtooth pattern along the beachface, and a strong longshore current is developed just off the beachface. (Upper Left) View looking north along St. Joe spit in northwest Florida showing waves approaching the beach obliquely out of a southerly direction. This high-tide picture shows the swash of the waves across the beachface. Arrows indicate the direction of the longshore currents generated by the obliquely approaching waves. Note waves breaking on the offshore bar with strong longshore currents generated in the trough on the landward side of the bar. Photograph taken in April 1973. (Upper Right) Rhythmic topography along the shore of Matagorda Bay, Texas. Arrows indicate direction rhythmic bulges of sediment are moving through time, also under the influence of southeasterly waves. The water was relatively calm on the day this photograph was taken in August 1979. The white sediments are coarse-grained shell material (mostly oyster shells).

is the creation of a triangle-shaped sand body, called a **tombolo**, in the lee of the island at the point where two opposing sediment transport directions converge.

When waves come ashore at an angle, the wave uprush generated by the breaking wave flows at an angle to the slope of the beachface, which allows any sand grains moved by the shallow flow of the wave uprush to move obliquely up the beachface. On the other hand, the water in the backwash moves in a perpendicular sense directly down the slope of the beachface under the influence of gravity. Therefore, with each wave, sand is moved along the shore a certain distance, depending upon the size of the wave and its angle of approach (bigger waves and higher angles increase the distance of movement of the sand grains). Thus, the sand grain moves along the shore in a sawtooth pattern called beach drift (Lower Left diagram in Figure 36). That type of sediment transport is restricted to the beachface itself. At the same time, the obliquely approaching wave is piling up water in the surf zone that flows away from the direction of the approaching wave, thus generating a current that runs parallel to the shore (called the longshore current). On many beaches, the longshore current is somewhat restricted to a trough that separates the beachface from a nearshore bar (Figure 36 – Upper Left). These longshore currents can be quite strong, commonly exceeding a knot or two, and, in concert with the beach drift, they play a major role in transporting sediment along the shore. Such currents can also carry swimmers far away from where they entered the water. [NOTE: A knot is equal to one nautical mile per hour. A nautical mile = 6,076 feet. Therefore, a knot is 15% faster than 1 mile per hour.]

Determinations of the volume of sediments that move along the shore in the longshore sediment transport system can be done in several ways, including: 1) Measuring volumes of sand trapped by breakwaters or jetties; 2) Determining volumes of sand dredged from harbors or other collection sites; and 3) Calculations using formulas based on wave and sediment data. For the Columbia River littoral cell on the Oregon and Washington Coasts, Ruggiero et al. (2006) reported that 1.3 million cubic yards of sand per year are transported from south to north.

Wang, Ebersole, and Smith (2002) determined that plunging waves create much greater sediment suspension in the breaking zone than spilling waves, as well as more sediment transport in a wider and more energetic swash zone. This explains the general rule that beaches are more apt to accrete (build out) when the waves are spilling than when they are plunging.

Another mode of longshore sediment transport occurs when evenly spaced bulges of sand along the beachface move down the shore like the sinusoidal loops one might throw along a rope. Those bulges of sand, called **rhythmic topography** (see photograph in the Upper Right of Figure 36), are typically spaced 10s to 100s of feet apart. They are created by waves that approach the beach at an oblique angle. The side of the sand bulge facing directly into the wave approach direction orients perpendicular to that direction, with the beachface being aligned parallel with the approaching wave crests, the most stable configuration possible. However, the distance the sediment on the exposed beachface of the bulge can extend offshore is limited by water depth in most situations. Sand is therefore either lost from the outer edge of the bulge into deeper water or is transported to its lee side. The result of this process is that the exposed face of the bulge erodes while sand accumulates on its sheltered side. Therefore, the bulge moves along the shore in the downdrift direction (direction in which the individual sand grains are moving along the shore within the longshore sediment transport system), as indicated in Figure 36 (Upper Right). Such "bulges of sand" move at different rates depending on the size and angle of approach of the waves. When these types of features were first described, some of the early authors used the term "cusp-type sand waves" in recognition of the fact that the individual sand bulges protrude out from the beach somewhat in the same fashion as the very common, smaller perpendicular protrusions off the beach called beach cusps (Figure 37). Those two features do not form in the same way; thus, these days, the more definitive term, rhythmic topography, is used for the larger of the two features. Rhythmic topography occurs on the sand beaches of Oregon and Washington with some degree of regularity.

Rip currents (Figure 38) form most commonly along coasts where waves come straight on shore (not at an angle usually, but there are some exceptions to this general rule). Also, there is typically some degree of reflection of the waves off the beachface. These currents, which typically flow in a perpendicular direction straight off the beach, tend to be very regularly spaced. Based on extensive research, most rip currents owe their formation to the occurrence of edge waves – standing waves with their crests normal to the shore, according to Bowen

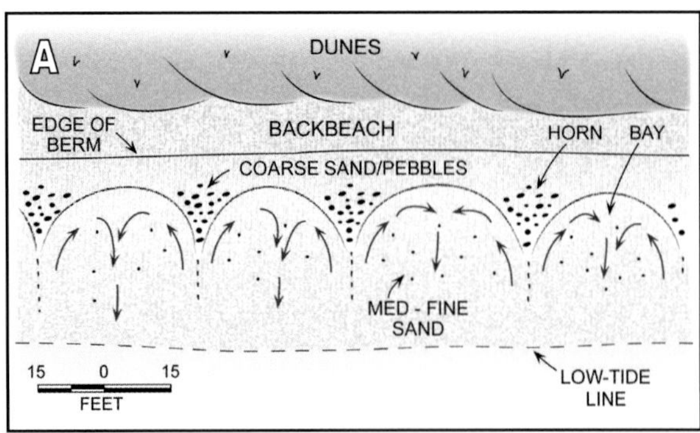

FIGURE 37. Beach cusps. (A) The general model. (B) Cusps on Point Reyes Beach, CA on 18 August 2008.

and Inman (1969) and Komar (1976). Therefore, the direction of motion of the water within these edge waves is parallel to the shore and their crests are perpendicular to the shore. Furthermore, they are "trapped" against the beach, decreasing in amplitude exponentially in the offshore direction. If, indeed, they are standing waves, it would follow that the rip currents would be positioned at equally spaced distances, which they typically are, where the waves have their maximum amplitude, not at the nodal points.

Rip currents may also be associated with some types of nearshore bars and off certain man-made structures (e.g., groins, that are discussed later), but the most common ones appear to be related to wave reflection and edge waves. At any rate, one of the authors of this book, Hayes, swears he has seen edge waves with his own eyes off a steep beach in Ghana, but Michel is a non-believer in edge waves, because she says "I have never seen one."

Beach cusps are relatively small-scale features (a few tens of feet maximum) that develop on the scalloped seaward margin of a mound of sand deposited on the intertidal beach called a berm (Figure 37A). They consist of a regularly spaced sequence of indented bays and protruding horns. They are thought to be primarily depositional features (i.e., on the seaward flank of a prograding beach) that form as a result of the interaction of shore-normal standing waves (edge waves) with shore-parallel, reflective waves, the same general process that produces rip currents (but at a smaller scale). That mode of origin was originally proposed by Guza and Inman (1975). There are other theories that challenge the edge-wave solution, believe it or not.

Both rip currents and beach cusps are very common on the Pacific Northwest Coast, and even a casual observer touring along the coastal roads will see some. If you want to study them at home, check out the imagery of several of the long, straight sand beaches in Oregon and Washington (on Google Earth) and you will see plenty.

Komar (1997) has shown that, where strong rip currents create a deep channel perpendicular to the

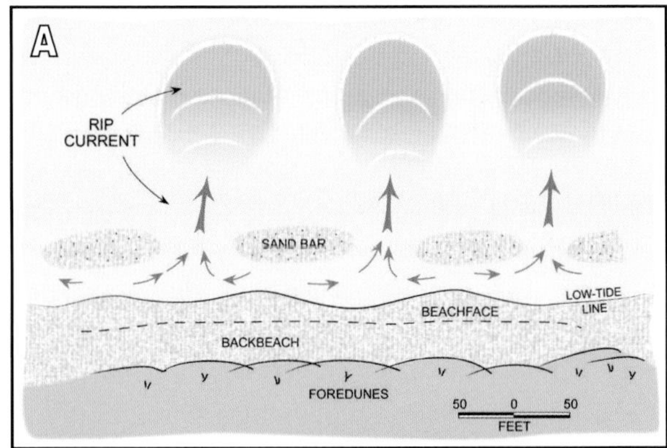

FIGURE 38. Rip currents. (A) General model. (B) Rip currents along the shore of Cape Cod, MA in April 1972. Note the onshore welding of sand bars in the space between the rip currents.

beach, winter storm waves can travel up the channel and break right on the beach, causing localized beach erosion "hot spots." Such erosion is most serious after previous storms have already eroded the beach and narrowed the dune field.

Rip currents can be very dangerous. If by chance you ever get caught in one while swimming off the beach, make sure you swim parallel to the beach so as to escape the current, which has the potential of carrying you far offshore. KMOnews.com made the following report on 5 August 2014:

WARRENTON, Ore. (AP) - Alarmed by the number of people swept into the ocean during the past two months along the Oregon and Washington Coasts, the Coast Guard is warning beachgoers to be aware of possible dangers. Since July 3, four of those cases have ended in fatalities. Cmdr. Bill Gibbons is chief of response for Coast Guard Sector Columbia River. He notes that those caught in NW ocean currents are often visitors to the area who don't realize the dangers. Gibbons says in one case, a victim was only in water up to his knees when he was knocked down by a wave and pulled out into the ocean.

Waves on the Coasts of Oregon and Washington

The daily sea conditions on most coasts are related primarily to global wind patterns. Inman and Jenkins (2005a) made the following points regarding the interaction of global wind patterns with the World Ocean:

1) A tremendous amount of kinetic energy is expended through momentum exchange between the atmosphere and the ocean (at the sea surface);

2) The prevailing wind systems cause all large bodies of water to have windward (prevailing wind blowing onshore) and leeward (prevailing wind blowing offshore) coasts;

3) The Pacific Coasts of the Americas are, in general, windward coasts and those on the Atlantic are leeward coasts; and

4) As illustrated in Figure 39, the general wind flow in the Northern Pacific Ocean region is clockwise around a semi-permanent, mid-latitude area of high pressure called the North Pacific High. Consequently, when these winds reach the Pacific Northwest Coast, they are blowing out of the west or northwest. In the Southern Hemisphere, the wind flow is counterclockwise. Consequently, when these winds reach the coasts of Chile and Peru, they are blowing out of the southwest.

These wind patterns are augmented significantly by coastal storms in determining the average size of the waves, which have a major input in shaping the coast. Coastal storms during *El Niños* usually generate by far the largest waves that occur along the Pacific Northwest Coast. As noted by Ruggiero et al. (2005), *during the 1997/1998 El Niño, U.S. Pacific Northwest beaches experienced monthly mean winter wave heights up to 1.0 m* (3.3 feet) *higher than usual, and wave directions having a more southwest approach than typical.*

Before reviewing more specific information on the Northwest Pacific waves, a new term, **significant wave height**, should be defined. This measure portrays the average height of the waves that comprise the highest 33% of waves in a given sampling period. As it turns out, an experienced person making visual observations will most frequently report heights equivalent to the average of the highest one-third of all waves observed. Furthermore, it is the larger waves that are of the most concern to sailors, coastal engineers, and the like.

Tillotson and Komar (1997) reported that, for the Pacific Northwest Coast, significant wave heights were 4.1 to 5.7 feet (with periods of 5-10 seconds) in summer and 6.6 to 9.8 feet (with periods of 10-20 seconds) during winter. During major winter storms, significant wave heights in deep water were 20 to >23 feet, which would translate into waves 30 to >33 feet high when breaking on the shore.

Allan and Komar (2006) analyzed data from offshore wave buoys over the period 1976-2002 and found that the winter significant wave heights have increased in Washington, Oregon, and northern California, but not in south-central California. Figure 40 shows the data for the wave buoys off Washington and Oregon. These data were used to determine that over the 25-year period, at the Washington buoy: 1) the annual average wave height increased by 2 feet; the average winter wave height increased by 2.6 feet; the five largest winter wave heights increased by 7.9 feet; and the maximum winter significant wave height increased by almost 9 feet. The reason for that increase is still a matter of debate, with ocean-wide oscillations and increased water temperatures in the last quarter of the century being possibilities. But, as Inman and Jenkins (2005a) pointed out, it is unclear what portions of those increased wave heights are also associated with global warming (Graham and Diaz, 2001).

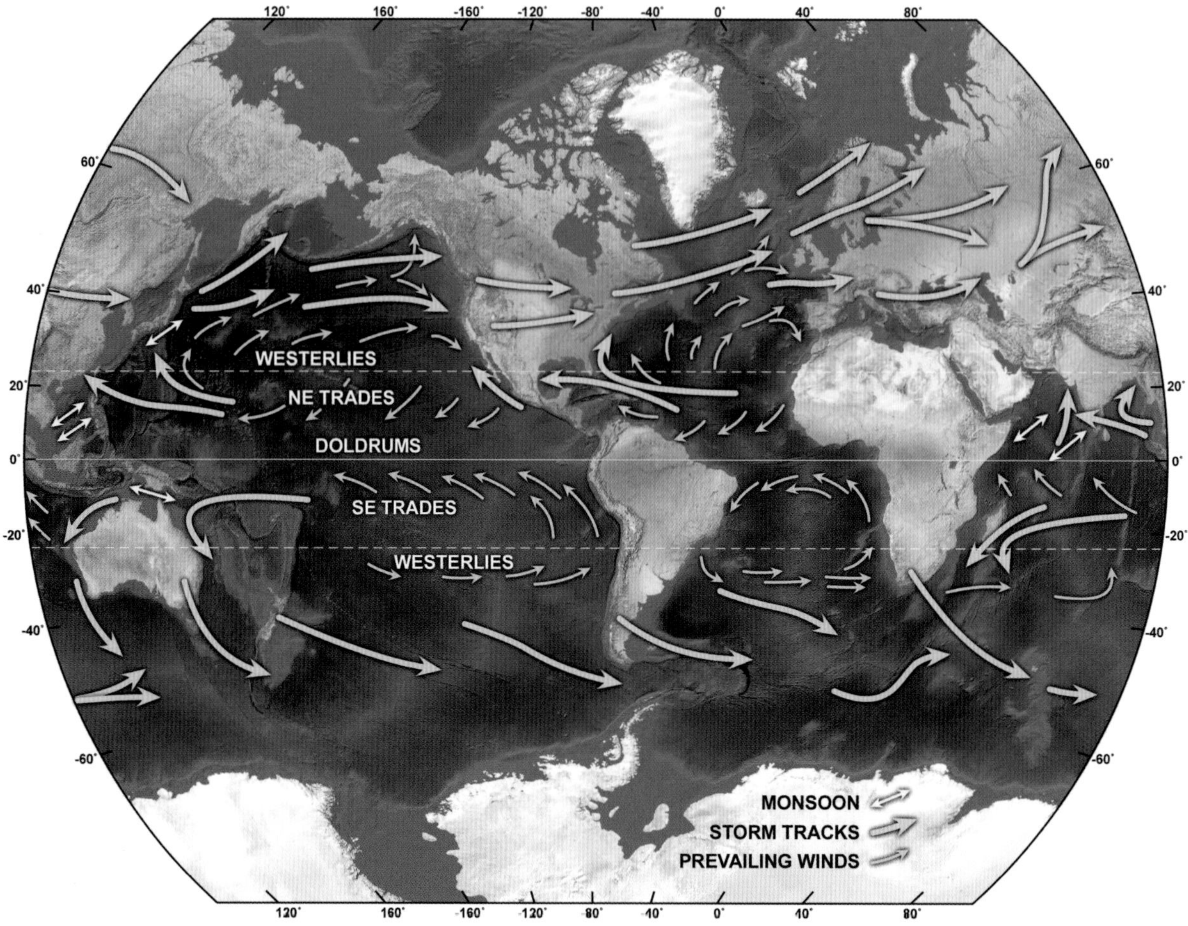

FIGURE 39. Prevailing winds and storm tracks for the World Oceans (modified after Inman and Jenkins, 2005a).

As those data indicate, the wave climate offshore from the Oregon Coast is one of the more extreme in the World, with winter storm waves commonly ranging from 15-30 feet and extreme heights of 50 feet (Allan and Komar, 2002; 2006). These large waves are because the storm systems coming from the North Pacific travel over fetches that are typically a few thousand miles in length and are also characterized by strong winds, the two factors that account for the development of large wave heights and long wave periods. Those storm systems originate near Japan or off the Kamchatka Peninsula in Russia, and typically travel in a southeasterly direction across the North Pacific towards the Gulf of Alaska, eventually crossing the Pacific Northwest Coast (Allan, Geitgey, and Hart, 2005.)

Much of the data just discussed are from buoys located several miles offshore. Ruggiero et al. (2006) summarized wave records for two inshore buoys at Grays Harbor and the Columbia River showing similar but less intense wave conditions. Mean summer wave heights were between 3.3 and 6.6 feet and from the west-northwest. Mean winter wave heights were between 6.6 and 13.2 feet and from the west-southwest.

MAJOR COASTAL STORMS

In the North Pacific Ocean, the most intense wave action results from cyclogenesis (development or intensification of a low-pressure center, or cyclone) over the poleward-flowing western boundary current, the Kuroshio Current, that is matched by its twin, the Gulf Stream Current, along the western boundary of the Atlantic Ocean. As noted by Inman and Jenkins (2005a), *these warm water jets carry equatorial heat to higher latitudes and produce strong temperature gradients that spin-up intense storms.* In winter in the North Pacific, cold and dry Siberian winds blow out of the west across the Kuroshio Current. And, *these winds produce a series*

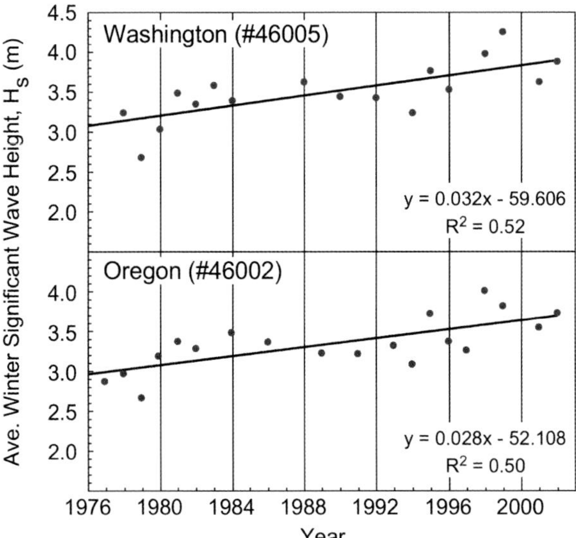

FIGURE 40. Average winter significant wave heights.

of cyclonic cold fronts that collectively have long fetches and generate the high waves for the Pacific Coast of North America. When they are far out in the ocean, those storms produce swell along the coast from California to Washington, and once they approach the shore, they generate waves that strike the coast along their tracks.

The largest storms by far occur during the *El Niños* that take place when the typical weather patterns discussed above and illustrated in Figure 39 are abruptly and severely modified on a global scale. Conditions during *La Niñas* are quite the opposite from those during *El Niños*. Those two globally coupled, ocean-atmosphere phenomena are complex and the subject of intense research at the present time for obvious reasons. A brief outline of the characteristics of the two events is given below based on Inman and Jenkins (2005b) and Griggs, Patsch, and Savoy (2005).

First, the *El Niño*-Southern Oscillation (ENSO), the technical name of the *El Niño*:

1) The name is from the Spanish for "the little boy," in reference to the Christ child, because its appearance is usually noticed by the arrival of unusually warm water off the coast of Peru, that tends to occur around Christmas time;

2) They are signaled by anomalies in the pressure fields between the tropical eastern Pacific and Malaysia. Those anomalies yield an atmospheric signature called the Southern Oscillation Index (that reflects the monthly or seasonal fluctuations in the air pressure between Tahiti and Darwin, Australia). If the index is negative (i.e., lower pressure at Tahiti and higher pressure at Darwin), then an *El Niño* or warm ENSO event occurs;

3) Another associated condition that develops during *El Niño* events is a generally high-pressure anomaly over the northern half of North America and a huge low-pressure anomaly over the eastern North Pacific Ocean that hugs fairly close to the North American Coast (mostly in winter). This forces storm-track enhancement along the southern edge of the low that aims the storm tracks towards the Pacific Northwest Coast.

4) Three factors—the relatively low pressure over the Pacific Ocean, the thermal expansion of warm water, and a weakening of the trade winds (during normal conditions, these strong winds generate a very strong westerly flow of the surface waters to the west in the equatorial region)—cause an increase in water level on the Pacific Northwest Coast during *El Niños* (as much as 8-12 inches; see Figure 31). Such an increase in water level leads to significant coastal flooding and sea cliff erosion;

5) During those events, unusually heavy rains typically accompany the increased number of winter storms; and

6) Major *El Niño* events occurred in the years 1790-93, 1828, 1876-78, 1891, 1925-26, 1982-83, and 1997-98 (Davis, 2001), and more recently 2009-2010 and 2015-2016. The three strongest *El Niño* periods on record were in 1982-83, 1997-98, and 2015-16.

On the other hand, some of the characteristics of the *La Niñas* include:

1) *La Niñas* generally show effects opposite to those of the *El Niños*. For example, a decrease in water levels along the Pacific Coast is possible. Rainfall is greatly diminished and the trade winds are usually strengthened;

2) Another associated condition that develops during *La Niña* events is a generally high-pressure anomaly over the eastern North Pacific that also hugs fairly close to the North American Coast. This forces storm tracks to come ashore further north than those that occur during *El Niño* events, leading to a higher incidence of large magnitude storms impacting the Pacific Northwest; and

3) There have been strong *La Niña* episodes during 1973-74, 1975-76, 1988-1989, 1999-2000, 2007-08, and 2010-11.

Inman and Jenkins (2005b) noted *that hypotheses explaining the causes of unusual ENSO events are numerous and range from the extrusion of hot lava on the seafloor to variations in intensity of the sun's radiations. Most climate modelers favor a complex set of changes in the interaction of the atmosphere and ocean that cause fluctuations in trade winds, monsoon intensity, and sea surface temperature.* No kidding?! Well, enough is now known about the two systems that it is possible for weather forecasters to predict their arrival and relative intensity. However, there is much uncertainty about how climate change may affect *El Niño* intensity, with studies that show it will intensify, weaken, or have no change.

The impacts of *El Niño* and *La Niña* storms on the Pacific Northwest Coast were summarized in some detail by Komar et al. (2000) and Allan and Komar (2002). Some of their major points include:

1) The most significant climate event in terms of its erosion impacts has been the occurrence of a major *El Niño*;

2) During the "super" *El Niños* of 1982-83 and 1997-98, there was extensive beach erosion along the entire West Coast, primarily occurring in hot-spot zones;

3) The hot-spot erosion at the south ends of the littoral cells to the north of the headlands, that is depicted in Figure 41, occurs because *unusually large quantities of the beach sand within the littoral cells are transported northward during an El Niño winter*;

4) Particularly noteworthy is that an *El Niño significantly elevates the measured tides, on average by about 0.30 m* (1 foot; see Figure 31) *but achieving a maximum difference of about 0.60 m* (2 feet) *between the measured and predicted tides*; and

5) The 1997-98 *El Niño* erosion was significantly expanded during the following winter of 1998-1999, *when several storms generated waves that exceeded what then had been projected to be the 100-year extreme event [later to be reanalyzed and increased to a 16-m* (52.8 feet) *significant wave height]*. The March 1999 *La Niña* storm was one of the most severe storms to hit the Pacific Northwest since 1962 (Allan and Komar, 2002). Komar (pers. comm., 2018) noted that the heights of the individual largest measured waves would be on the order of 1.5 to 1.9 times larger than the significant wave height, placing the highest waves at 74-93 feet.

The 2015-2016 *El Niño* was likely the strongest in the last 145 years; off the California and Oregon Coasts, significant wave heights ranged from 26-36 feet and maximum wave heights ranged from 39-62 feet (Barnard et al., 2017). Wow! They also reported that the amount

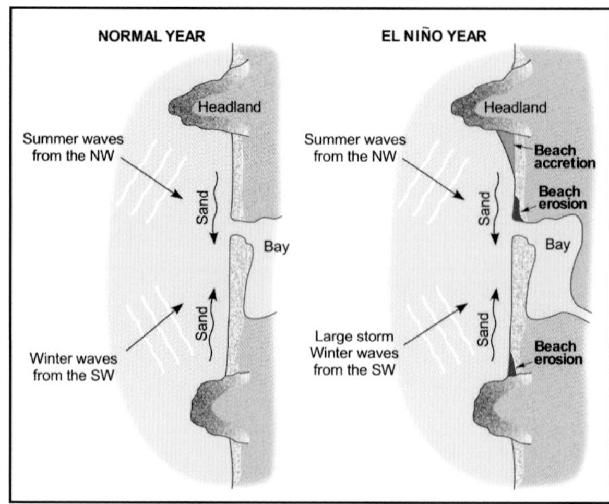

FIGURE 41. Illustration after Komar (1998b) that shows how longshore sand movements within Oregon's littoral cells during a normal year contrast to that during a major *El Niño*, when strong winds from the southwest generate northward transport of the sand along the beach along with the development of 'hot spot' erosion sites (shown in red).

of winter shoreline erosion during this *El Niño was the highest on record, with erosion 76% above the normal winter shoreline retreat, 27% higher than any other winter and easily eclipsing the El Niños of 2009-2010*. They noted that such high erosion did not result in erosion of dunes and cliffs along the Pacific Northwest Coast however, likely because of less erosion during the previous two mild winters so there was more beach sand to counteract the erosion.

Scientists who are studying how climate change might affect the frequency and intensity of such storms in the future say that these storms could shift north, which would increase the impacts of winter storms in Oregon and Washington (National Research Council, 2012). There are also hints that the largest waves have been getting higher and the winds have been getting stronger over the last few decades. In December 2018, a new study was published that clearly shows that global warming will enhance both the amplitude and frequency of eastern Pacific *El Niño* events, with widespread climatic consequences (Ham, 2018). So, such storms are likely to cause more beach erosion, landslides, and flooding in the future, unless efforts to reduce greenhouse gases are successful, soon!

7 MAJOR LANDFORMS OF THE COAST

RELATIVE ABUNDANCE OF COASTAL HABITAT TYPES

The distribution of the different coastal habitat types of Oregon and Washington has been mapped in detail by our group (NOAA/RPI, 2015). The coastal types are ranked on a scale of 1-10 based upon their sensitivity to oil spills and other considerations. The scale we use is called the Environmental Sensitivity Index (ESI) (Table 2). On this scale, exposed rocky shores are ranked 1 and

TABLE 2. Coastal habitat types mapped for the Oregon and Washington coasts (NOAA/RPI, 2015).

ALL COMPARTMENTS - CALIFORNIA BORDER TO CANADIAN BORDER			
ESI VALUE	**ESI DEFINITION**	**LENGTH (miles)**	**% ABSOLUTE LENGTH**
1A	Exposed Rocky Shores	170	5.6%
1B	Exposed, Solid Man-Made Structures	16	0.5%
2A	Exposed Wave-Cut Platforms in Bedrock	255	8.4%
2B	Exposed Scarps and Steep Slopes in Clay	1	0.0%
3A	Fine- to Medium Grained Sand Beaches	359	11.8%
3B	Scarps and Steep Slopes in Sand	12	0.4%
4	Coarse-Grained Sand Beaches or Bars	59	1.9%
5	Mixed Sand and Gravel Beaches or Bars	164	5.4%
6A	Gravel Beaches or Bars	182	6.0%
6B	Riprap	136	4.5%
6D	Boulder Rubble	35	1.1%
7	Exposed Tidal Flats	566	18.6%
8A	Sheltered Rocky Shores and Sheltered Scarps	86	2.8%
8B	Sheltered, Solid Man-Made Structures	13	0.4%
8C	Sheltered Riprap	90	2.9%
8F	Vegetated, Steeply-Sloping Bluffs	1	0.0%
9A	Sheltered Tidal Flats	819	26.9%
9B	Vegetated Low Banks	68	2.3%
10A	Salt- and Brackish-Water Marshes	970	31.9%
10B	Freshwater Marshes	303	10.0%
10C	Swamps	330	10.9%
10D	Scrub and Shrub Wetlands	150	4.9%
	Total ESI Length	4,341	142.9%
	Absolute Shoreline Length	3,039	100.0%

wetlands are ranked 10. This system, which was first created by us during a study of Lower Cook Inlet, Alaska in 1976 (Hayes, Michel, and Brown, 1977), is now used worldwide for oil-spill contingency planning purposes.

The following is a description and brief discussion of some of the more common habitat types that occur on these coasts:

1) Exposed rocky shores (ESI = 1A): The term exposed relates to the size of the waves breaking onshore, which in this case, is quite large because of the exposure of the outer coasts in these areas to the prevailing westerly winds that blow across the open Pacific Ocean and to common large swell waves as well as fairly numerous *El Niño* storms. Those shores are mostly vertical rock cliffs carved by waves (Figure 42A) and occur along 5.6% (170 miles) of the coastal area surveyed.

2) Exposed wave-cut platforms in bedrock (ESI = 2A): These are also highly exposed coastal habitats where the waves have carved out a flat, eroded platform (typically rocky) that is usually uncovered in the lower intertidal zone during low tide (Figure 42B). Such platforms occur along 8.4% (255 miles) of the coastal area surveyed.

3) Fine- to medium-grained sand beaches (ESI = 3A): See Figure 6 for the grain-size scale. These types of beaches are common along the open ocean shores (Figure 43A). They occur along 11.8% (359 miles) of the coastal area surveyed.

4) Mixed sand and gravel beaches (ESI = 5): In order to be placed in this class, a mixed sand and gravel beach must be composed of at least 20-30% gravel (Figure 43B). This habitat type occurs along 5.4 % (164 miles) of the coastal area surveyed.

5) Gravel beaches (ESI = 6A): A relatively uncommon coastal habitat (Figure 43C). These beaches occur along 6.0% (182 miles) of the coastal area surveyed.

6) Exposed tidal flats (ESI = 7): Being exposed to both wave action and strong tidal currents, these tidal flats are typically composed of sand (Figure 44). They occur along 18.6% (566 miles) of the coastal area surveyed.

7) Sheltered tidal flats (ESI=9A): Because these tidal flats are sheltered from wave action and strong tidal currents, they are usually composed of mud (Figure 45A). They occur along 26.9% (819 miles) of the coastal area surveyed. They typically occur in the upper reaches of the larger bays. They represent such a large length because of the many tidal channels in the upper parts of the bays.

8) Salt- and brackish-water marshes (ESI = 10A): Like the sheltered mud flats, these habitats are not typically exposed to significant wave action (Figure 45B). They occur along 31.9% (970 miles) of the length of the coastal area surveyed. Also like the sheltered mud flats, they are fairly common at the heads of the larger bays where many tidal channels contribute to the total shoreline length.

More detailed coverage of the geographic occurrence of the different habitat types is presented in the discussion of the four Compartments presented in Section III.

ROCKY COASTS

Introduction

Rocky coasts are some of the most breathtakingly beautiful places in the World, with those on the Pacific Northwest Coast taking a backseat to few others. The combination of exposed rock cliffs and wave-cut rock platforms occur along 14 % of the shores of Oregon and Washington. According to Davis and Fitzgerald (2004), rocky shores make up 75% of the World's coast, a number that may be a little high, in our estimation. Whatever the exact number, no doubt this is a common coastal type, a fact that people living on trailing-edge coasts, like those in the southeastern U.S., might be surprised to learn.

First, we will start with some excuses as to why coastal geomorphologists do not know more than they do about rocky coasts:

1) The changes that take place on some rocky coasts may be almost imperceptible over a human lifespan, unlike sand beaches where significant changes may take place within a few minutes. However, noticeable changes may occur rapidly during a major *El Niño*, especially if the rock cliff is composed of easily erodable rocks or is subjected to a landslide;

2) Therefore, the most significant changes usually occur during intense coastal storms;

3) Difficulty in making wave measurements and diving observations, because of the rough terrain and the necessity of making such measurements during storms; and

4) Difficulty of access to the high and precipitous cliffs.

FIGURE 42. Rocky coasts. (A) Exposed rock shore at Cape Flattery, WA in May 2013. Credit: NOAA OAR 2013 Photo Contest. Photo by Sophie McCoy, CC by 2.0. (B) Wave-cut rock platform at Birdpoint near Tillamook Head, OR on 1 June 2011. Credit: Oregon ShoreZone, CC by SA.

If, after reading this section, you still hunger for more knowledge about rocky coasts, we refer you to books by A.S. Trenhaile, *The Geomorphology of Rock Coasts* (1987) and T. Sunamura, *Geomorphology of Rocky Coasts* (1992), the only two books we know about devoted entirely to that topic.

Rock Cliffs

Nearly vertical rock cliffs mark the landward boundary of the shore along most rocky coasts. With a few exceptions, the cliffs are created by wave erosion and they erode at rates of less than an average of 3 feet/year globally, according to Stephenson and Kirk (2005). Usually, erosion rates differ spatially along the shore, with the erosion typically taking place in brief spurts during storms.

Assuming a relatively constant wave size and storm

FIGURE 43. Oregon beaches. (A) Example of a medium- to fine-grained sand beach (Cannon Beach). Photo taken on 6 August 2012 by Walter J. Sexton. (B) Sand and gravel beach named Sporthaven, on the south side of the jetty at Brookings. Photo taken on 1 August 2014 by Walter J. Sexton. (C) Example of a gravel beach at Cape Arago, south of Coos Bay, that shows subtle imbrication of the individual clasts (mostly boulders) as well as armoring. Photo taken on 7 July 2014, courtesy of Hugh Shipman (gravelbeach.blogspot.com).

FIGURE 44. Oblique aerial view of the entrance to Netarts Bay, OR in June 2011. Note the exposed sandy tidal flats. Courtesy of Oregon ShoreZone, CC by SA.

frequency along a given stretch of shore, as is the case for the Pacific Northwest Coast, the rate of retreat is dependent for the most part on the type of rocks that make up the cliff. The principal physical process involved in creating rock cliffs is wave-induced erosion, which acts in the following ways:

1) A process called wave hammer, the high shock pressures generated against the cliff by the breaking waves;

2) At high tide during storms, the breaking waves may trap air between the cliff face and the oncoming water. *As the wave collapses, extremely high pressures are instantaneously produced and the air is compressed on the rock surface .... This process is particularly important when air pockets are compressed into the crevices of rocks, leading to the enlargement of the crack and ultimately to a shattering of the rock* (Davis and FitzGerald, 2004);

3) Quarrying, a process whereby pieces of rock, which may vary in size from tiny fragments to large blocks, are removed from the cliff face by the breaking wave;

4) If the water in the breaking waves contains a lot of suspended sediments, ranging in size from sand to pebbles, or perhaps even up to small cobbles, the result is intense abrasion of the rocks in the cliff, especially if they are soft sedimentary rocks; and

5) Because the waves tend to break further offshore during low tide, the zone of greatest wave erosion at the cliff edge is during neap and spring high tides, and particularly during the even higher water levels generated during winter storms.

The rock type along the coast is an important factor in cliff formation and erosion by wave attack. The more resistant volcanic basalts form headlands, whereas the weaker mudstones, siltstones, and terrace sandstone erode at higher rates and tend to be fronted by beaches. The beach acts as a buffer to cliff erosion, except during high wave conditions, when the beach is eroded and the waves break directly on the cliff face.

An important process of cliff formation and retreat is a process called mass wasting, which is defined as a mechanism whereby weathered materials are moved from their original site of formation to lower lying areas under the influence of gravity. Examples of mass wasting include: a) When pieces of rock tumble down the rock scarp (called rock fall); and b) Deep-seated landslides, that require suitable geological conditions to occur (Griggs and Trenhaile, 1994; to be discussed in more detail later). The results of such mass wasting are debris piles of soil and fragments of rocks up to boulders in size at the base of the cliffs. That accumulation of material,

FIGURE 45. Mud flats and marshes. (A) Mud flat. (B) Salt marsh photo courtesy of Oregon ShoreZone, CC by SA.

sometimes referred to as talus, is typically moved either down the beach by longshore sediment transport or offshore during big storms.

Once such material is removed from the base of the cliffs by wave action, the waves can impact the cliffs directly during later storms. Obviously, for cliffs where such talus or a wide beach typically is present, the retreat rates will be slower than that experienced by cliffs composed of similar material but without such protection. Furthermore, mass wasting will be accelerated where the waves cut an eroded notch into the base of the cliffs.

As the rock cliffs retreat, a dazzling array of erosional remnants can be left behind on the rock platforms offshore of the cliffs, namely sea stacks and sea arches. Sea stacks are isolated pinnacles of rock that are typically taller than they are wide with tops at a somewhat lower elevation than the top of the adjacent rock cliff (Davis and FitzGerald, 2004). The stack is usually composed of rocks that are significantly more resistant to wave erosion than their neighboring rocks in the formerly existing cliff. Figure 46 shows examples of sea stacks along the outer coasts of Oregon and Washington. It is important to point out that sea stacks are one of the favorite nesting and roosting sites for birds, most notably cormorants and common murres, as well as haulouts for sea lions and seals.

The presence of joints in the rocks (plane surfaces along which the rocks have split apart, usually during one or more of the ever present tectonic movements along this coast), especially vertical joints oriented perpendicular to the shore, favor the formation of sea stacks. Joints also speed up the erosion process along the cliff face.

Sea arches and sea caves are most commonly formed in flat-lying sedimentary rocks, or layered volcanic deposits, because the separate rock layers commonly have different degrees of resistance to wave erosion. In a situation where the top layer is more resistant than some underlying layers, parts of the underlying layers may be removed by wave action to form a sea cave (Figure 47A). In the situation where a narrow peninsula composed of layered rocks with a more resistant top layer projects offshore, wave refraction around the headland may focus the wave erosion such that two caves eroded into the opposite flanks of the peninsula meet, forming a sea arch. As the waves continue to erode the peninsula and enlarge the sea arch, eventually the top of the arch

FIGURE 46. Sea stacks. (A) The Needles at Cannon Beach, OR 16 April 2013. Photo by Tiger635, CC BY-SA. (B) Sea stacks near La Push, WA. Photo by Joe Holmes on 18 June 2008. (C) Medium- to fine-grained sand beach (with offshore sea stacks) located approximately 2 miles north of Pistol River, OR (between Brookings and Gold Beach). Photo by Walter J. Sexton on 2 August 2014.

FIGURE 47. Sea caves and sea arches. (A) Sea Lion Caves near Haceta Head, OR. Photo by Art Bromage, CC BY-SA, 29 July 2007. (B) Natural arch at Second Beach, WA. Photo by Joe Holmes, 18 June 2008.

collapses, creating a sea stack. Sea arches are common on the outer coasts of Oregon and Washington (Figure 47B).

Rock Platforms

As the eroding rock cliff retreats, a flat rock platform can be left behind, some of which have sea stacks. In the geological literature, several pleas have been entered to call these features shore platforms, because when non-conforming, unrefined field people like ourselves call them wave-cut rock platforms, we are implying that we know how they originated.

We have been doing this (calling these things wave-cut platforms) during mapping projects along many coasts since 1976, and the only place we felt uncomfortable about calling them "wave-cut" was for the very wide platforms in some of the relatively sheltered portions of Lower Cook Inlet, Alaska, where it is probable that the waves are augmented considerably in the erosion process by the abrasion of the rocks by sediment-bearing ice layers (during the winter time). A good case can be made that rock platforms are formed, at least in part, by other processes in the coldest polar regions and in the tropics. However, a discussion of those features in this book seems unwarranted.

Besides the obvious fact that the waves have to be pretty big to erode rock platforms, it is also necessary that sea level remains fairly constant for a significant amount of time. How significant? That depends on how fast the cliff is retreating and what processes may account for the downcutting of the platform (also mostly related to wave action, in our opinion).

Raised Marine Terraces

Raised marine terraces on the Oregon and Washington coasts are located landward of presently eroding rock cliffs, resembling giant stairsteps up the side of the mountains (Figure 48). Those flat or slightly seaward dipping terraces usually are at least a few hundred yards wide, and they are flanked on their landward sides by a somewhat degraded, former wave-cut rock cliff.

A process for the formation of elevated terraces was suggested by Davis and FitzGerald (2004) and Griggs, Patsch, and Savoy (2005), based on studies in New Guinea by Chappell (1983). According to that hypothesis, the following steps are required for the formation of a series of raised platforms on shores subject to episodes of uplift like those on the U.S. Pacific Coast (see general model in Figure 49):

1) A period of stable sea level occurs during which the waves erode the rock cliffs, the cliffs retreat, and a wide rock platform is created. During the relatively stable highstand of the past 5,000 years, rock platforms up to a few hundred yards wide have been carved in some places;

2) Sea level drops to a lowstand for thousands of years, as has happened at least four times during the major glaciations that have occurred within the past 650,000 years;

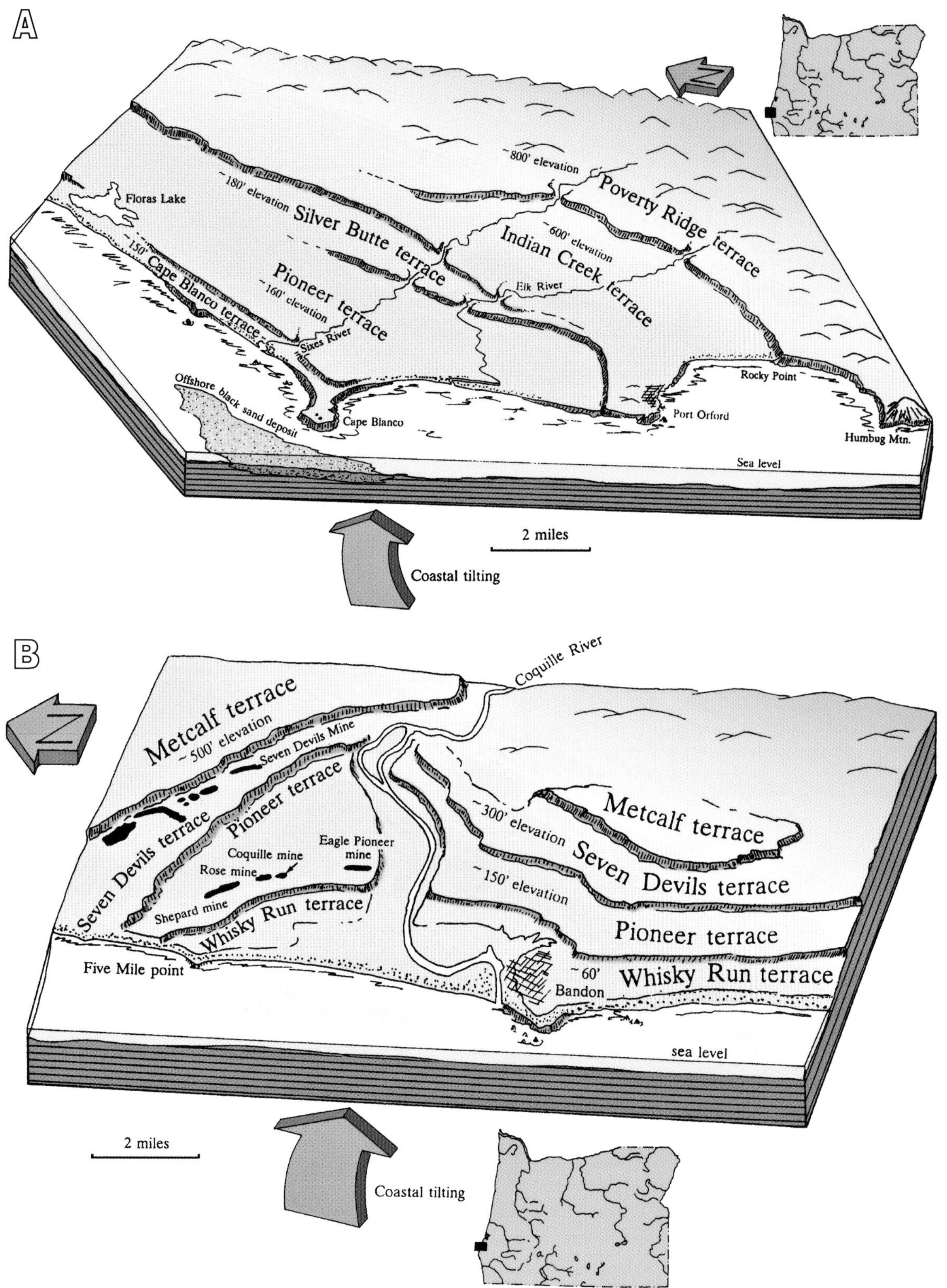

FIGURE 48. Raised marine terraces in the vicinity of: (A) Cape Blanco, OR, and (B) Bandon, OR. Modified after Orr and Orr (2000).

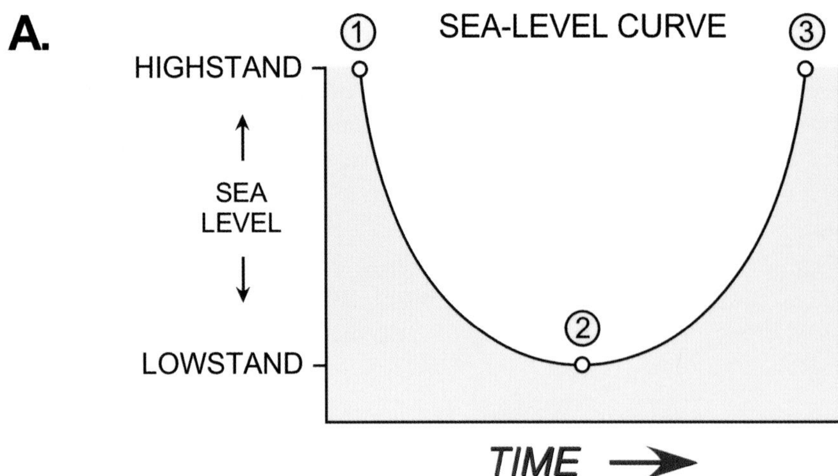

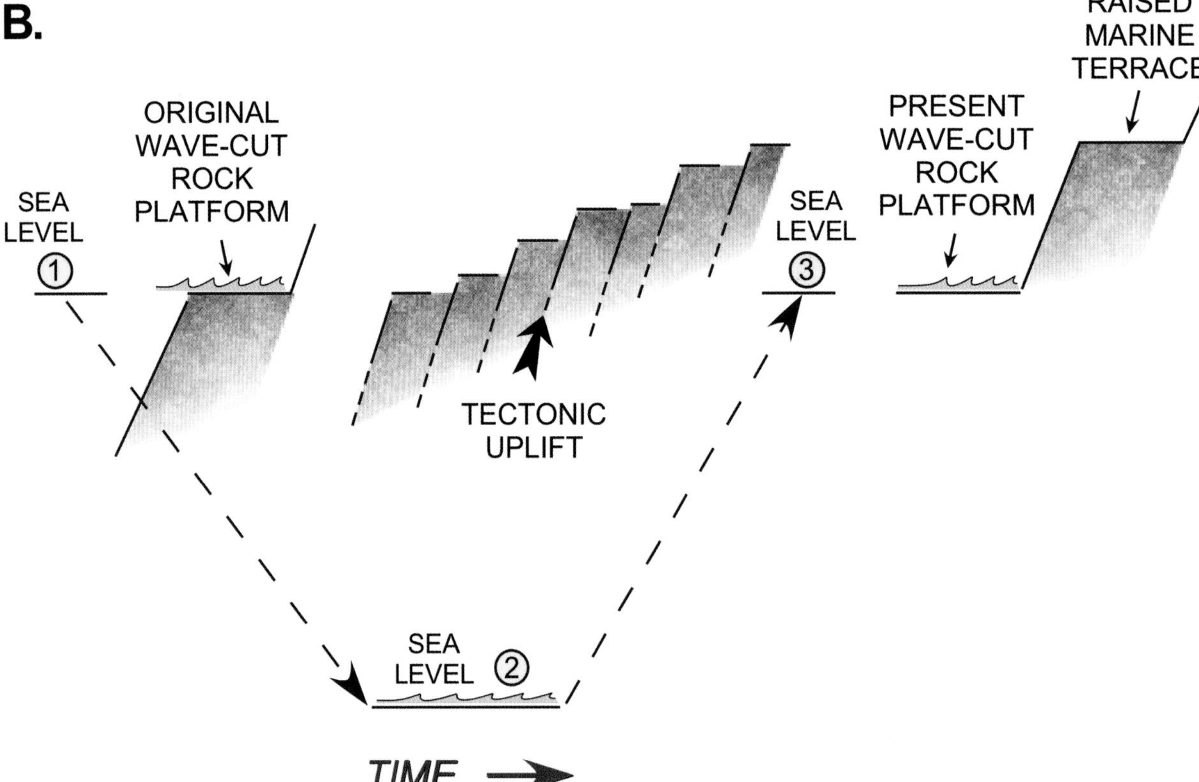

FIGURE 49. Model of the formation of raised marine terraces (after Davis and FitzGerald, 2004). (A) Typical sea-level curve showing the change from a highstand to a lowstand and back to another highstand through time. (B) The different steps required for the formation of a raised marine terrace, in this case on the Pacific Coast of the U.S. During the first highstand (1), a wide wave-cut rock platform is carved by the waves. The land is slowly uplifting, but the downcutting action of the waves compensates for it. During the interval between the two highstands, while the sea level is lower, eventually retreating all the way to the lowstand level (2), the land continues to rise, either steadily or sporadically. By the time the second highstand occurs (3), the land has uplifted a considerable vertical distance (200 feet is common on the Oregon Coast). Finally, a second wave-cut rock platform develops a few hundred feet or so below the former one that has by then been converted to a raised marine terrace.

3) For a systematic arrangement of raised marine terraces to develop as illustrated in the general model in Figure 49, it would be necessary for the adjacent land to continue to rise at a fairly consistent rate (and faster than sea level), during both lowstands and highstands. As shown in Figure 25, the Oregon Coast between Crescent City and Coos Bay is rising, relative to mean sea level, thus marine terraces only occur in this region;

4) When sea level comes back during the next highstand, the sea level at that time is considerably lower relative to the adjacent land mass than when the first rock terrace was created, because the landmass has continued to rise. By that time, the original wave-cut rock platform has been raised as much as a hundred feet or maybe considerably more, and it is located landward of this later highstand shoreline. If we use 1.5 feet/thousand years, obviously if sea level were to be down 100,000 years before it comes back up, the elevated terrace would be 150 feet above the new shoreline, assuming that the sea comes back to the same level it was at before. The youngest of the raised terraces in Oregon are at the elevation of about 60 feet (Figure 48B); and

5) This cycle can be repeated over and over as long as sea level continues to go up and down in a somewhat regular fashion, as it did during the Pleistocene epoch and, of course, the land continues to rise at some relatively consistent rate.

Komar (2010) noted the following about raised marine terraces on the Oregon Coast:

The long-term net tectonic uplift of the Oregon Coast, together with the cycles of sea level rise and fall during the Ice Ages, has given rise to marine terraces that in places form stairways up the flanks of the Coast Range. Along the south coast the highest and oldest terrace in the series reaches elevations up to nearly 500 m (1,650 feet). *The lowermost terrace, extending along much of the coast, dates back about 80,000 years. The Pleistocene terrace sands (former beaches and dunes), together with the underlying Tertiary mudstones and siltstones, are being eroded by the waves, forming sea cliffs that back sand beaches. Many of Oregon's coastal communities are situated on this nearly level terrace; cities such as Cannon Beach, Lincoln City and Newport have suffered property losses as the cliffs progressively retreated.*

Orr and Orr (2000) discussed raised marine terraces along the southern Oregon Coast between Coos Bay and Port Orford that reach a maximum height of 1,600 feet. Two of those areas are illustrated in Figure 48. The Cape Blanco terrace, dated at 80,000 years, covers much of the point at Cape Blanco, while the extensive Pioneer terrace, dated at 103,000 years, continues on the landward side of the Cape toward Port Orford. Dates for some of the other terraces shown in Figure 48 include Whisky Run at 83,000 years, Seven Devils at 124,000 years, and Metcalf at 230,000 years.

Regarding these terraces, Orr and Orr (2000) remarked that *terrace development is not a case of simple step-like uplift of the coast with stationary stages during which the terraces were cut. The terracing reflects both the eastward tilting of the Coast Range block as well as the rise and fall of the sea during Pleistocene time.* They also pointed out that the highest and oldest terraces are *thoroughly dissected by erosion making them difficult to trace laterally along the coast. By contrast, lower, younger terraces are in much better condition and form natural flat surface ideal for human cultural use.*

SAND BEACHES

A River of Sand

Beaches will form on any shore where clastic particles somewhat resistant to wave abrasion are available and there is a site for sediment accumulation protected from extreme daily erosive wave action, such as wave reflection off rocky cliffs. On the coasts of Oregon and Washington, those clastic particles range from sand-sized fragments of rocks with a wide range of composition up to boulders. The discussion in this section will focus on sand beaches.

A sand beach is an accumulation of unconsolidated sand-sized sediment transported and molded into characteristic forms by *wave-generated* water motion. The landward limit of the sand beach is the highest level reached by average storm waves, and the seaward limit is the lowest level of the tide. In Oregon and Washington, the beach's landward limit typically is either a line of wind-blown sand dunes or at the base of a rock scarp. As is indicated in Table 2, the approximately 3,039 miles of the coast mapped by our team (including all the tidal channels in marshes), the landward side of beaches (including beaches of all sediment classes) abuts against a man-made structure (vertical concrete seawalls and riprap) along about 254 miles of it (8.4%).

Most sand beaches are a virtual "river of sand" with the sand being in constant motion up and down the beachface and along shore. We call the zone within which the sand moves back and forth the zone of dynamic change (ZODC), that is illustrated in Figure 50 for areas where the beach is backed by a line of wind-blown sand called foredunes. The foredunes act as a storage area for sand between storms. Thus, the landward limit of the zone of dynamic change is the point at which the foredunes are eroded during "normal" storms (not extreme *El Niño* storms). Accordingly, during those "normal" storms, the sand formerly deposited on the intertidal beach, as well as up to several feet of the foredunes, is usually transported into the subtidal area, which, in some areas, contains an offshore bar (see Figure 50; the red zone represents the portion of the beach/dune sand that potentially could be eroded during a typical, moderate storm). The seaward limit of the ZODC is what engineers call the closure point, the most seaward distance for sand to move in an offshore direction under relatively *normal adverse conditions* (again, not during extreme *El Niño* storms!!). As indicated in Figure 50, all of the sand grains in motion do not stay within the zone of dynamic change at a specific geographic site on the beach for an infinite amount of time, but most likely will eventually "escape" from the zone by one of three primary types of "leakage:"

1) Alongshore out of the zone, under the influence of longshore sediment transport by wave-generated currents as illustrated in Figure 36;

2) Offshore beyond the point of closure during major storms, such as during extreme *El Niño* events; and

3) Onshore by wind action, beyond recovery by offshore winds or by erosion during storms.

Another way sand grains may "escape" from the river of sand for a long period of time is for the shore to build out seaward to the extent that more lines of foredunes develop until they are located far enough landward to

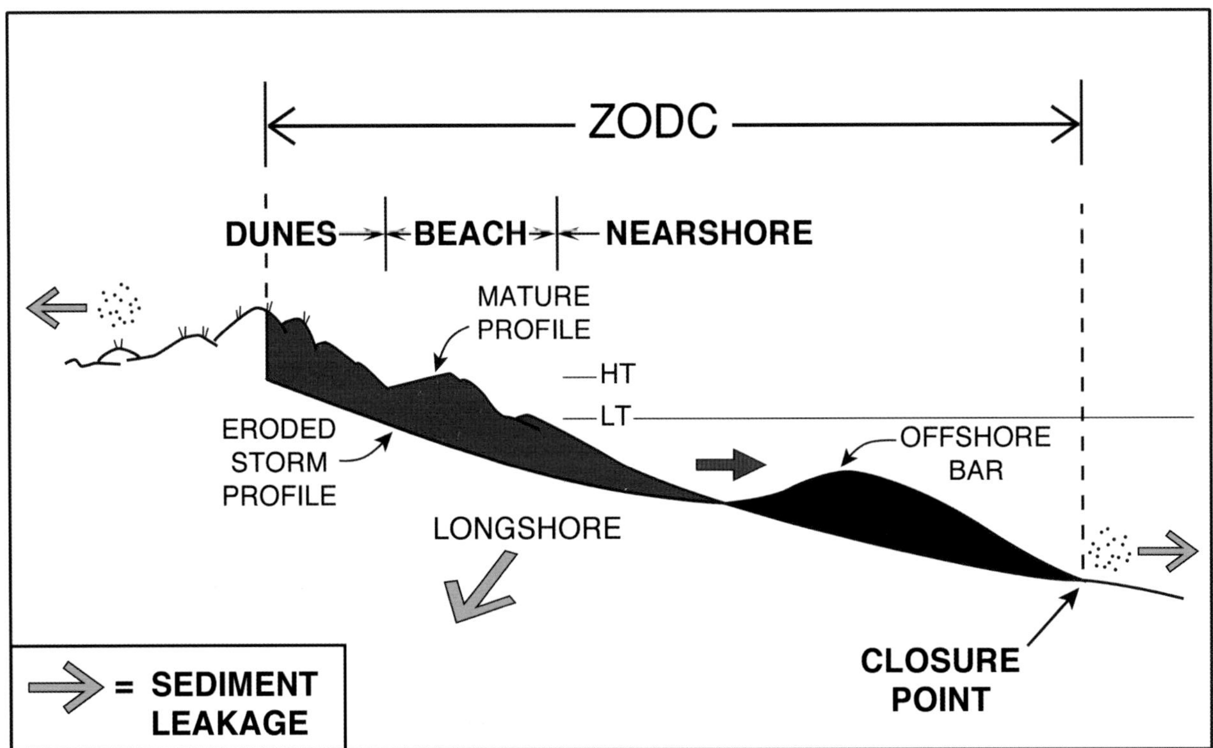

FIGURE 50. Zone of dynamic change (ZODC). The area in red depicts sand that may be eroded from the beach area during storms. Much of this sand is deposited on the offshore bar (area in black). In between storms, much of the eroded sand is returned to the beach/dune area. The orange arrows indicate the three mechanisms by which the sand might be lost: 1) Carried offshore beyond the closure point during large storms, such as hurricanes or extreme *El Niños*; 2) Blown landward out of the foredune area during periods of high wind activity; and 3) Transported alongshore by beach drift and longshore currents as illustrated in Figure 36.

be out of the reach of normal storm waves. For example, North Beach Peninsula that forms Willapa Bay, is over 2 miles wide on its south end, and it was built by successive beach ridges with sand from the Columbia River in the last 5,000 years or so, since sea level stabilized. However, shores of that type are relatively rare on the coasts of Oregon and Washington.

The Beach Profile

When coastal geologists or engineers study sand beaches to determine such issues as their potential for erosion, they usually measure a topographic profile across the beach surface using a variety of surveying techniques. Some studies use a remote sensing system called LiDAR (Light Detection and Ranging) to produce topographic images of the beach and nearshore zone. That system is more expensive than simple horizontal-leveling surveys carried out by two or three junior surveyors; therefore, they are usually spaced at least six months apart. Others use aerial photography to track the position of the shoreline over time.

Such beach surveys are usually repeated over a period of months or years to determine the cycle of erosion and/or deposition on a beach. Studies of that type have been carried out along many of the coasts of the World, including the coasts of Oregon and Washington (e.g., Aguilar-Tunon and Komar, 1978; Ruggiero et al., 2005; Allan, Geitsey, and Hart, 2005).

For purposes of discussion, the different morphological components of the intertidal sand beach need to be defined.

1) Berm: A wedge-shaped sediment mass built up along the shore by wave action. It typically has a relatively steep seaward face and a gently-sloping landward surface (2-3 degrees). A sharp peak (berm crest) usually separates the two oppositely slopping planar surfaces on the top of the berm. There frequently are two berms present during all tidal phases except during unusually high spring tides, a high berm, the most landward, oldest berm, and an active berm, the most seaward and most recently activated berm (Figure 51A).

2) Beachface: The zone of wave uprush and backwash during mid- to high-tide. It usually slopes seaward at an angle of 5-10 degrees and is commonly the seaward face of the berm. However, the berm may be washed away during a storm. In that situation, the beachface would be a flattened-out, eroded surface subject to the uprush and backwash of the waves.

3) Low-Tide Terrace: A relatively flat surface that is located seaward of the toe of the beachface and slopes offshore at a small angle (2-3 degrees). The landward margin is usually near mean sea level and the seaward margin near mean low water. The sand that composes the terrace is commonly water saturated throughout, essentially representing the intersection of the water table with the beach surface. However, during periods of recovery of the beach following an episode of erosion, intertidal bars migrate across the low-tide terrace in a landward direction. An intertidal trough is commonly located on the landward side of the elevated intertidal bar.

Komar (1976) presented a generalized profile of the littoral zone intended to represent the outer coasts of Oregon and Washington (illustrated in Figure 51A). In this general model, the backshore is typically composed of depositional berms. An erosional scarp typically lies on the seaward edge of the berms. The foreshore area contains the beachface, whereas the inshore zone contains an alongshore trough and an alongshore bar.

A key aspect of any beach profile is the slope of the beachface. Beach research as early as during World War II focused on that issue, due to its relevance to beach landings by amphibious watercraft. Those early researchers found that the slope of the beachface in any given area will attain an equilibrium profile (i.e., a nearly constant slope of the beachface) under what they referred to as "steady" conditions, that is, relatively unchanging, day-to-day wave characteristics.

During storms, sand beaches are commonly eroded down to a flat surface that essentially parallels the water table in the beach with sand typically being transported off the beach to an offshore bar (as suggested in Figure 51B). When sediment returns to the beach after the storm, the depositional profile attained by the beachface is steeper than the beach was immediately following the storm. Komar (1976) explained why:

Due to water percolation into the beachface and frictional drag on the swash, the return backwash tends to be weaker than the forward uprush. This moves sediment onshore until a slope is built up in which gravity supports the backwash and offshore sand transport.

Several field studies and wave tank experiments have shown that coarser-grained beaches, especially those composed of gravel, have steeper slopes (of the beachface) than those composed of finer-grained

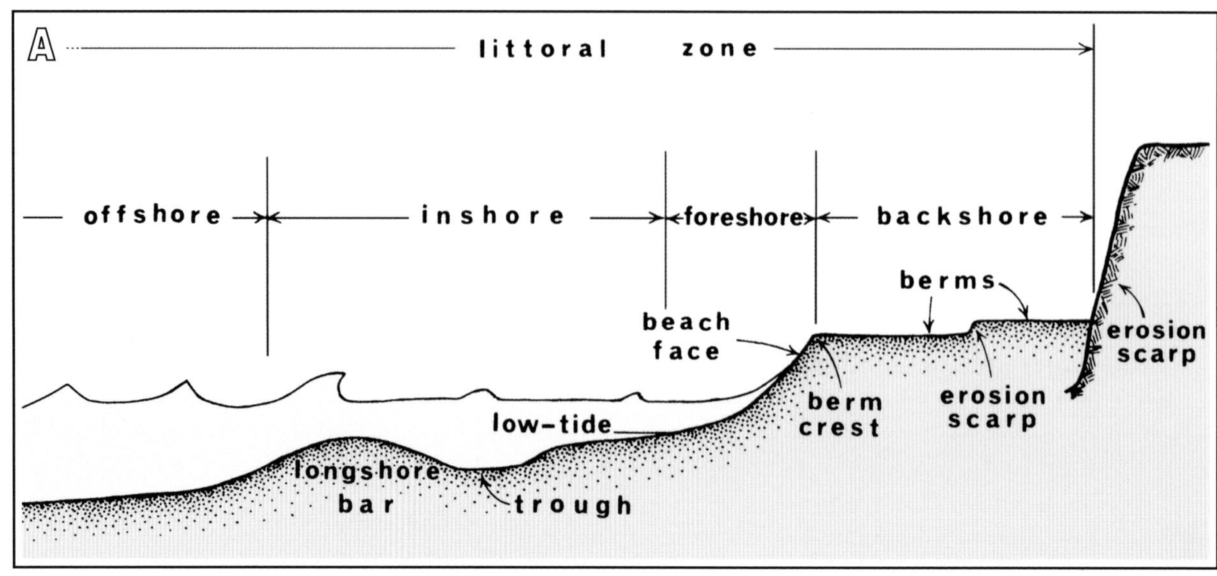

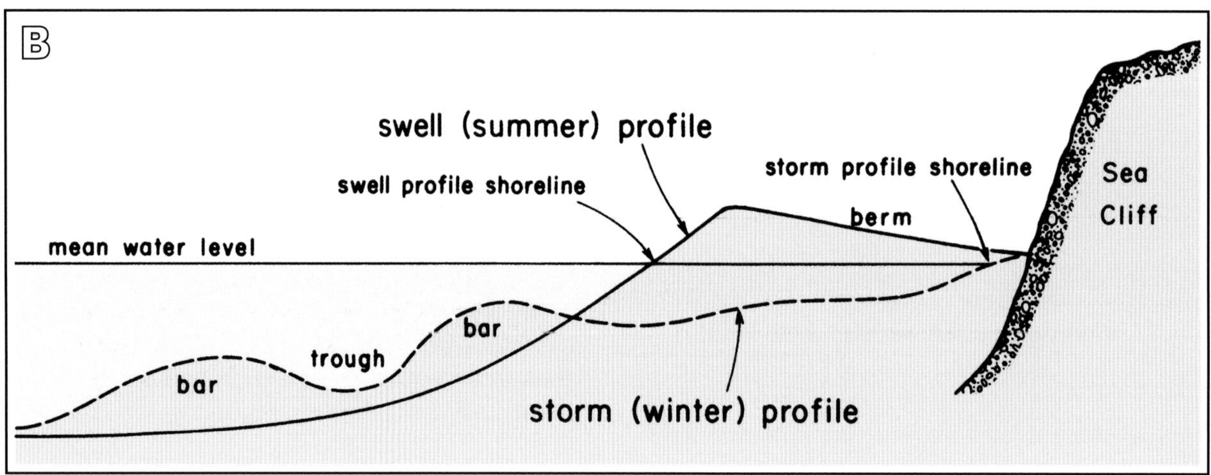

FIGURE 51. (A) Terminology used to define aspects of the beach after Komar (1976). As noted by Allan, Geitgey, and Hart (2005), the backshore is typically composed of some combination of a foredune, a foredune backed by a dune field, or a bluff. The erosion scarp typically lies on the seaward edge of the foredune or bluff. (B) General pattern of seasonal changes in beach profiles related to seasonal variations in wave conditions. Modified after Komar (1976).

sediments (Wiegel, 1964). Coarser-grained beaches allow greater percolation of the water brought to the beachface in the wave uprush than finer-grained beaches, thus reducing the strength of the return of the backwash. Consequently, a steeper beachface is required to maintain an equilibrium profile on coarse-grained beaches (gravity-supported transport). On sand beaches of comparable grain sizes, those beaches consistently exposed to large waves tend to be flatter than sheltered beaches exposed to smaller waves, because larger waves lose relatively less of their total volume to percolation into the beach sediment than smaller waves, therefore they produce stronger backwash, which tends to flatten the profile.

The Beach Cycle or "Normal Seasonal Changes"

The concept of the cyclical change of beaches from a flat, erosional profile in winter to a wide, depositional berm in summer had been well ingrained in both the popular and scientific literature for many years following research carried out in the 1940s. The concept originated from detailed studies on the Southern California Coast (e.g., Shepard, 1950b; Bascom, 1954). The idea was that, generally speaking, erosional storm waves are more common in the winter, and flatter, depositional swell waves are more common in the summer – hence the terms "winter beach" and "summer beach" were commonly used. Beach profiles measured at Scripps pier in La Jolla in the 1940s commonly showed wide beaches in the late summer and fall and narrow beaches in the winter and spring.

One of the more clearly expressed discussions of the winter/summer beach profile concept was presented as follows by Komar (1992) in a treatment of the sand beaches of Oregon:

Beaches respond directly to the seasonal changes in wave conditions. Then he introduced a diagram that illustrates this process (Figure 51B). *The beach is cut back during the winter months of high waves, when sand is eroded from the shallow underwater and from the beach berm (the nearly horizontal part of the beach profile that is above the high-tide line). This eroded sand moves to deeper water, where it then accumulates in offshore bars, approximately in the zone where the waves first break as they reach the coast. Sand movements reverse during the summer months of low waves, moving back onshore from the bars to accumulate in the berm.* From there, he went on to note that there may be exceptions to this pattern at times, making the following statement. *Although this cycle between the two beach-profile types is approximately seasonal due to changing ocean waves, the response is really one of high storm waves versus low regular swell waves. At times, low waves can prevail during the winter, and the beach berm may actually build out, although not generally to the extent of the summer berm. Similarly, should a storm occur during the summer, the beach erodes.*

Despite its early acceptance regarding West Coast beaches, the winter/summer beach profile concept has not withstood the test of time for certain other parts of the World. In 1969, co-author Hayes and his colleague Cy Galvin observed a striking contrast between beach cycles on the U.S. East Coast and those presented in the literature for the West Coast. Eroded winter profiles and strongly depositional summer profiles are not a dominant theme on northern Atlantic beaches, particularly in the New England area. The beach cycle in that area is controlled by the passage of individual storms. This cycle also appears to reflect, at least to some degree, less severe wave climate and relatively smaller seasonal changes in wave climate on the northern Atlantic coast. Mean wave heights and periods are 19 inches and 6.9 seconds along the northern Atlantic coast, considerably smaller values than those for the West Coast. These dissimilarities in wave climate result to some extent from differences in their locations on western and eastern edges of oceans with predominantly westerly winds, with the winds blowing predominantly onshore on the West Coast and offshore on the East Coast. However, Ed Clifton (pers. comm.) commented that *local winds are less a factor than the fact that the largest storms (typhoons) develop on the western side of the ocean, and that every large weather system in the western Pacific (and the Gulf of Alaska as well) generates long period swell which arrives on the U.S. West Coast.* As alluded to earlier, the largest waves that occur along the West Coast are generated in one of two ways: 1) When major *El Niño* storms cross the coast; or 2) When some of the largest swell from especially huge offshore storms reach the coast.

In the 2002 Coastal Engineering Manual, published by the U.S. Army Corps of Engineers, the following statement appears: *the seasonal onshore-offshore exchange of beach and nearshore sediments ... is now known to apply best to swell-dominated coasts, such as the U.S. West Coast, where wave climate changes seasonally.* Therefore, the general consensus today is that the terms "winter beach" and "summer beach" should not be used with reference to erosional and depositional cycles on most beaches, although many published discussions of West Coast beaches still use these terms, or else refer to "normal seasonal changes," which presumably means more-or-less the same thing. Paul Komar, working on the Oregon Coast, suggested the terms **storm profile** (flat beach) and **swell profile** (well-developed berm).

However, a study by Ruggiero et al. (2005) on the outer Washington Coast favored the concept of winter and summer beaches. The area they studied is called the Columbia River Littoral Cell, extending from Seaside, Oregon to Grenville, Washington. In their discussion of seasonal climate and beach change variability, they noted that *the seasonal cycle in waves and water levels along the* Columbia River Littoral Cell *forces a seasonal cycle in beach morphodynamics, with offshore and northerly sediment transport resulting in beach erosion during the winter and onshore and weak southerly sediment transport dominating beach recovery in the summer months.* In their conclusions, they gave some specific numbers, pointing out that winter conditions resulted in beach lowering on the order of 1.7 feet while the shore retreats horizontally between 33 and 132 feet, with recovery taking place in the summer months.

Smaller-Scale Physical Sedimentary Features

Because of the relatively large tides on the coasts of Oregon and Washington, when you visit the sand beaches at low tide you may be able to walk across an intertidal zone that contains an abundance of small-scale physical sedimentary features that will no doubt arouse your curiosity. Although geology has long been considered to be primarily a field science, the greatest breakthrough in the understanding of these sedimentary features was the result of laboratory flume studies carried out by hydraulic engineers (e.g., Simons and Richardson, 1962). Those workers demonstrated that the sediment bed of a laboratory flume (box-like trough with a sand bottom and a mechanism to generate water flow across the sand bed) could be made to pass through the following sequence of features (called **bedforms**) by simply increasing the velocity of the water flowing across the sand bed or by changing the depth of the water:

flat bed → small ripples → large ripples (megaripples) → washed-out megaripples → flat bed → antidunes

While doing this work, they came up with an empirical relationship that dictates which of these different bedforms are active at any given time. It is called the Froude number, which equals:

$$\textit{Froude number} = \frac{V}{\sqrt{gD}}$$

V = velocity
g = gravitational constant
D = water depth

Therefore, as the water velocity increases, the Froude number also increases, and as depth increases, the Froude number decreases. Incidentally, in flowing water, sand grains on the underlying sediment surface start to move down current at velocities of about 10 inches/second.

The first part of the sequence up to the washed-out megaripples occurs when the value of the Froude number is <1.0. That part of the sequence is called the lower flow regime. By increasing the Froude number to values >1.0, by either increasing the velocity of flow or decreasing the water depth (see formula), the sand bed goes into the upper flow regime, where either flat beds or antidunes are formed.

The lower flow regime bedforms are asymmetrical, with the sand grains moving in discreet steps by rolling or bouncing up the gently sloping upcurrent side of the bedform and sliding down the steeply dipping surface on the downcurrent side (called the slip face). Thus, these forms move along in the direction the current is flowing, sometimes several feet during a single tidal cycle. On a beach, it is rare to see any bedforms larger than **ripples** (spacing between crests less than 2 feet) and they usually occur in the intertidal troughs on the landward side of the intertidal bars, where the water depths are greater than elsewhere on the beach (Figure 52A). **Megaripples** (spacing between crests 2-18 feet) are seen more commonly in tidal channels, where the water is deeper and the currents are stronger (Figures 52B and 52C). The mode of sediment transport over these types of bedforms is illustrated in Figure 53. Larger bedforms called **sand waves** (spacing between crests greater than 18 feet) do occur on this coast, but they are mostly found in deep tidal channels or on sand bodies in tidal inlets.

As you walk across the beach and tidal sand flats at the coast, you will notice some dramatic changes in the nature of the lower flow regime bedforms (at the scale of ripples), even though the changes in the topography are quite subtle. As the diagram in Figure 54 shows, there is a progression of the three-dimensional shapes from straight-crested to undulatory ripples at conditions of relatively low flow strength. As flow strength increases, the ripples take on a cuspate shape. That shape is very common in the intertidal troughs on the beach. As the flow approaches the upper flow regime (Froude number >1.0), the ripples convert to a planed-off rhomboid shape just before the bed goes flat. The photographs in Figure 55 illustrate those bedform types. You will undoubtedly see the pattern illustrated in Figures 54 and 55 repeated time and time again during your walks over the intertidal bars during low tide. The pattern will almost always be present where the high-tide waters were flowing down

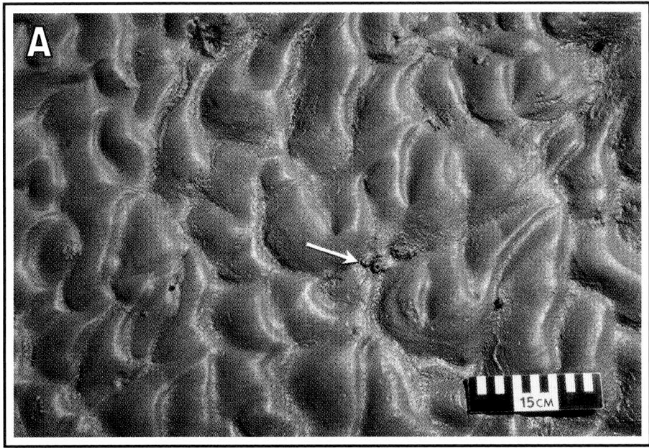

FIGURE 52. Ripples and megaripples. (A) Ripples. Arrow points to exposed ghost-shrimp burrow. Scale is ~ 6 inches. (B) Megaripples. The ebb current that created these megaripples was flowing from the lower left to the upper right. (C) Close-up view of a megaripple shown in photograph B. Note the high angle of dip of the beds created by the migration of the bedform by the mechanism illustrated in Figure 53. The machete is about 2.5 feet long. The ebb current that created this bedform was flowing from left to right.

very subtle slopes on the sand surface.

The most common bedform type under upper flow regime conditions is the **flat bed**, over which the flow of the sand grains is more-or-less continuous, streaming in long streaks across the sediment surface. This is the most common type of sand surface occurring on sand beaches, where water depths are typically small and wave-generated water flow is generally fast. Thus the Froude number is typically >1.0 (see formula again). Under conditions where the velocity is faster than normal, for example, when the water flows down a subtle slope (e.g., over the crest of an intertidal bar or down a more steeply sloping high beachface), a feature called an **antidune** is formed. Under that condition, the sand surface develops a sinusoidal shape and a relatively thin layer of water flows in phase across the top of these features, that are linear and parallel with a typical spacing of about 15-20 inches.

As shown in Figure 56A, the water flowing down the slope on the downcurrent side of the sinusoidal antidune picks up sand grains from that relatively steep slope only to drop them out again as the water flows up the slope on the upcurrent side of the next antidune in line. A slice through a preserved antidune, shown in Figure 56B, illustrates the deposited sand on the steep upcurrent side of the antidune. At first glance, as you watch antidunes in motion, you experience an optical illusion that the sand grains are moving upstream against the current. Actually, you are viewing the individual forms of the antidunes that are moving against the current, because of erosion on the downcurrent side and deposition on the upcurrent side. Although they move in a series of stops and starts, the sand grains always move in the direction the water is flowing.

If you look carefully, you will almost always see antidunes forming on the Oregon and Washington beaches in one of three places:

1) Where water draining from intertidal trough outlets cut through and across the intertidal bar located just seaward of the trough;

2) Anywhere along the upper or lower part of the beach where the backwash flow gains velocity as it moves back down the beach slope; and

3) Where the outlets of small streams flow through shallow channels cut through the upper portion of the beach (best seen during low spring tides).

It has been our experience that students walking

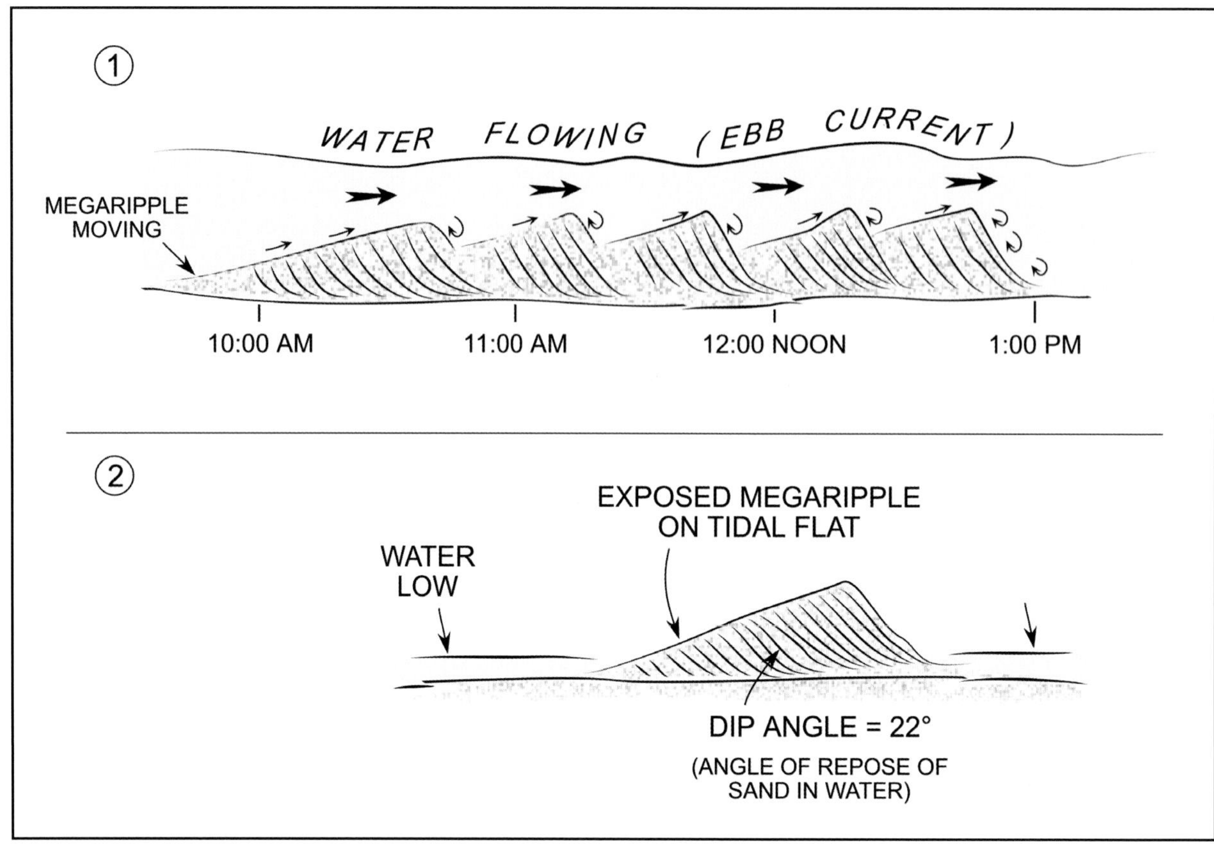

FIGURE 53. Mechanism for the movement of megaripples. Compare the implied slip faces of the bedform as it moves along (in the sketch) with the ones exposed in the photograph in Figure 52C.

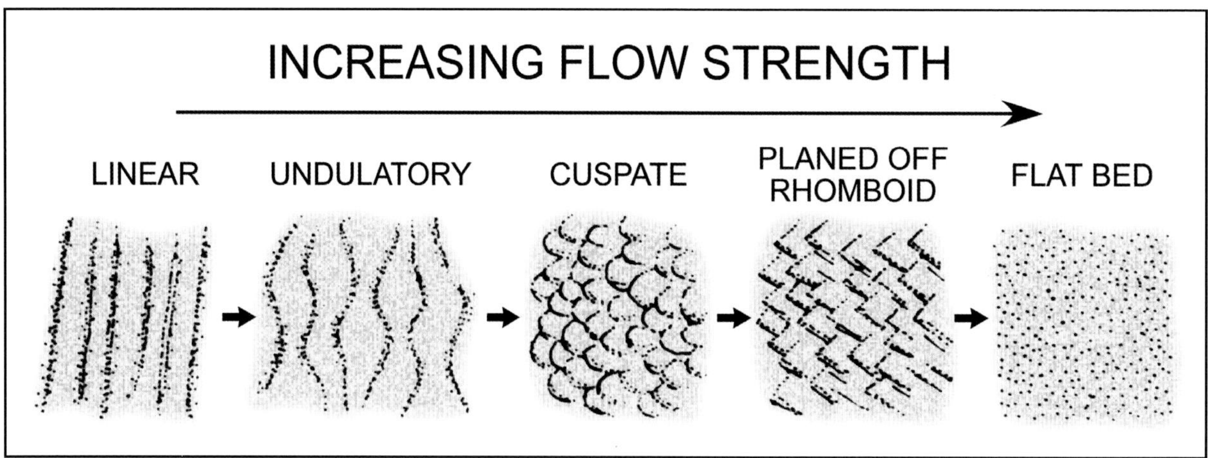

FIGURE 54. Changes in linearity of crests and 3-dimensional shape of ripples under conditions of increasing flow strength. Based on data of Allen (1968), Boothroyd (1969), and numerous field observations by our group on the intertidal bars and sand flats around the World. Water current that creates the ripples flows from left to right.

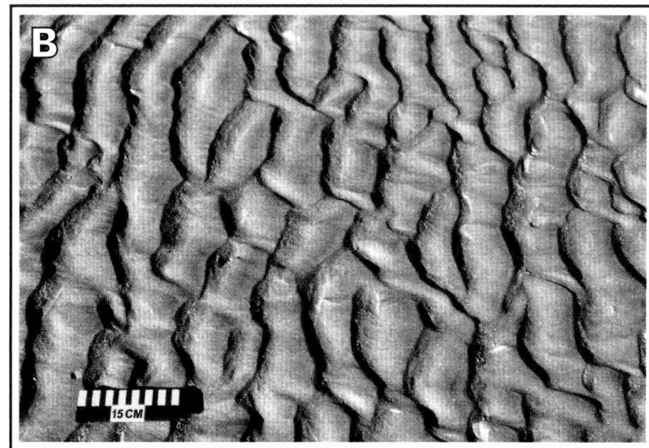

FIGURE 55. Photographs of the ripple types illustrated in Figure 54. (A) Linear ripples. Scale is one foot. The ebb current was flowing from left to right. (B) Ripples with undulatory crests, formed by an ebb current flowing from left to right. Scale is 15 centimeters (~ 6 inches). (C) Cuspate ripples, formed by an ebb current flowing from the upper right to the lower left. (D) Small rhomboid ripples. The current was flowing from left to right. Scale is 1 foot. This ripple form is the last to develop before the bed passes to the flat-bed stage in the transitional zone between lower and upper flow regime.

in the swash zone get pretty excited when they feel the antidunes move past their feet. Sand moving uphill!! Not exactly, but close.

The shallow, rapid flow of water over antidunes, or flat beds for that matter, tends to sort the sediment by composition, moving out the larger and lighter quartz and feldspar grains and leaving behind more dense and smaller grains (Figure 57A). These more dense minerals, called **heavy minerals**, are darker in color than the lighter minerals, hence they may leave linear traces that mark the position of the antidunes in which they were deposited (see Figure 57B). Another place where you will see relatively thick deposits of heavy minerals is at the base of an eroding dune scarp at the back of the beach, where the wind has carried the lighter grains away and left the heavier ones behind. Occasionally, the high part of the beachface is covered with a layer of them, usually after a series of storms have eroded the beaches. According to Komar and Wang (1984), heavy mineral deposits in Oregon (both on modern beaches and in the old beach deposits on the uplifted Pleistocene terraces) have been mined for gold and chromium; heavy mineral concentrations can reach 96%. You may also see such heavy mineral deposits in the dune fields landward of the beach.

A variety of minerals occur in these black layers on the Oregon and Washington beaches and dunes. Clemens and Komar (1988) collected sand samples from *26 beaches extending from immediately south of the Columbia River to Cape Blanco on the south-Oregon coast. In total, 18 different minerals were identified in microscopic analyses of the samples, but the major constituents are augite, hornblende, hypersthene and garnet.*

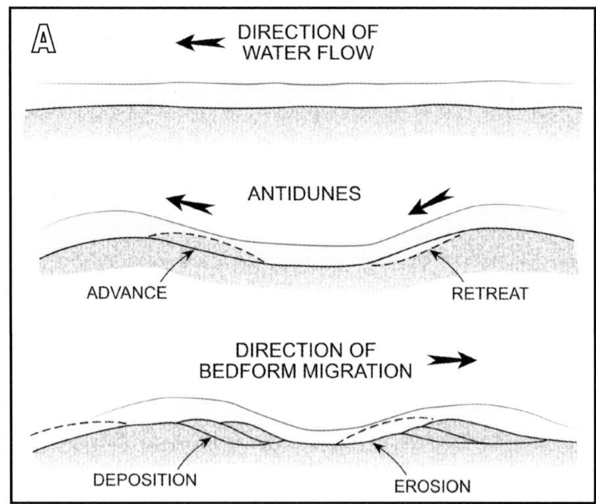

FIGURE 56. Antidunes. (A) Formation and growth. As the shallow sheet of water flows across the linear form of the antidune, water flowing down the downcurrent side of the antidune tends to erode sand from that slope. As it flows up the slope of the next antidune down current, the water slows down and sand is deposited on that slope, causing the feature to migrate upcurrent. This bedform is called an antidune, because it moves in the opposite direction of the current in which it forms. Lower flow regime ripples, on the other hand, move in the same direction as the current (compare this illustration with the diagram in Figure 53). However, the sand grains themselves move downstream in both situations. (B) Trench through antidunes on the beach on Seabrook Island, SC. Water was flowing from right to left when the antidunes were forming. Compare the internal sand layers in the preserved antidune with the sketch in diagram A. Scale is 1 foot.

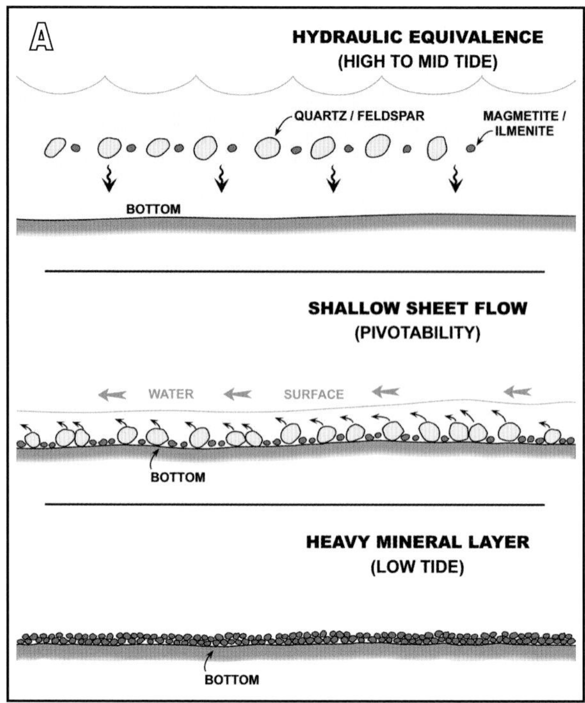

FIGURE 57. Surface layers of heavy minerals. (A) The process whereby surface layers of heavy minerals are created. (B) Bands of black heavy minerals on the surfaces of antidunes on Kiawah Island, SC. Scale is 15 cm (~ 6 inches).

[NOTE: In the usual mix of heavy and light minerals on a beach, the heavy minerals will be smaller in diameter than the lighter minerals, because of their hydraulic equivalence (e.g., if dropped into a cylinder of water, they would all reach the bottom of the cylinder at nearly the same time). That is, as they fall through the water in a breaking wave, the smaller, heavier minerals fall at the same rate as the larger, light minerals, because the buoyancy of the relatively dense water the grains are falling through creates more resistance on the larger grains than on the smaller grains. Once the sand grains are laying on a flat beach, the larger, lighter mineral grains, such as quartz and feldspar, will be transported away more readily by the shallow sheet flow of the water in the wave swash than the smaller heavy mineral grains, such as magnetite and garnet, because they protrude further up into the water than the smaller heavy minerals (a process called **pivotability** to be discussed in more detail later). Over time, this action may leave behind a surface layer of the heavy minerals because they do not stick so far up into the water column as the large lighter grains, which are carried away (see Figure 57A). Ed Clifton suggested a different process for the formation of heavy mineral layers on the "back edges of beaches" based on his observations on Oregon beaches. He commented that *sand transport during wave uprush is mostly in the form of suspension, whereas during the backwash, it moves in a flowing bed along the bottom where the gains are colliding with each other in something akin to grain flow. In this flow, "collisional sorting" causes the largest particles to move to the top and front of the flow. The small size of the heavy minerals promotes their transport in suspension up the beach. But during the modified grain flow under backwash, the small size and their density work together to inhibit transport. So the heavy grains are carried in suspension with the large light grains to the back of the beach and the larger, lighter grains are selectively carried back down the beach, leaving the heavies behind.*]

In some ways, we are both saying the same thing, except we did not distinguish between suspended sediment transport and bottom creep in the uprush and backwash. So what is more important, grain flow or pivotability of the individual sand grains? Do both processes work, maybe at the same time? Good questions to be addressed with future research.

Grain flow is the movement of sediment under gravity where the sediment is moved by direct grain-to-grain contact. This differs from turbidity flow, where the sediment moves under the influence of gravity in a turbulent flow of water.

GRAVEL BEACHES

We define gravel according to the Wentworth (1922) scale, which includes four classes – *granules* (median diameter = 2-4 mm), *pebbles* (median diameter = 4-64 mm), *cobbles* (median diameter = 64-256 mm), and *boulders* (median diameter >256 mm) (see Figure 6). The general term *gravel clast* is used in this discussion when referring to gravel particles irrespective of their size class. Gravel beaches composed of all these size ranges are present on the Pacific Northwest Coast (see example in Figure 43C). These types of beaches, in combination with sand and gravel beaches, occur along 11.4% of the length of the outer coasts of Oregon and Washington, with pure gravel beaches occurring on 6%.

For waves to form a gravel beach, clearly the shore has to be either exposed to a significant fetch and/or subject to relatively large swell waves. However, the size of the waves is not necessarily the deciding factor. The presence of a source for the gravel is even more essential. Hayes (1967a) and Davies (1973) recognized that the common occurrence of gravel in continental shelf and shore zone sediment budgets in the higher latitudes (>40 degrees north and south) is primarily the result of glaciation at those latitudes, either under present conditions or during the Ice Ages of the Pleistocene epoch. Glaciers have the competence to carry huge volumes of coarse material to the coast, where it is redistributed along the shore by wave action.

A general model of highly exposed (to wave action) coarse-grained gravel beaches based on our field experience on open-ocean coasts is presented in Figure 58. By coarse-grained, we mean beaches made up primarily of cobbles and boulders (Figure 6). The most landward portion of those beaches commonly consists of an elevated, linear pyramidal-shaped mound referred to as the **storm berm**, that is composed typically of coarse pebbles and cobbles. That berm is only activated during major storms and, unless the storm is unusually severe (e.g., during a major *El Niño* storm), the berm is built up even higher during the storm event and not eroded away as is common for depositional berms on most sand beaches during storms (Shepard, 1950a; Hayes and Boothroyd, 1969).

The issue of the landward transport of the gravel clasts to form the storm berm has been discussed at some length in the literature. According to Lorang (1991; 2000), the height of the crest of the storm berm is controlled by the incident breaker height and period, combined with the gravel size. The larger the waves and the coarser the gravel, the higher the berm. Furthermore, it is generally agreed that, on a gravel beachface, the influence of infiltration of water into the sediments is more extensive during the backwash than during the uprush, considerably reducing the strength of the backwash in comparison (Packwood, 1983). This weaker **flow competency** (a term that accounts for the

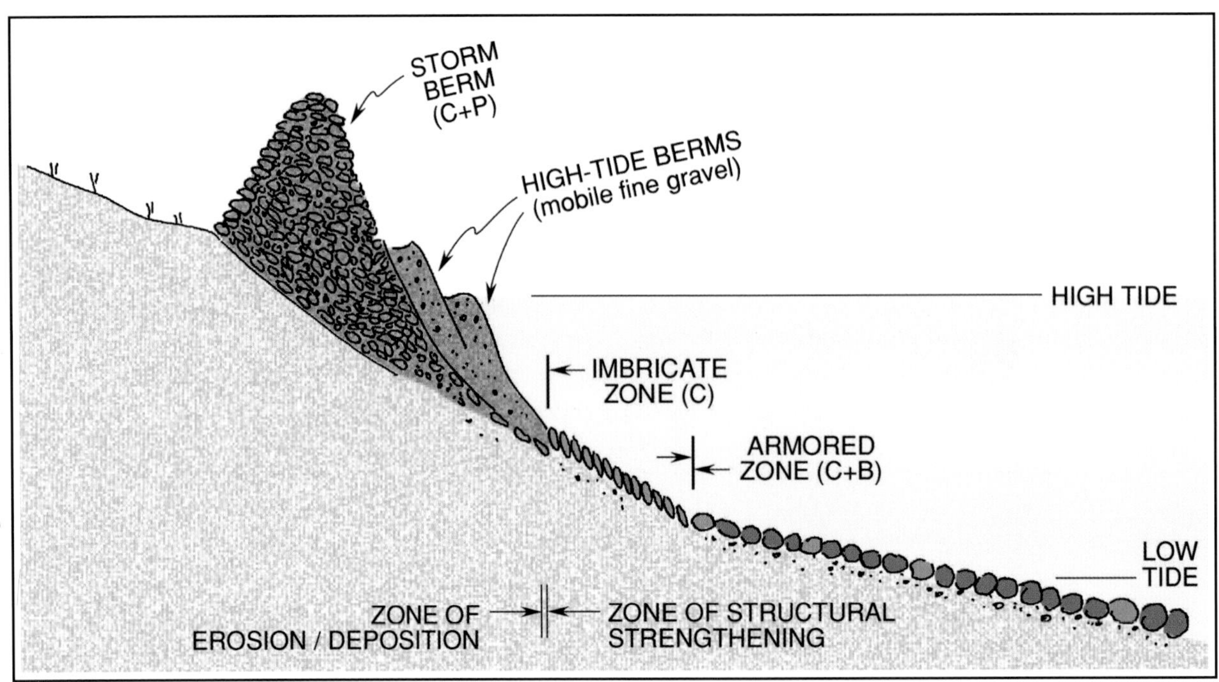

FIGURE 58. Typical topographic beach profile and sediments of an open ocean (exposed) coarse-grained gravel beach. The landward half of the profile is primarily depositional, with the storm berm being relatively stable during storms and the minor, high-tide berms eroding during storms and reforming after they pass. The seaward half of the profile is an area where structural strengthening occurs by either clast imbrication or by the formation of a surface armor of coarse gravel. B = boulders; C = cobbles; P = pebbles.

ability of the moving water to transport sediment) of the backwash relative to the uprush brings about a net landward transport of the gravel clasts up the beachface, stranding particles too large to be transported back down slope by the weaker backwash. Also, the loss of the volume of the backwash through percolation aids in the generation of the characteristic steep slopes of gravel beachfaces. In discussing the slope of the beachface on sand beaches, Komar (1976) explained that, because of the discrepancy between the competence of the uprush and backwash to transport sand, the uprush moves sand onshore until a slope is built up in which gravity supports the backwash and offshore sand transport. That is, a steeper slope thus created compensates for the decreased water volume in the backwash (relative to the uprush). A similar explanation must also apply to gravel beaches, which means that at some point in its evolution, the beachface on a gravel beach attains an equilibrium slope considerably steeper than those on even coarse-grained sand beaches. The greater permeability of the gravel allows a greater loss of water into the beachface than would occur on a sand beach subject to the same size of waves.

Numerous papers have pointed out that disc-shaped clasts are common on gravel beaches, especially on the storm berm. The gravel beaches we have studied show a very distinct segregation of gravel shapes along the beachfaces of storm berms – discs on the top, spheres and blades further down, and spheres and rollers at the base. The different gravel shapes, that you will no doubt see the entire spectrum of if you spend much time on the gravel beaches of Oregon and Washington, are illustrated in Figure 59. The just-described pattern of the distribution of gravel shapes along the surface of gravel beaches develops because once the uprush of the wave has moved the pebbles or cobbles up the beach, disc-shaped ones remain preferentially up high on the storm berm for two reasons: 1) Discs stay in suspension longer than the other more compact shapes because of their greater maximum projection areas (Bagnold, 1940; Folk, 1955) and, therefore, they can be carried further landward during the uprush than more spherical (or compact) clasts of equal volume that settle quicker; and 2) Discs have a lower pivotability than the other

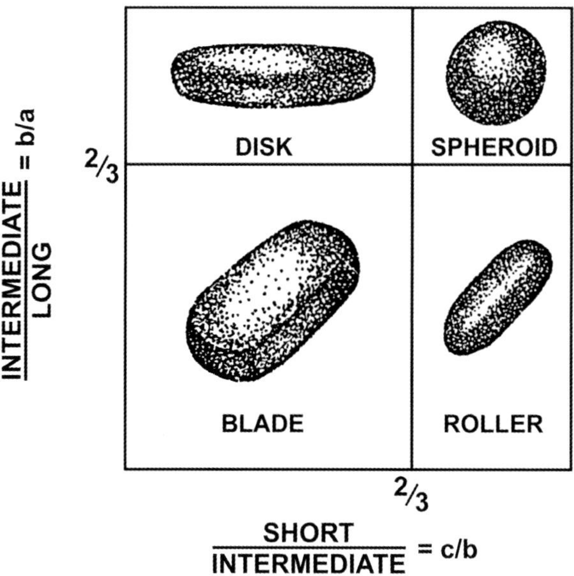

FIGURE 59. Gravel shapes (from Zingg, 1935).

shapes, so they are less likely to roll back down slope. You may wonder why the heck we are elaborating so much on shapes of pebbles, cobbles, and boulders. Well, if you spend some time wandering up and down gravel beaches during your excursions to the coasts of Oregon and Washington, you will be amazed how well the waves have sorted the shapes of the pebbles and cobbles during storms.

Komar and Li (1988) discussed the selective transport of sediment particles by water (called entrainment) of varying sizes within a sediment accumulation. A key factor determining whether a particle would be moved is its relative projection above the bed, called its *pivotability*. The fact that coarser-grained sand particles composed of light minerals, such as quartz and feldspar, are readily transported over beds of placer deposits composed of finer-grained heavy minerals is well studied (Komar and Wang, 1984; Slingerland, 1977), and was discussed in some detail earlier (Figure 57A). Sand beaches with a variety of sand sizes may experience a similar size sorting, with the medium- to coarse-grained sand being preferentially transported offshore during storms, leaving behind a flat, fine-grained post-storm beach composed of finer-grained quartz as well as an increase in the abundance of heavy minerals. As Komar and Li (1988) so eloquently put it, *"first movement occurs when the fluid forces of drag and lift overcome the grain's immersed weight, pivoting it over the underlying particles so that it is rotated out of resting position."* Therefore, on a storm berm composed of a mixture of Zingg's (1935) basic shapes (Figure 59) of relatively equal volumes – discs, blades, rollers, and spheres – the discs will be moved last of all the shapes by a backwash-generated current because of their lower pivotability. Therefore, once you reach the top of that storm berm on your favorite gravel beach, there should be a lot of discs up there. However, on some Oregon and Washington beaches, the base of an eroded cliff at the backside of the beach may only contain an ill-sorted array of coarse cobbles and boulders, rather than a well-sorted storm berm, because many of those clasts are deposited as piles of talus that are only rarely reworked by the waves. Also, the reflection of waves off the rock cliffs during storms may disrupt the size sorting.

Rollers or rods, with two short axes and one long axis (Figure 59), are particularly susceptible to the downward push of the returning waters of the backwash and the pull of gravity because of their high pivotability (as well as relatively large maximum projection areas; the largest outline of the cobble or pebble, that is the portion of the clast most subject to push by running water). Therefore, they tend to congregate at the base of the storm berm, assuming such a variety of clast shapes are present on the beach. Spheres will move easily down the beachface slope as well. In some cases, of course, the shape originally inherited from the source material may not contribute such a diverse combination of shapes.

As shown by the general model in Figure 58, one or more normal high-tide berms composed of mobile fine gravel typically occur on the lower half of the storm berm. The highest of those high-tide berms will have been deposited during a spring tide; lower high-tide berms are deposited as the tide falls from spring to neap conditions. During storms, high-tide berms can be eroded and redeposited over a time frame of minutes to hours. However, in some instances, the coarser gravel in those berms will be added to the storm berm proper. During the post-storm recovery period, the finer gravel returned from down the beach or from offshore is redeposited as normal high-tide berms again. On mixed sand and gravel beaches, a sand zone may form seaward of the storm berm, such that during periods of maximum sand deposition, only the upper portion of the storm berm itself will be exposed (see example in Figure 60).

FIGURE 60. A common type of beach on the Oregon Coast with lower intertidal zone composed of a sand beach and the upper intertidal zone composed of armored gravel (cobbles and pebbles). Also note the classic example of a sea cave. Site is located approximately 1 mile north of Arch Cape in Hug Point State Park. Photo taken on 6 August 2012 by Walter J. Sexton.

LITTORAL CELLS

Introduction

The term **littoral cell** is one that comes up frequently in both written and verbal communications about the sand beaches on the Pacific Coast. It is a concept first suggested by Inman and Chamberlain (1960) as a way of defining a sand budget for the sand beaches of the Southern California Coast. As Inman (2005) noted, the boundaries of the cells *delineate the geographical area within which the budget of sediment is balanced.* The budget is "balanced" if the amount of sand leaving the cell is the same as the amount that comes into it. Obviously, if more sand is lost from the beaches than comes into the cell, beach erosion problems will most likely ensue.

Littoral cells have notably different characteristics among the diverse coasts of the World (e.g., trailing-edge coasts, polar regions, etc.), with those on the leading edges of continental plates, illustrated in Figure 61, having the following defining characteristics (Griggs, 1986; Inman, 2005; Griggs and Patsch, 2006):

1) The so-called updrift, or up coast, side of the cell, the direction from which the sand comes, is typically composed of a rock headland around which very little sand passes to the beaches located downdrift, or down coast, from it;

2) The sand that reaches the beaches located within one particular littoral cell may be derived from streams, eroding rock cliffs, and eroding bluffs composed of less resistant material than rocks (and of course by artificial beach nourishment, but that is a relatively new entry into the equation);

3) Once the sand is on the beach, it moves alongshore in response to prevailing wave conditions by the process of longshore sediment transport illustrated in Figure 36; and

4) The sand within a cell can be removed from it primarily by one or all of three natural processes: a) Funneled down submarine canyons; b) Blown landward into dune fields; and c) Carried so far out onto the continental shelf during *El Niño* storms that it cannot be returned by normal onshore sediment transport. Some sand might also be lost into major bay/estuarine systems.

According to Inman (2005), *the littoral cell now plays*

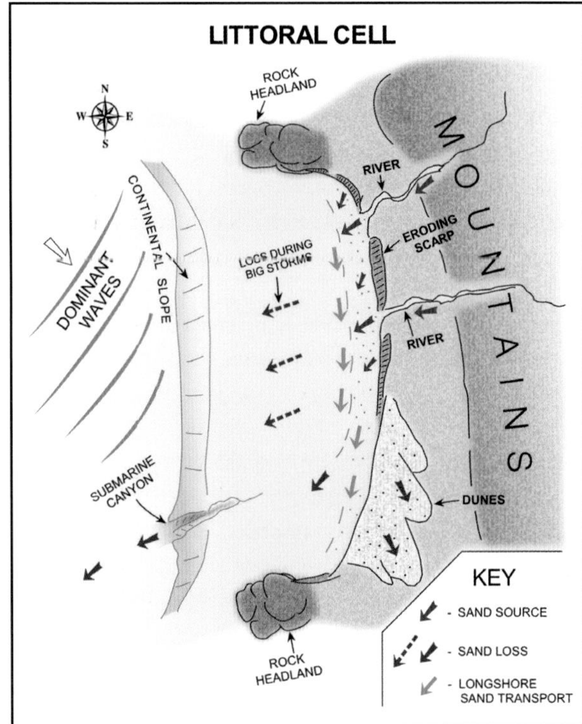

FIGURE 61. General model of a littoral cell, a concept first suggested by Inman and Chamberlain (1960) as a way of defining a sand budget for the sand beaches of Southern California. Sand is delivered to the cell by rivers and streams or by the erosion of bluffs along the shore. It is lost down submarine canyons, into coastal sand dunes, and possibly, in some cases, onto the outer reaches of the continental shelf during storms.

an important role in the U.S. National Environmental Protection Act (1974) and it has become a necessary component of environmental impact studies. In the realm of public policy and jurisdictions, the littoral cell concept has led to joint-power legislation that enables municipalities within a littoral cell to act as a unit (Inman and Masters, 1994).

Littoral Cells in Oregon and Washington

Figure 62 shows the littoral cells along the Oregon and Washington Coast. Note that the littoral cells in Oregon occur as isolated pocket beaches, whereas the ones in Washington are dominated by the Columbia River Littoral Cell. Allan, Geitgey, and Hart (2005) made the following summary statement about the littoral cells on the Oregon Coast:

There are at least 18 major littoral cells identified on the Oregon Coast, with the majority of the shoreline (72%) consisting of dune-backed sand beaches, while the remaining 28% of shore is comprised of a mixture of bluff-backed beaches, rocky shores, and coarse grained (gravel) beaches. Because the headlands extend into deep water, wave processes are generally regarded as unable to transport beach sediment around the ends of the headlands. As a result, the headlands essentially form a natural barrier for sediment transport, preventing sand exchange between adjacent littoral cells. Thus, a littoral cell is essentially a self-contained compartment, deriving all of its sediments from within that cell.

Some of these cells in Oregon are quite short, with the three longest ones being the Coos Cell at 59 miles, the Bandon Cell at 55 miles, and the Newport Cell at 27 miles.

Those features are natural study sites in which to learn about key issues, such as alongshore sediment transport, trends in beach morphology and sedimentation, and so on. Shih and Komar (1994) did a study of sediments, beach morphology, and sea cliff erosion within the 14.4-mile-long Lincoln City littoral cell on the central Oregon Coast. That cell has headlands on either side that prevent bypassing of beach sands, in effect serving as a large pocket beach. Some of their conclusions follow:

1) There is a seasonal reversal in the sand transport along this cell, but the net littoral drift is zero;

2) Despite the "extreme wave energy," with 22 feet significant wave breaker heights, this "energy" is uniform along the cell;

3) *In spite of this uniformity, there is a marked longshore variation in grain sizes of the beach sediments and an accompanying change in beach morphology from dissipative to reflective beaches;*

4) Cliff erosion, which introduces coarse sediment (up to gravel in size) into the beach sediments, did not begin or was insignificant *until about 300 years ago, at which time a major subduction earthquake caused subsidence of this portion of the coast and rapid cliff recession.* We assume they are talking about the earthquake that generated *"The Orphan Tsunami of 1700"* that was discussed earlier; and

5) *The longshore variations in beach sand grain sizes and accompanying beach morphology are playing an important role in the continuing sea cliff erosion with cliffs backing the coarse-grained reflective beaches eroding more rapidly than the cliffs buffered by the fine-grained dissipative beaches.*

These are interesting findings that further emphasize the diversity of the geomorphology and sedimentation processes on this complex coast.

[NOTE: **Reflective beach**: A beach that reflects a major part of the incoming wave. Reflective beaches have a convex beach profile, without nearshore bars, which is steeper than the wave it reflects and is typically caused by accretion from swell waves. Reflection may generate edge waves on these beaches. Reflective beaches are typically steep in profile with a narrow shoaling and surf zone, composed of relatively coarse sediment, and characterized by surging breakers. Coarser sediment allows percolation during the swash part of the wave cycle, thus reducing the strength of backwash and allowing material to be deposited in the swash zone. They are generally steep and coarser-grained.

Dissipative beach: Dissipative beaches are wide and flat in profile, with a wide and shoaling surf zone, composed of finer sediment than nearby reflective beaches, and characterized by spilling breakers. Multiple intertidal bars may be present.

Actually, it is a bit more complicated than that, but we hope you get the idea.]

Back to littoral cells, our earlier discussion of Komar's (1998a;b) erosional "hot spots" showed where these features are located within a littoral cell (Figure 41). Also, the information on winter/summer beach profiles presented by Ruggiero et al. (2005) and discussed earlier was carried out in the Columbia River Littoral Cell,

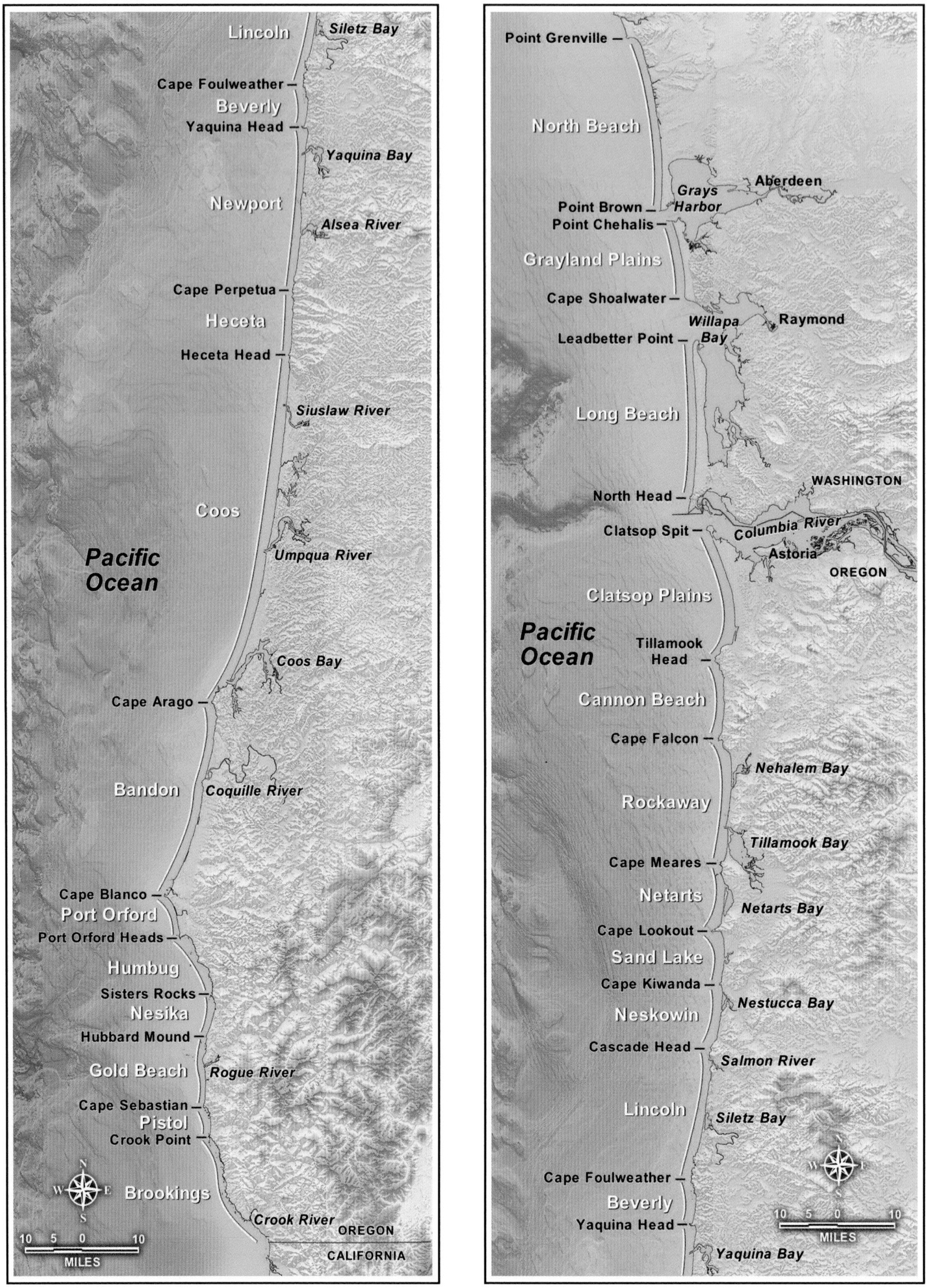

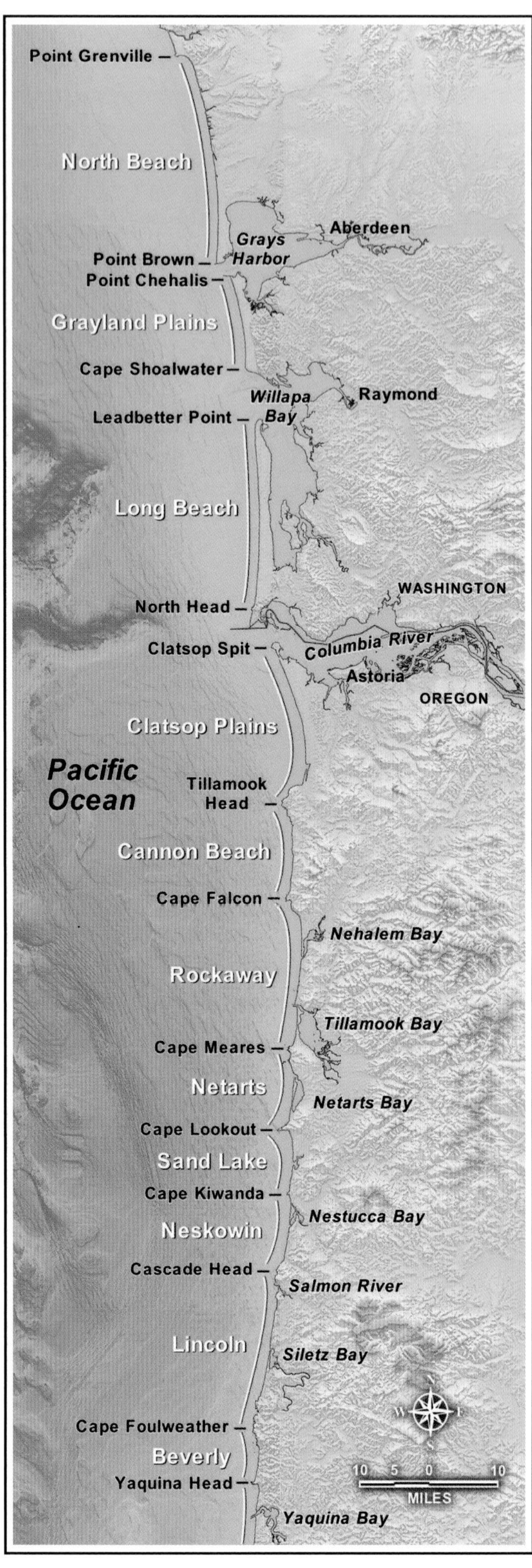

FIGURE 62. The major headlands defining the principal littoral cells along the coast of Oregon and southwestern coast of Washington. Modified after Komar et al. (2013), and Kaminsky et al. (2010).

which is a high-energy, dissipative littoral cell.

CRENULATE BAYS

The Pacific Coast of North America contains a number of morphological features known as crenulate bays (see general model in Figure 63). Those features typically have a fishhook shape, with the shank of the hook pointing away from the dominant approach direction of the larger waves and swell along the coast. The sand beaches in the crenulate bays are most commonly located between two segments of rocky shore. The shape of the bay is the result of refraction/diffraction of the dominant waves around a rock headland area located at the end of the bay facing the wave approach direction. This general pattern of wave-front bending is illustrated in Figure 63. Note that the wave crests are almost parallel to the shore as the waves approach the shore. The waves arriving at the shore refract and diffract into the shadow zone in the lee of the up-coast bedrock headland, creating longshore currents that remove sediment from that zone and transport it toward the opposite end of the bay (except for some reversals, minor in some bays and major in others, in which the finer-grained sand moves back toward the updrift headland), eventually producing a stable equilibrium beach with a crenulate, or fishhook, shape.

Crenulate bays are relatively scarce on the Oregon and Washington Coasts, with only two of them containing clearly depositional features: 1) Crescent Bay near Port Angeles on the Strait of Juan de Fuca (Figure 64); and 2) At the very north end of the Oregon Coast near Seaside. There are two shore segments with the shape of crenulate bays on the northwest coast of Washington that appear to be erosional features that contain little in the way of depositional components: 1) The north end of Grenville Bay (Figure 65); and 2) Northeast of Alexander Island.

Crenulate bays are present all around the World on coasts with patchy bedrock distribution, and they are one of the most definitive clues of long-term sediment transport directions that exist. When we did the overview study of the leading-edge coasts of North and South America, we saw literally hundreds of them while studying the Google Earth images along those coasts. One striking contrast we observed was the difference in the orientation of the crenulate bays in the northern hemisphere, particularly those under the influence of the dominant northwesterly winds, from those in the southern hemisphere, which are mostly under the influence of dominant southwesterly winds. The shanks of the northern hemisphere bays point toward the southwest, and the shanks of the southern hemisphere bays point toward the northwest, a very striking pattern, one of the most arresting trends we saw during that survey.

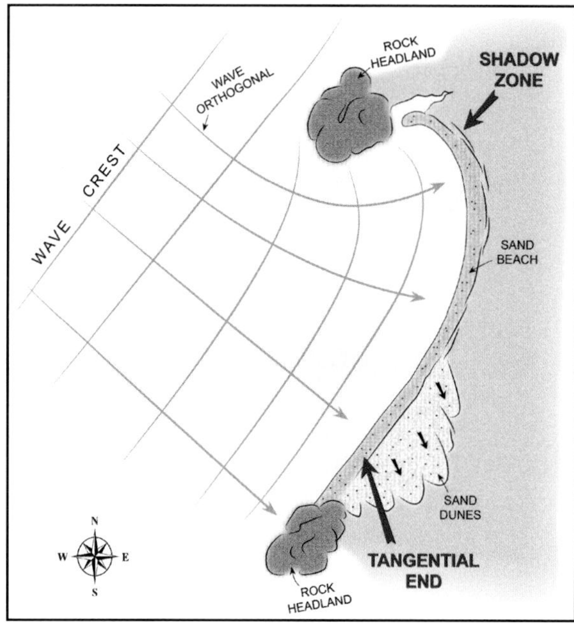

FIGURE 63. General model of a crenulate bay. The shape and orientation of the bay is the result of the bending (refraction/diffraction) of the dominant waves and swell around a rocky headland located at the end of the bay that faces in the direction from which the dominant waves approach the bay.

TIDAL INLETS

Introduction

On a coast dominated by barrier islands, such as along the East Coast and parts of the Gulf of Mexico, tidal inlets are usually defined as major tidal channels that intersect barrier islands, or separate two barrier islands, usually to depths of tens of feet (see general model in Figure 66). The tidal inlets are the focus of the most dynamic changes that occur on any component of those barrier-island systems. Because tidal inlets connect the open ocean with more sheltered backbarrier settings, including some potential harbor sites, engineers have long been engaged in efforts to control them, commonly

FIGURE 64. Crescent Bay near Port Angeles on the Strait of Juan de Fuca, an example of a crenulate bay. Dominant waves are from the northwest. Compare it with the general model of such bays in Figure 63. NAIP 2013 imagery, courtesy of Aerial Photography Field Office, USDA.

by constricting their migration with a set of two parallel jetties.

When a storm surge floods a barrier island or barrier spit, water flowing through breaks in the dunes is capable of scouring channels across the island or spit. This is one mechanism by which many new tidal inlets are formed in barrier islands and barrier spits (Hayes, 1965, 1967b; Shepard and Wanless, 1971). However, most of these new channels are closed off by wave-transported and wind-blown sand within a few weeks after the storm. Needless to say, some survive if they are deep enough and other conditions, such as altered tidal current patterns, promote the longevity of the inlet.

The ultimate morphology, or shape, of a tidal inlet through a barrier island (or barrier spit) is the product of the constant interaction, or contest if you will, between tidal and wave-generated forces. Longshore currents and beach drift generated by wave action move sediment along shore in such volumes that would normally fill the inlet in the absence of tidal currents. Sediment carried into the inlet throat by the wave-generated currents is constantly swept away by tidal currents that scour the deep inlet throat. Sediment transported from the scoured-out throat of the inlet by flood currents may be deposited as a lobe of sediment on the landward side of the inlet called the **flood-tidal delta**, and that carried seaward by the ebb currents may form a seaward lobe called the **ebb-tidal delta**.

FIGURE 65. Shore with a crenulate-bay shape at the north end of Grenville Bay, WA. NAIP 2013 imagery, courtesy of Aerial Photography Field Office, USDA.

The general model of tidal inlets in Figure 66, that shows relatively equal volumes of sediment in the two tidal deltas, is rare in nature. More typically, in microtidal, wave-dominated areas such as the Texas Coast, the flood-tidal delta is much larger than the ebb-tidal delta, because of the influence of wave action in both moving sediment to the inside of the inlet and in erosion of the outer lobe (among other factors).

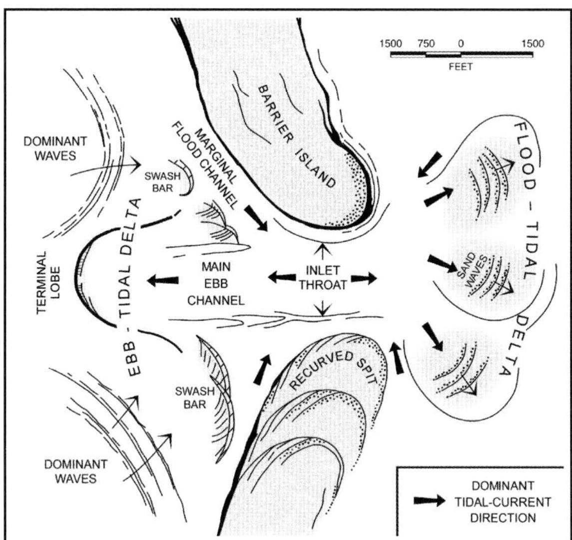

FIGURE 66. Morphology of tidal inlets. General model.

Because of the big waves and relatively small tides and relatively few bays with large tidal prisms, except for the one off the mouth of the Columbia River, massive ebb-tidal deltas are rare on the coasts of Oregon and Washington. A major reason for the paucity of inlets such as those on the East Coast is that Oregon's Coast has a number of bedrock shores, with no barrier islands per se (but it does have some barrier spits), primarily because of its location on the tectonically active, western coast of the North American Plate. The issues of shipping, dredging, and sediment supply that revolve around the ebb-tidal delta off the mouth of the Columbia River will be discussed in some detail in Section III.

Oregon's Tidal Inlets

In this discussion, we broaden the definition of tidal inlets a bit to include any channel-like opening that connects the sea with the mainland. According to our newly proposed definition, there are 65 tidal inlets along the Oregon Coast. Most of those inlets do not fit the morphological patterns based on the barrier island shores on the East Coast that were discussed earlier, with a few exceptions. For example, the two inlets shown in Figure 67, Netarts Bay Inlet and the inlet at the entrance to Sand Lake, both have large flood-tidal deltas and very small ebb-tidal deltas presumably because of the relatively small tidal range, little freshwater flow, and extra large waves along the outer beaches.

The following are the most common types of tidal inlets on the Oregon Coast (based on our broader definition):

1) Entrances to rivers of significant size. Ten inlets are of that type (e.g., at the mouths of the Columbia and Rogue Rivers);

2) Outlets and entrances to bays, harbors, and/or large lakes. Fifteen inlets are of that type (e.g., entrances to Nehalem and Yaquina Bays);

3) Jettied inlets. Ten of the originally natural inlets on the Oregon Coast have been stabilized with jetties, that are defined as two parallel, linear structures extending into a body of water (in this case the Pacific Ocean) that are designed to provide access to a harbor or marina. They are usually composed of large pieces of boulder-sized rock called riprap. Komar et al. (1977) analyzed all of the Oregon Coast jetties, showing that beaches built out on both the north and south sides of the jetties. Figure 68 shows the jetties at the mouth of the

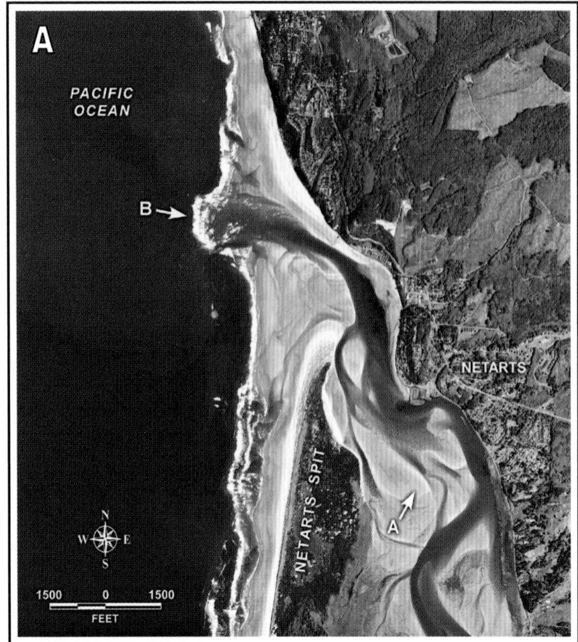

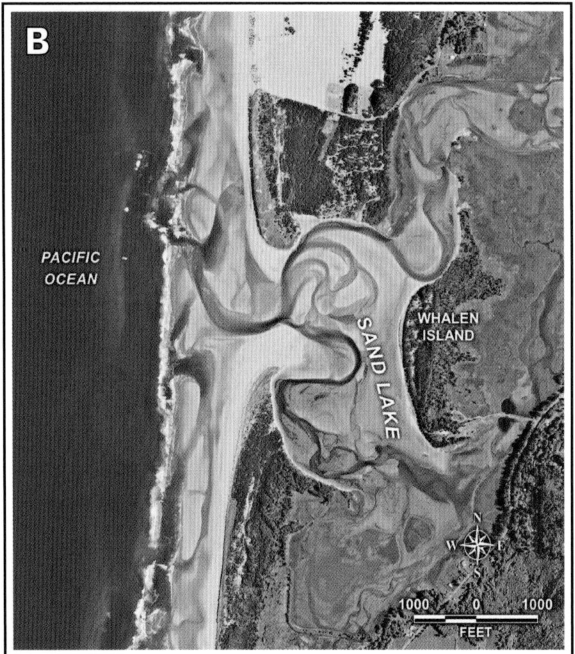

FIGURE 67. Tidal deltas. (A) Netarts Bay Inlet, OR, which conforms fairly closely to the general model in Figure 66, except for the relatively small size of the ebb-tidal delta. Arrow A points to the flood-tidal delta and arrow B points to the ebb-tidal delta. NAIP 2014 imagery, courtesy of Aerial Photography Field Office, USDA. (B) Tidal inlet at the entrance to Sand Lake, OR. NAIP 2009 imagery, courtesy of Aerial Photography Field Office, USDA.

Columbia River; and

4) Stream-mouth/bar systems. By far the majority of tidal inlets in Oregon (30 out of 65) occur at the mouths of somewhat small, relatively steep-gradient streams. The stream mouth usually has a sand or sand and gravel spit built across it (see Figure 69). The spit, which is typically a few tens of yards wide, is frequently overwashed during high spring tides and storms, and it may be removed or highly modified during the larger floods on the stream. Minor wetlands commonly occur along the stream channel landward of the spit.

[NOTE: The term **spit** is defined as a linear projection of sediment across an embayment, stream outlet, and so on that builds in the direction the longshore transport system is moving the sediment. Some spits grow in spurts, with the welding beach berm on the downdrift end curving most of the way around the extension. After a period of little growth, the next spit that follows projects beyond the end of the previous one, creating a series of the recurving spit ends. That type of spit is called a recurved spit, but most of the spits in the stream-mouth/ bar systems on the Oregon Coast are relatively straight rather than recurved (see example in Figure 69B).]

Tidal Inlets on the Outer Shore of Washington

In this discussion, we will follow the broader definition of tidal inlets that we used for those on the Oregon Coast. Also, this discussion will be focused on the 33 inlets on the outer coast of Washington and the Strait of Juan de Fuca, specifically those occurring in Compartments 3 and 4 (Figure 1). Significant ebb-tidal deltas are rare, probably because of the large waves that occur along this coast (as well as the relatively small tidal ranges). Flood-tidal deltas are also fairly rare. The long northward projecting spit to the north of the mouth of the Columbia River (North Beach Peninsula) does indicate a consistent northward transport of littoral sediments in that area (Komar, 1992; as verified by Ruggiero et al., 2006).

Of course, Washington's Coast has a number of bedrock shores, no barrier islands per se, but some barrier spits, primarily because of its location on the tectonically active, western coast of the North American Plate. The following are the most common types of tidal inlets on the outer coast of Washington:

1) Entrances to rivers of significant size. Eight inlets are of this type (e.g., those at the mouths of the Lyre and

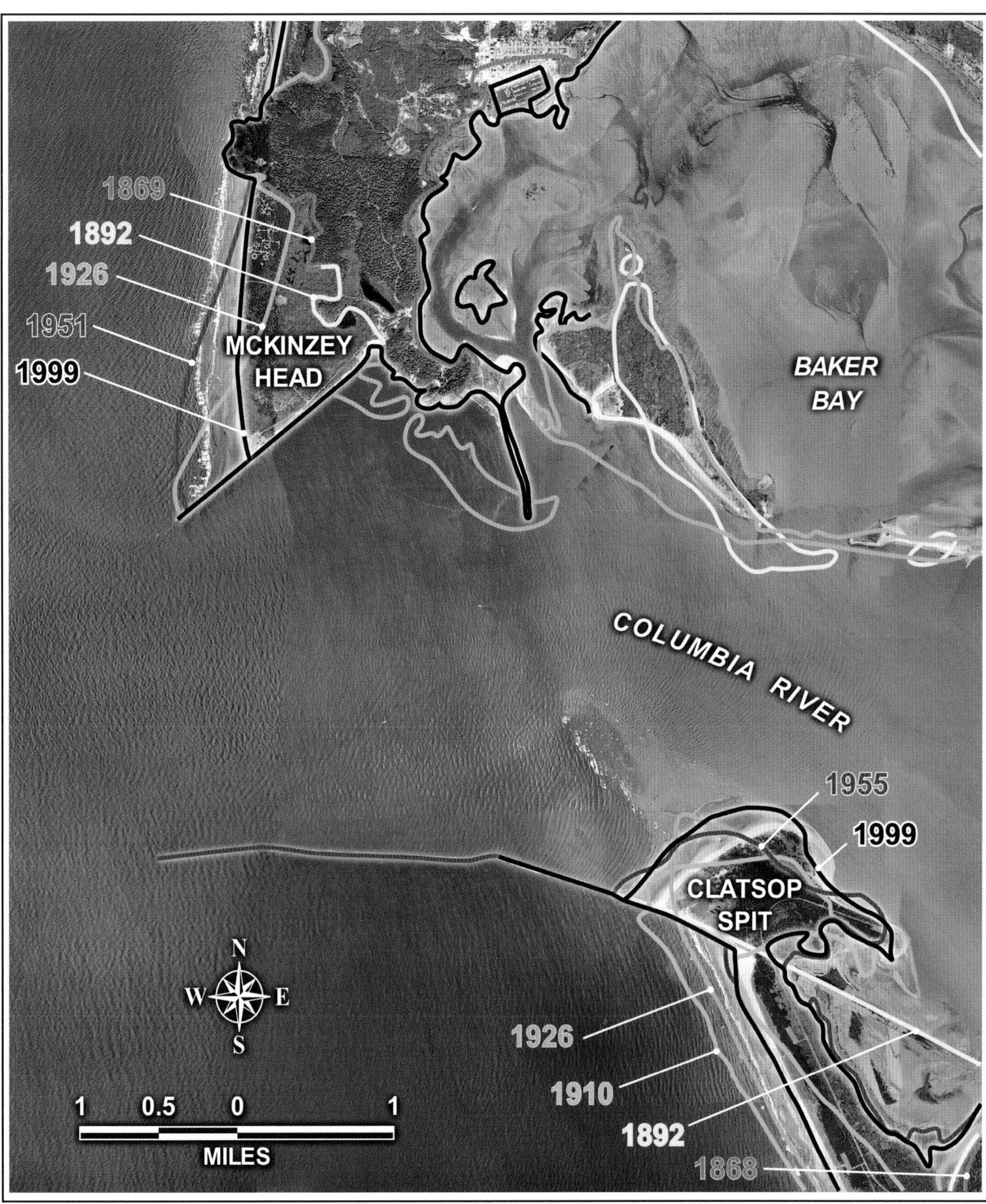

FIGURE 68. Shoreline changes at the Columbia River mouth due to construction of the jetties from 1885 to 1917. Note the accumulation of sand on both the north and south jetties. (Kaminsky et al., 2010). NAIP 2011-2013 imagery, courtesy of Aerial Photography Field Office, USDA.

FIGURE 69. Stream mouth bars at (A) Big Creek, north of Florence, OR and (B) Hunter Creek south of Gold Beach, OR. Note the northward building spit at the creek mouth. NAIP 2011 imagery, courtesy of Aerial Photography Field Office, USDA.

Elwha Rivers);

2) Outlets and entrances to bays, harbors, and/or large lakes. Five inlets are of this type (e.g., entrances to Willapa Bay and Grays Harbor);

3) Jettied inlets. Only three of the originally natural inlets on the outer Washington Coast are confined by jetties: 1) The mouth of the Columbia River (Figure 68); 2) The entrance to Grays Harbor (Figure 70A); and 3) The mouth of the Quillayute River at La Push (Figure 70B); and

4) Stream-mouth/bar systems. A little over half of the tidal inlets on the outer Washington Coast (17 out of 33) occur at the mouths of somewhat small, relatively steep-gradient streams. The stream mouth usually has a sand or sand and gravel spit built across it. The spit, which is typically a few tens of yards wide, is frequently overwashed during high spring tides and storms, and it may be removed or highly modified during the larger floods on the stream. Minor wetlands commonly occur along the stream channel landward of the spit.

The construction of the jetties at the mouth of the Columbia River between 1885 and 1917 resulted in the accumulation of sand on both the north and south jetties. In Figure 68, the pre-jetty shoreline position is indicated by the orange 1868-69 lines overlain on the 2011-13 satellite imagery. To the south, Clatsop Spit built out and attached to the south jetty by 1910, which acted as an artificial headland and created a sub-littoral cell. To the north, Benson Beach was created by the southerly transport of sand. Kaminsky et al. (2010) provided detailed explanation of all the processes that contributed to the patterns of sand accumulation and erosion after the construction of the jetties and the damming of rivers in the Columbia River watershed.

COASTAL DUNE FIELDS

It has been estimated that sand dunes are present along 45 percent of the Oregon Coast and 31 percent of the Washington Coast (Cooper, 1958), mostly adjacent to sand beaches. In order for such dunes to form, there must be a fairly large supply of sand along the coast and a topographic area landward of the beach that is relatively low. The dunes in Oregon compose the largest coastal dune system in North America. The Oregon Dunes National Recreation Area extends for 42 miles from south of Florence to north of Coos Bay. Dunes there are up to 600 feet high and have a wide variety of constantly changing dune types. According to Komar, Allan, and Ruggerio (2013), coastal dunes have increased in elevation since the 1930s due to the planting and expansion of European dune grass.

Most of the sand transported by the wind into the dune fields is moved by a process called saltation, whereby the sand grains bounce off the sand bed and "fly" for some distance at heights usually not more than a couple of feet (illustrated in Figure 71). Some of the grains roll slowly along the ground (a process called creep).

The four morphologic types of dunes that occur in these dune fields – transverse, blowout (usually lobate),

FIGURE 70. Jetties. (A) Grays Harbor, WA; (B) Mouth of the Quillayute River at La Push, WA. NAIP 2009-06 imagery, courtesy of Aerial Photography Field Office, USDA.

barchan, and parabolic – are sketched in Figure 72 (see also more detailed definitions in the glossary). Inspection of the aerial images of those dune fields indicates that transverse and barchan dunes are the most common types present. Photographs of some of the different types of types of dunes are presented in Section III. Where the dunes are vegetated, blow-out dunes, or possibly more commonly parabolic dunes, are the most usual type. The other two types are present depending on variables such as sand supply, consistency of the wind, viability of the vegetation, and so on.

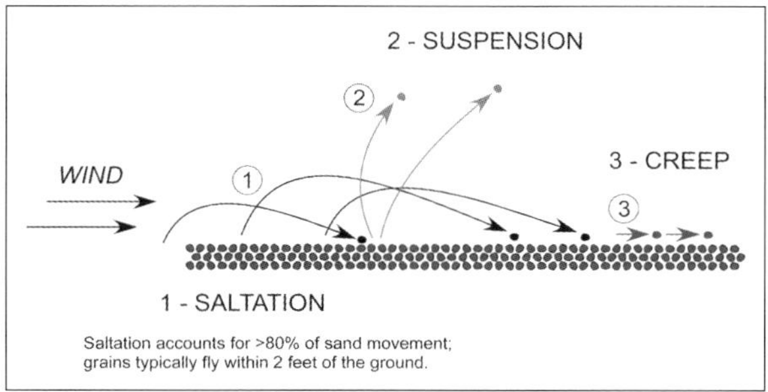

FIGURE 71. The mechanisms by which sand grains are transported by the wind – saltation, suspension, and creep. From Guadalupe Restoration Project – Dunes Center Manual.

Obviously the sand is moved by blowing wind, so it is worthwhile to examine the wind patterns on the Oregon Coast. As shown by the wind roses for Portland, Oregon in Figure 73, the winds show a bimodal pattern. Close analysis shows that the winds blow primarily from the northwest during May to September (the "dry season") and from the southeast during November to February (the "wet" season). Thus, one would expect to see an abundance of dunes formed by the northwest winds during the drier and warmer months of summer. An inspection of aerial images, such as the one shown in Figure 74, shows that to be the case; note the abundance of transverse dunes on the image that are oriented perpendicular to the northwest winds that formed them. Cooper (1958) discussed what he called "transverse ridges" that are similar to what we are calling transverse dunes, saying *the transverse-ridge pattern occurs in nearly all the major dune localities on the Oregon Coast, reaching its finest development on the Coos Bay dune sheet*. However, there are some exceptions to this trend (i.e., dunes not oriented perpendicular to the NW wind direction), depending upon different wind directions

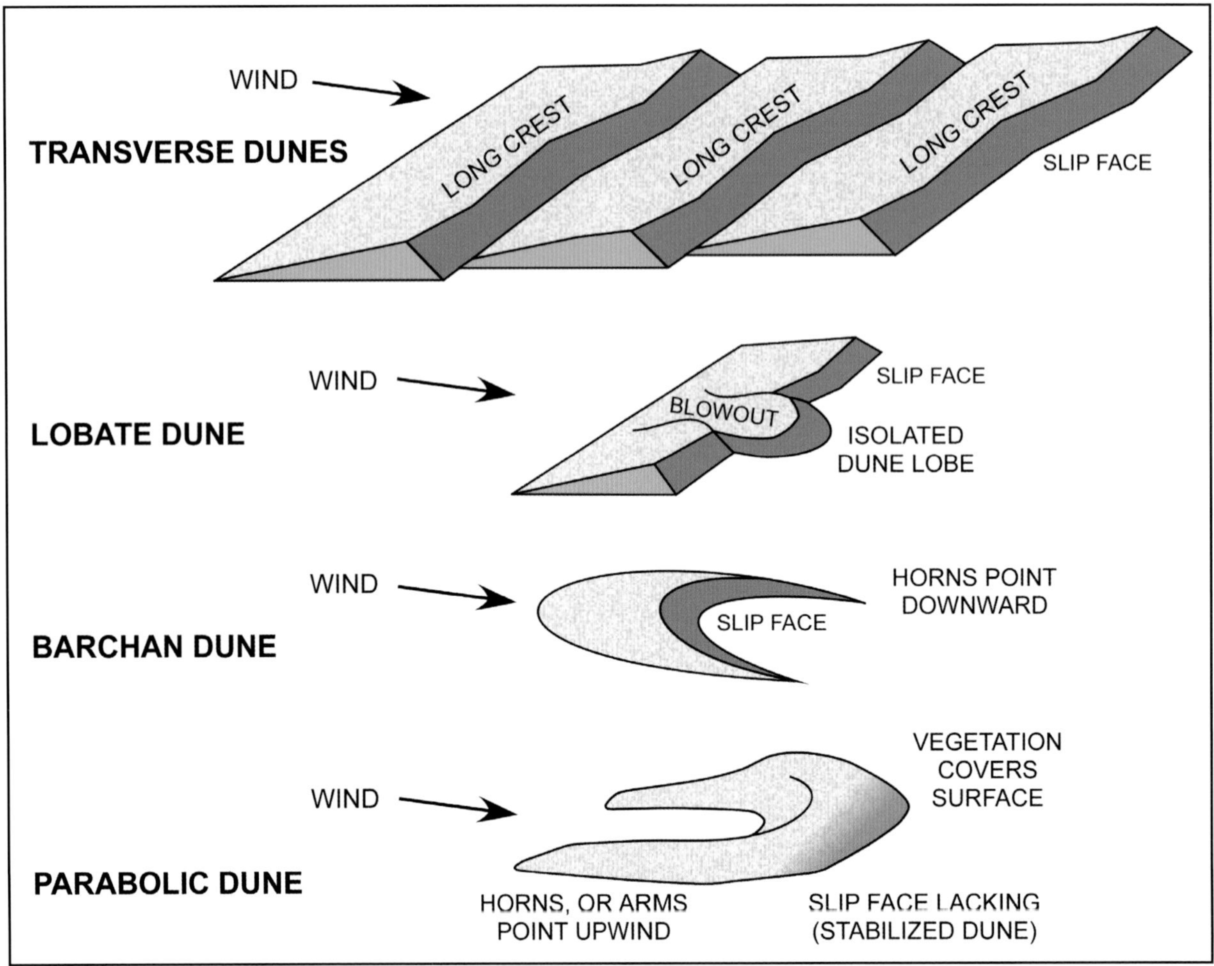

FIGURE 72. Four sand dune types typically present in the coastal dunes of the Oregon and Washington Coasts. From Guadalupe Restoration Project – Dunes Center Manual.

and other factors.

A key question is where did all this sand come from? Geotimes (2004) commented on that question as follows. The many rivers flowing westward from the Coast Range … *delivered a large proportion of the sand that is now incorporated into the dunes. Scientists, however, are uncertain about the mechanisms that brought the sand together to form the dunes. One theory, says Darren Beckstrand, a geologic consultant who wrote his master's thesis on the Oregon Dunes while at Portland State University, is that retreating beach sand caused by sea-level rise in the Holocene accumulated as it moved inland and formed the dunes. The competing hypothesis, favored by Beckstrand, is that the sand was blown inland from the continental shelf during periods of low sea level in the Pleistocene.*

RIVER DELTAS

An irregular bulge of the shore into a standing body of water at the mouth of a sediment-laden stream is normally referred to as a delta. Many deltas have an inverted pyramid shape, or perhaps more appropriately, an inverted version of the upper class Greek symbol delta. The distinction between deltas and estuaries is not readily apparent in some situations, as will be disclosed in more detail later in the discussion of estuaries.

In their classic summary paper on deltas, Coleman and Wright (1975) discussed over 50 parameters that have an impact on river delta morphology. Factors such as characteristics of the drainage basin, river slope, and coastal climate were acknowledged. Most present-day workers, however, simplify matters by focusing on

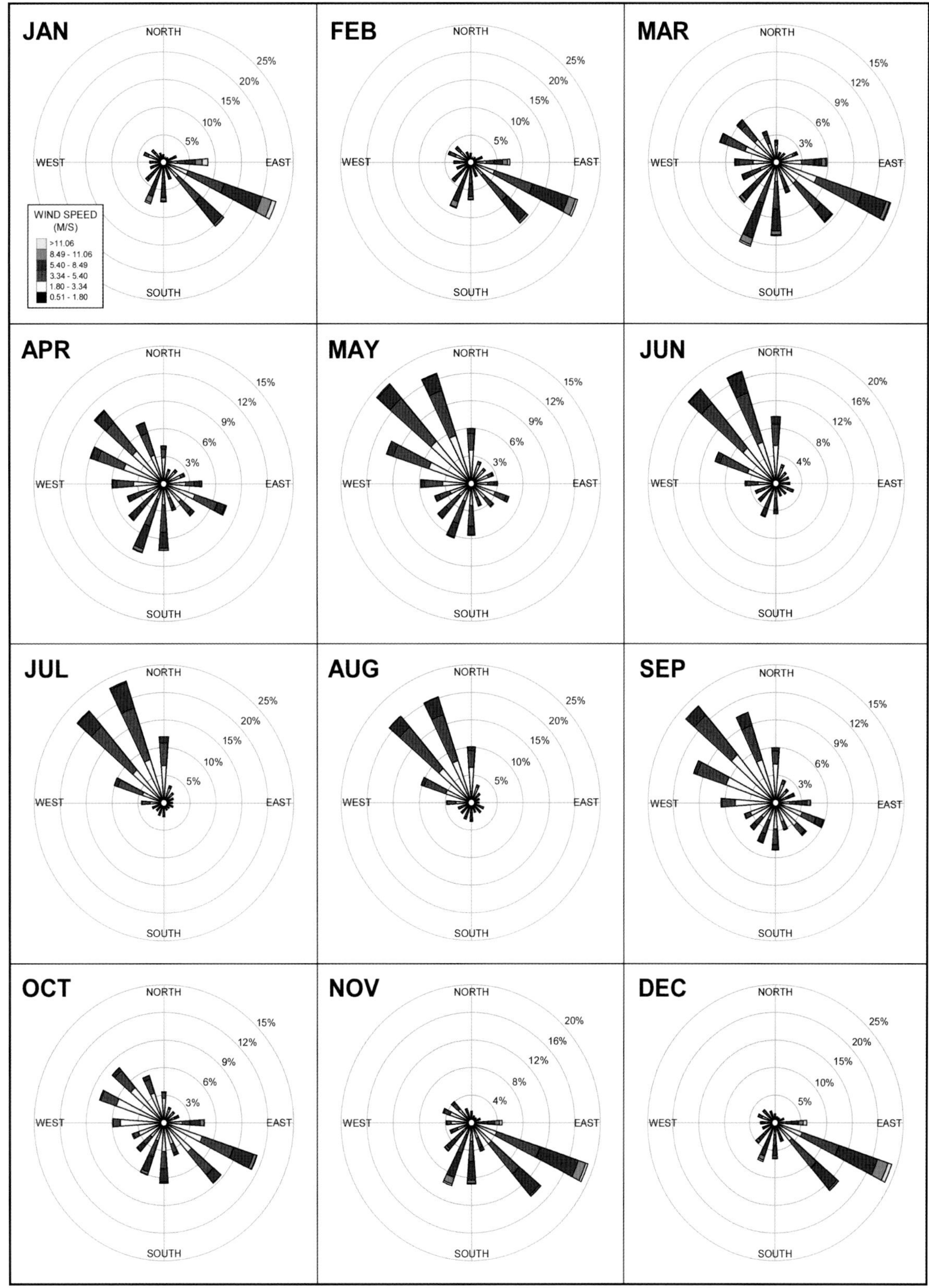

FIGURE 73. Wind roses for Portland, OR, derived from hourly measurements from 1961-1990. Source: National Water and Climate Center, Natural Resources Conservation Service, USDA.

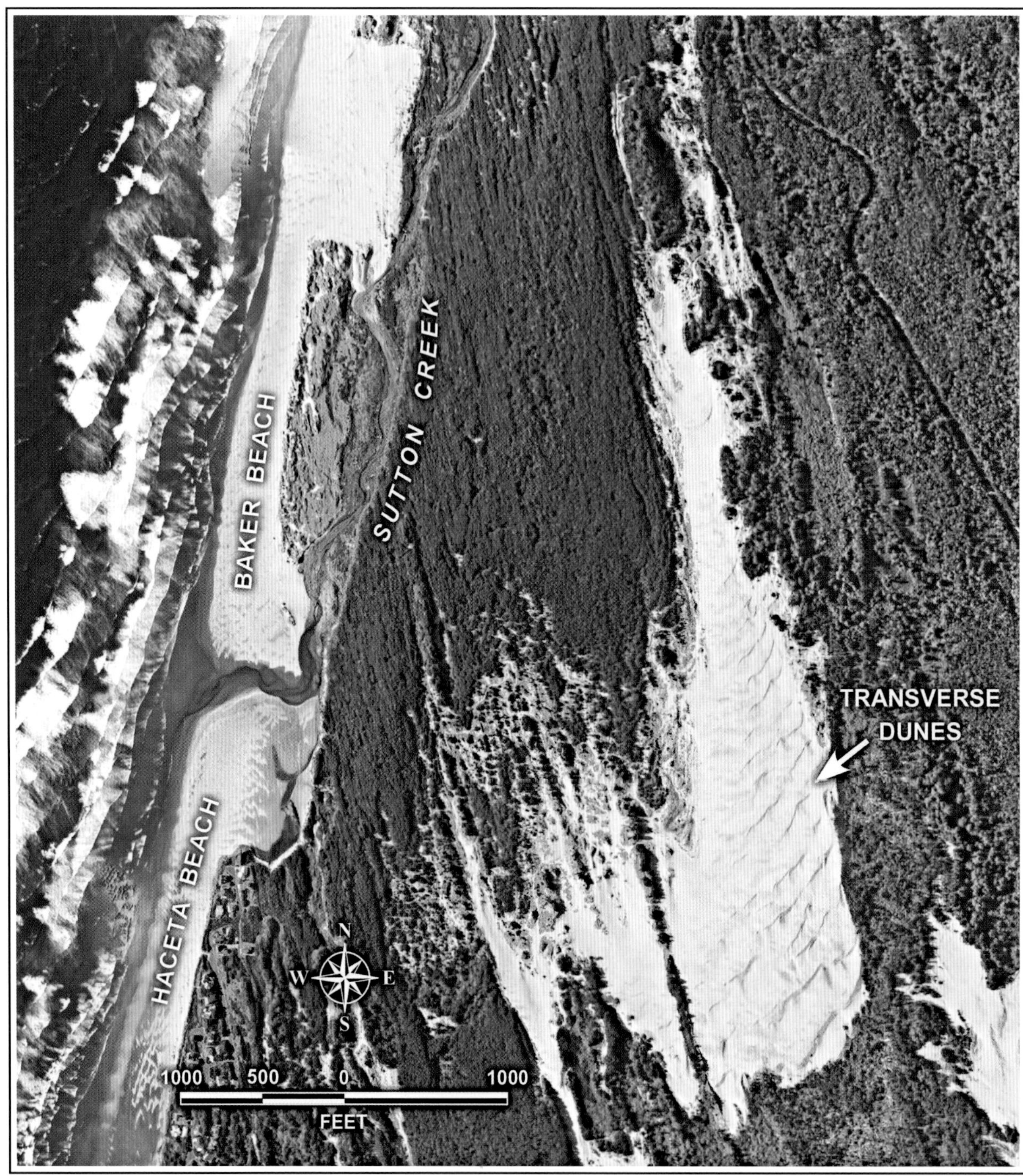

FIGURE 74. Barren sand dunes just to the north of Florence, OR. Note the transverse dunes oriented perpendicular to the northwest winds. NAIP 2014 imagery, courtesy of Aerial Photography Field Office, USDA.

three basic controls – sediment supply (river-dominated delta), wave energy (wave-dominated delta), and tidal current energy (tide-dominated delta) – in their attempts to classify deltas. For example, river-dominated deltas have a relatively large supply of sediment that tends to overwhelm the dynamic coastal processes, namely waves and tides, that are active around the deltas margin. Such deltas consist of a prograding lobe of sediment with a number of outward fanning, distributary channels. On the other hand, wave-dominated deltas have a rounded outer margin that contains wave-built outer bar systems. Figure 75 shows the wave-dominated delta of the Elwha River.

This topic will not be pursued any further, because there are no significant river deltas on the outer shores of Oregon and Washington and two relatively small ones along the Strait of Juan de Fuca.

FIGURE 75. The wave-dominated Elwha River Delta located on the Strait of Juan de Fuca. NAIP 2013 imagery, courtesy of Aerial Photography Field Office, USDA.

8 HABITATS OF COASTAL WATER BODIES

ESTUARIES

Definition

There are seventeen coastal water bodies, commonly called estuaries, along the coasts of Oregon and Washington. As observed by Emmett et al. (2000), such estuaries have historically played and are presently playing vital roles in the economy and life of the residents along the West Coast. They are gateways for shipping, they provide flat land for agriculture, they are nursery areas for fish and shellfish, and they are important areas for recreation, aquaculture, and urban development. The area covered in this book presently has one estuary designated as a Land Margin Ecosystem Study Area by the National Science Foundation (Columbia River), two estuaries that are included in the Environmental Protection Agency National Estuary Program (Columbia River and Tillamook Bay), and one designated as a National Estuarine Research Reserve (South Slough).

According to the original definition, which was basically a description of Chesapeake Bay by Pritchard (1967), an estuary has three defining qualities:

1) Flooded river valley that was formed during the lowstand of sea level that culminated about 20,000 years ago;

2) Water body with a substantial freshwater influx; and

3) Water body subject to tidal fluctuations.

The second and third criteria are self explanatory, but the first requires some understanding of sea-level fluctuations during the ice ages of the Pleistocene epoch. As suggested earlier, when the sea level was low during the major glaciations, streams flowing across the mainland to the sea carved deep valleys that extended out onto the present continental shelf in some areas. In many cases, this type of erosion was confined to the same valleys during each of the four major drops in sea level. When sea level started to rise at the beginning of the Holocene epoch, about 12,000 years ago, those valleys were flooded as the shore of the rising sea advanced across the then-exposed continental shelf. When sea level essentially stopped this sudden rise around 5,000 years ago, the valleys were flooded with salt water for some distance inland, creating the original estuaries. In some places, the valleys were eventually filled with sediment and a bulge of sediment protruded out into the ocean away from the general trend of the shore, converting the original "estuaries" into what we now call "deltas." A second type of delta, that some call a bay-head delta, may form further up the flooded valley far short of the open ocean. Bay-head deltas formed in situations where the volume of sediment supplied by the river was not sufficient to completely fill the valley out to the open ocean shore within the time allowed. The head of Tillamook Bay is a good example of this type of delta (Figure 76).

Estuarine Characteristics

A key factor in defining the physical structure of an estuary is the way in which the salt and fresh water mix. Salt water is more dense than fresh water, resulting in vertical stratification within estuaries, the extent of which is depending on physical forces (tide, wind) within the estuary. As shown in Figure 77, a highly stratified estuary with a well-defined salt wedge occurs where a significant freshwater stream enters a flooded valley with a small tidal range, with the more dense salt water hugging the bottom of the channel during a rising tide (see Upper Diagram in Figure 77). With such a large river discharge, the estuary at the mouth of the Columbia River is a salt-wedge estuary. Mixing of the salt and fresh water is enhanced by an increase in tidal energy; therefore, with a larger tidal range, the two water

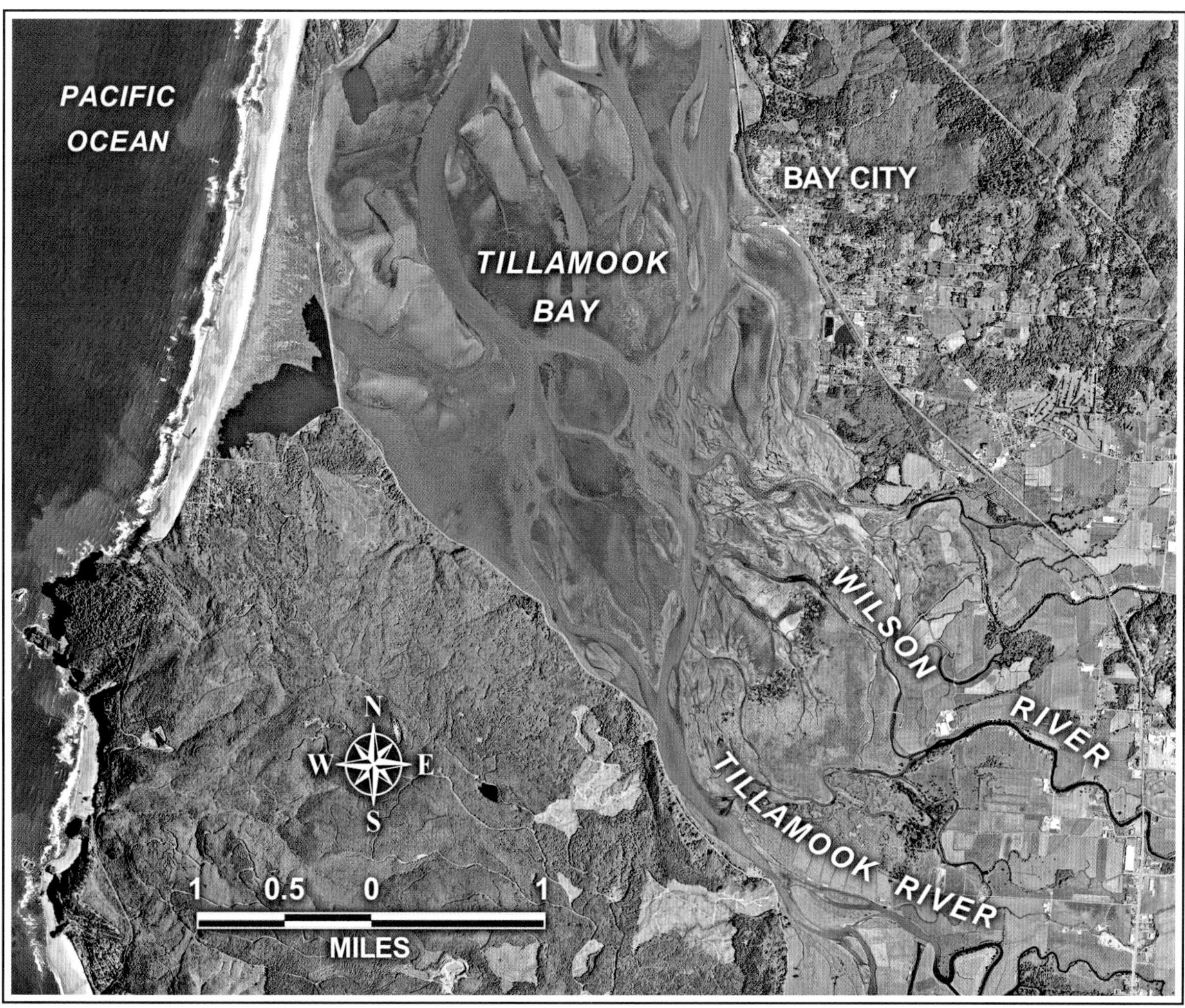

FIGURE 76. The bayhead delta in Tillamook Bay, OR. NAIP 2009 imagery, courtesy of Aerial Photography Field Office, USDA.

masses become partially mixed (as indicated by the arrows in the Middle Diagram in Figure 77). With an even larger tidal range, the water within the estuary may become homogeneous from top to bottom, completely fresh top to bottom at the head of the estuary and pure salt water from top to bottom at the entrance (Lower Diagram in Figure 77).

Figure 78A shows a plan view model of an estuary, and Figure 78B shows a cross-section illustrating the circulation within these typically partially mixed estuaries (middle type shown in Figure 77). We consider the most landward boundary of an estuary to be the place where the tide "stops going up and down," the so-called "head of tides." In the upper portions of the type of estuary illustrated in Figure 78, the primary hydrodynamic process is the river flow; whereas at the entrance, waves and tides dominate. The sediments in the middle and upper portion of the estuary are typically provided by the incoming river, but in some estuaries, a considerable amount of sediment comes into the entrance of the estuary from offshore under the influence of flood-tidal and wave-generated currents.

Considering the different components of the system as illustrated in Figure 78A, the upper estuary is a zone of relatively low tidal flow, and the river flow is still strong in a seaward direction. In the middle of the estuary, tidal flow is increased, the flood tide reverses the river currents when the tide rises, and a complex two-way, two-layer circulation occurs, creating a zone of mixing of the two water masses. In that zone of mixing, demonstrated in

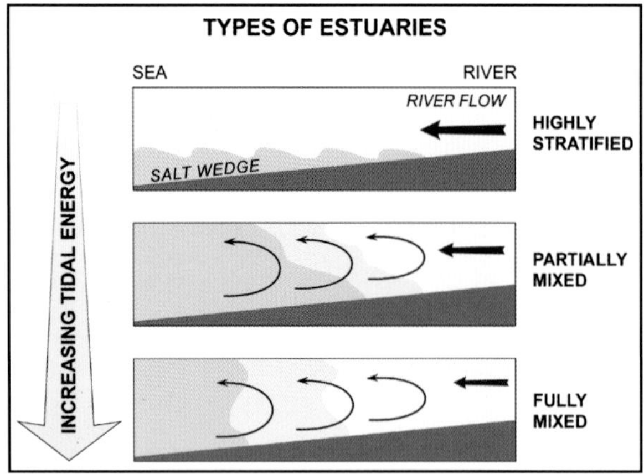

FIGURE 77. Effect of tides on the mixing of salt and fresh water in estuaries (modified after Biggs, 1978).

the cross-section in Figure 78B, an important process known as **flocculation** of the clay particles in suspension occurs. Flocculation takes place because the abundant clay particles in suspension brought in by the river slow down their seaward movement at the point of mixing as a result of the reversing of tidal flow directions twice a day. That allows an increase in the number of clay particles in the water column to the extent that they begin to collide with one another. Those colliding particles tend to adhere together under the influence of the increased concentration of cations, such as $Na+$, $Ca++$ and $Mg++$, derived from the seawater. As a result, the clay particles, which may have diameters as little as one micron, cluster into groups (**flocs**) that may have diameters measured in hundreds of microns. As a result, many of the flocs sink to the bottom at slack tide (Figure 78B). Once on the bottom, some of the flocs are resuspended by ebb- and flood-tidal currents. The net result is the formation of a zone in the middle reaches of the estuary that contains abundant fine-grained sediment in suspension known as the **zone of the turbidity maximum**. That zone of turbidity can migrate up and down the estuary over time, moving downstream during periods of flood on the river and upstream during low-flow periods. A side effect of this process is the creation of a "mud zone" on the tidal flats and in the tidal channels in the middle of the estuary (Figure 78A).

At the estuary entrance, conditions take on an entirely different aspect in that:

1) Tidal currents are more dominant than river currents with common differentiation between ebb-dominated and flood-dominated channels;

2) The waters are more mixed, with diminishing estuarine stratification;

3) Wave activity becomes a more important dynamic process; and

4) Moving of sediments along the bottom by wave- and tide-generated currents causes the tidal flats and channel bottoms to become much more sandy than those in the central portions of the estuary in the "mud zone."

Emmett et al. (2000) produced a map showing the location of the seventeen key "estuaries" on the coasts of Oregon and Washington. The ones in Oregon, going from south to north, include – Rogue River, South Slough, Coos Bay, Umpqua River, Suislaw River, Alsea River, Yaquina Bay, Siletz River, Netarts Bay, Nehalem Bay, Tillamook Bay, and Columbia River. Estuaries in Washington include Columbia River, Willapa Bay, Grays Harbor, Hood Canal, Puget Sound, and Padilla Bay. All of those systems are drowned-river-valley type estuaries, except for Netarts Bay, which they call a bar-built system, and Hood Canal and Puget Sound, which they call fjord systems.

Details on specific coastal water bodies are presented in Section III.

MARSHES

The coastal water bodies of the Pacific Northwest Coast are host to fairly extensive developments of salt, brackish, and freshwater marshes that are well documented as being the primary food sources for coastal and nearshore ecosystems. Marshes in these estuaries are flooded with water on a regular basis and occur in areas that were originally tidal flats where sediment accumulated in the early stages of evolution of the estuarine system. As the flat was built up to or slightly above the appropriate sea level (typically variable between mean sea level and neap high tide), marsh grasses began to take root. Once the grasses grew on the flat, the sedimentation process was accelerated because of the baffling effect of the plants on the tidal currents. The salt marshes that presently occur in the Oregon and Washington estuaries are, in effect, intertidal flats vegetated with halophytes (plants that are adapted to grow in salty soil; Basan and Frey, 1977). Marshes can expand very rapidly, up to a yard or so a year, if the slope is flat and sediment is abundant. Vertical growth is much slower because it depends upon the rate of sedimentation

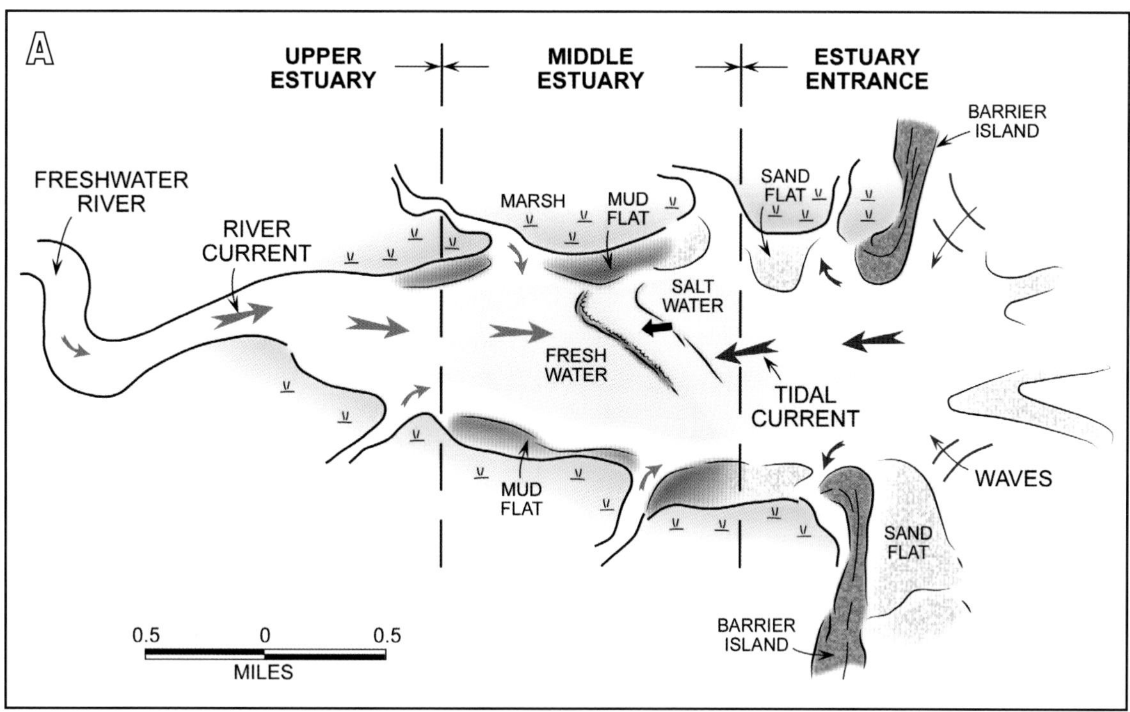

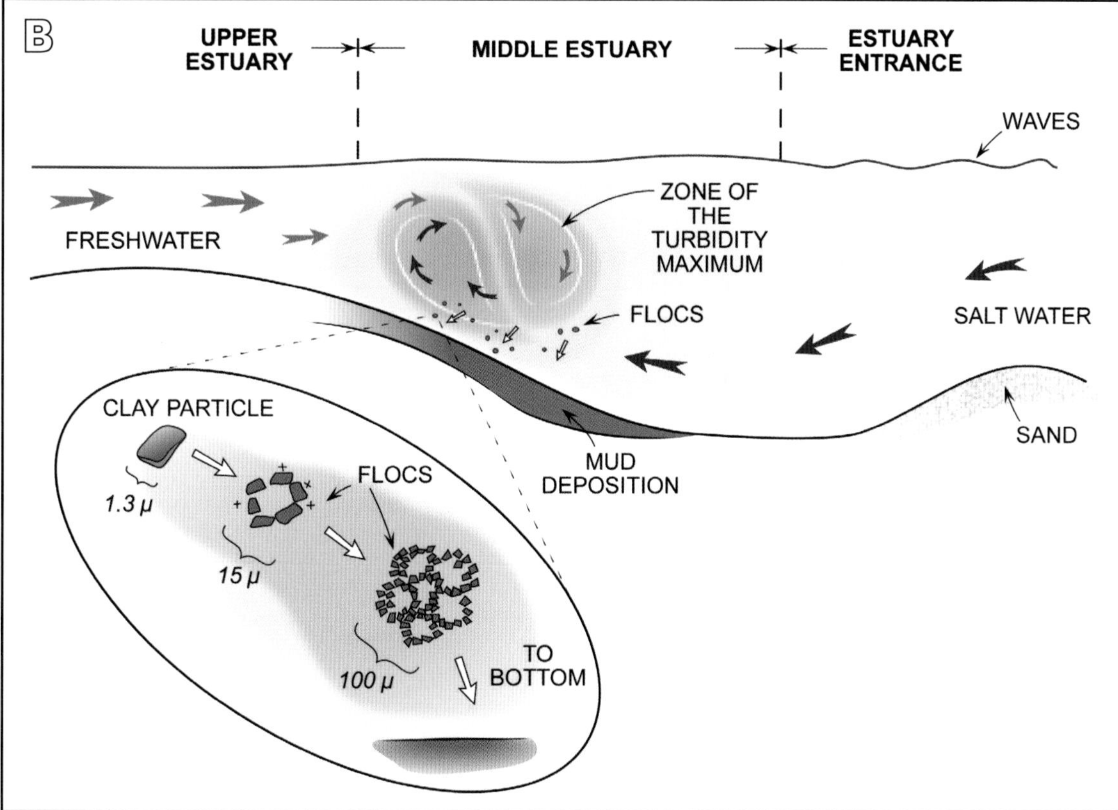

FIGURE 78. Estuaries. (A) Plan view of a typical partially mixed estuary. (B) Cross-section of the estuary illustrated in A showing predominant processes and circulation patterns. Especially noteworthy is the formation of a *zone of the turbidity maximum* where the clay particles (<0.004 mm in size; Figure 6) brought in by the river undergo a process called *flocculation.*

on the underlying tidal flats, which is typically quite slow.

Lateral salinity changes up and down the estuary, most notably the decrease in salinity toward the head of the estuary, have a striking impact on the plant communities. Freshwater marshes are most common in the upper estuary, where there are tides but very low salinities; brackish marshes occur most commonly in the middle estuary, where salinities generally average less than 15 parts of "salts" per thousand parts of water (ppt); and salt marshes occur in the lower estuary, where salinities range from 15 ppt to the low 30s.

Marshes are some of the most valuable coastal ecosystems, providing extensive ecosystem services, including nutrient and pollution filtration, protection from storms/erosion, carbon sequestration, nursery ground for marine species, bird habitat, habitat for commercially and recreationally harvested species. Unfortunately, along the Pacific Coast of North America, marsh habitats have been dealt serious blows by human development. *Since 1870, almost 24% of the Columbia River estuary's wetlands have been converted to diked floodplain, uplands, and non-estuarine wetlands (Thomas 1983). Willapa Bay lost 1,117 ha (hectares; 1 ha = 2.47 acres) between 1905 and 1974, while Grays Harbor lost 558 ha between 1916 and 1981* (Emmett et al., 2000).

The generalized profile of an Oregon intertidal marsh is shown in Figure 79, consisting of the following four subdivisions:

1) Low marsh indicators – arrow-grass (*Triglochin palustris*); pickleweed (*Salicornia virginica*); Lyngby's sedge (*Carex lyngbyei*);

2) High marsh indicators – tufted hair grass (*Deschampsia caespitosa*); common spike rush (*Eleocharis palustris*);

3) Transition species – silverweed (*Argentina anserine*); redtop (*Agrostis gigantean*); Baltic rush (*Juncus arcticus*); and

4) Upland – Sitka spruce (Picea sitchensis)

And finally, more bad news from Emmett et al. (2000). *Extensive freshwater swamp forests that once existed at the base of steep watersheds (such as in Tillamook Bay, Oregon) have been lost to diking.*

[NOTE: A **swamp** is a wetland forested with trees and shrubs (plants having a permanently woody main stem or trunk, ordinarily growing to a considerable height, and usually developing branches at some distance from the ground), whereas a **marsh** is a wetland inhabited by grasses (relatively short plants having jointed stems sheathed by long narrow leaves, flowers in spikes, and seedlike fruits).]

TIDAL FLATS

Definition

Tidal flats are nearly flat intertidal surfaces that are usually sheltered from direct wave attack. In most places, the sediment grain size decreases as you move landward, away from main tidal channels, because the sediments are carried to their final resting place by tidal currents that decrease in velocity away from the channels as the water flows up onto the flats. The depositional and erosional rates on tidal flats are much slower than on beaches on the open-ocean front, where the sediment is moved readily by wave-generated currents.

Exposed Sandy Tidal Flats

In the entrances to some of the larger estuaries, broad intertidal flats up to several hundred yards long and wide occur (Figure 44). Those flats are usually composed of sand, which indicates that tidal currents and waves are strong enough to mobilize the sediments from time to time, removing the muddy sediments. Even so, sedimentation rates are considerably slower than on the more exposed beaches. On the lower portions of some of the sandy flats, waves may have formed low ridges that migrate very slowly across the flat. Also, ripples and even larger bedforms may be present, either on the lower reaches of the flat or in the vicinity of tidal channels that cut across it. Despite this kind of sediment motion lower down, most of the sand and muddy sand on the upper half of these flats is relatively stable; therefore, it usually contains a huge population of infaunal organisms (those that burrow in the sediment).

Sheltered Mud Flats

For the most part, intertidal mud flats occur in the most sheltered portions of coastal water bodies, as well as on the most landward portions of wide tidal flats that are sandy on their lower reaches (see example in Figure 45A). Mud flats are important ecological habitats, and the abundant animals that live in the mud are key food sources for migratory birds and wintering waterfowl.

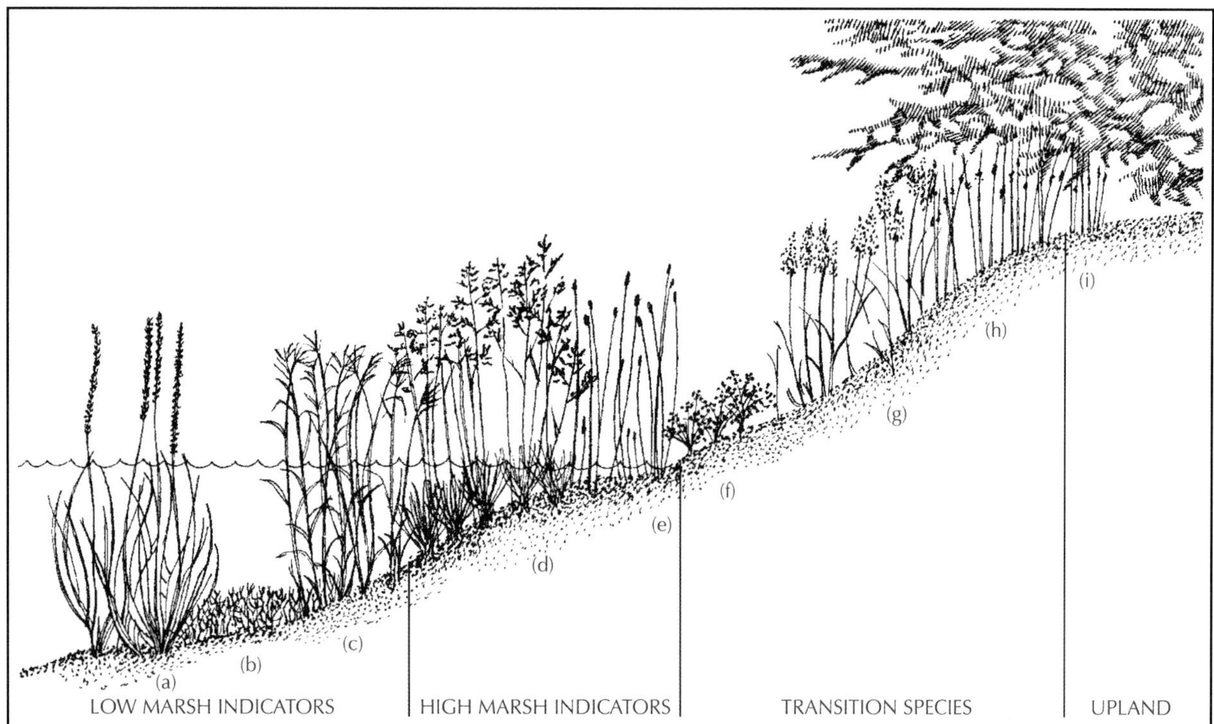

FIGURE 79. Generalized profile of an Oregon intertidal marsh. Although the plant species shown are prominent throughout their respective zones, they may also be found to some extent in the other marsh zones. Plants shown are: (a) arrow-grass, (b) pickleweed, (c) Lyngby's sedge, (d) tufted hairgrass, (e) common spike rush, (f) silverweed, (g) redtop, (h) Baltic rush, and (i) Sitka spruce. Modified from *Oregon's Salt Marshes*, an archived document maintained by the Oregon State Library.

The question of why large quantities of mud accumulate on the more sheltered tidal flats is an interesting one. Several Dutch workers have commented on that issue. Van Straaten (1950) and Postma (1967) concluded that the mud is deposited during the last stages of the flood flow, during slack water, or at ebb just before water is drawn away. Furthermore, a stronger current is required to pick the mud back up than the one that deposited it. The finest particles in the mud are composed of clay minerals, which have a thin, sheet-like shape (like a thin coin). This shape slows down the process of settling of the particles from suspension, which allows them to be transported landward by the waning flood currents beyond the point where more spherical particles with a similar mass would settle out. Also, once the thin, disc-shaped particles have settled on the bottom, their thin edges provide little resistance to the outflowing water because of their lower pivotability, as was explained for disc-shaped clasts on gravel beaches. Therefore, they are not easily picked back up and set into suspension. That process of permanent deposition is aided further by the fact that floccules of organic matter and suspended fine-grained sediment (see Figure 78B), as well as slime-secreting diatoms, create other particles (in addition to individual silt and clay particles) that readily settle out in the quiet water. Once that particular mud is deposited, it tends to remain in place as well because: 1) Diatoms move up through the mud and deposit slime; 2) The mud dries out somewhat during low tide; and 3) Burrowing organisms tend to stabilize the mud. In some cases, mud is trapped between plants on the flat.

Another factor of prime importance to mud deposition on tidal flats is the 'ecosystem engineering' that occurs due to the super abundance of filter-feeding organisms, such as oysters and clams. During the filtering process, those organisms compress the finely divided clay particles and bind them together in their intestines as fecal pellets that are excreted onto the flats. Another part of the suspended matter is coagulated in their gills and pushed back into the water as pseudo-feces. The feces and pseudo-feces are easily deposited, even in comparatively turbulent water.

Dominant Biogenic Features

The tracks, trails, and types of burrow structures of the animals that live on tidal flats and in the sediments (called biogenic features) have fascinated geologists for decades, because such evidence is useful for interpreting the depositional environments in ancient sediments (commonly used by geologists interpreting rock layers in their search for economic deposits, such as oil and gas). Knowledge on this topic has been greatly advanced by studies on the Georgia Coast, specifically by researchers at the University of Georgia (e.g., Frey and Howard, 1969; Howard and Dorjes, 1972). Generally speaking, the animals that live in the sediment on tidal flats, such as clams and burrowing worms of various kinds (called infauna), live in either vertical tubes or U-shaped burrows, that maintain less variable temperatures and salinities than exist on the surface of the flats. Figure 80 shows sketches of some of the more common burrow types in tidal flats. Except in lower reaches of some sand flats and tidal point bars, where highly mobile large-scale bedforms occur, the tidal flat sediments of the mid to upper estuarine regions of Oregon and Washington are highly burrowed throughout. One of the most common burrowers, according to Kozuka and Chaiudez (2008), is the ghost shrimp, *Neotrypaea californiensis*. Dense populations of ghost shrimp (Figure 81A) are found in the major estuaries of Washington and Oregon.

You may see some of the infauna and/or their burrows as you walk the intertidal regions of the coastal water bodies of the Pacific Northwest Coast (WITH SHOVEL IN HAND!!). These infauna provide an important food source for migratory birds and are the reason for large concentrations of waterfowl and shorebirds on tidal flats.

Speaking of shovels, looking in the right place one might use a shovel to dig up some of the delicious razor clams (Figure 81B). According to the Oregon Department of Fish and Wildlife (2014), *razor clams (Silqua patula) are found throughout Oregon's ocean beaches. Clatsop beaches (Columbia River to Seaside) have the most stable populations (because of beach stability), 95 percent of Oregon's razor clam digging occurs there. Other areas such as Agate Beach, Waldport Beach, Whiskey Run, Myers Creek, and other beaches along the coast also have razor clam populations* that tend to be less available than those on Clatsop beaches. In Washington, razor clam fishing is popular at Long Beach, Two Harbors Beach, Copalis Beach, Mocrocks Beach, and Kalaloch Beach. Check with state shellfish managers for when the razor clam fishery is open, as well as for information on shellfish safety.

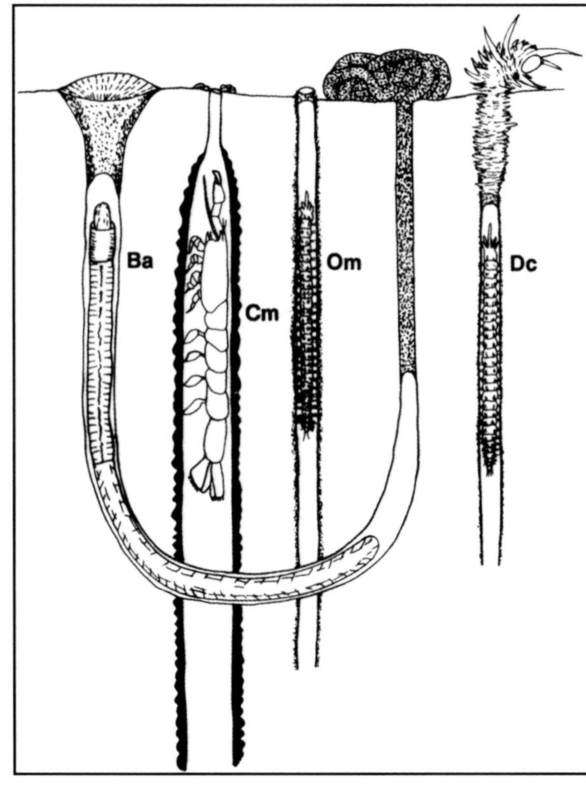

FIGURE 80. Burrowing characteristics of infauna on tidal flats.
Ba – The iodine worm (*Balanoglossus*), which has an open U-shaped burrow. Note the excreted sand pile, a conspicuous feature on the sandy tidal flats where this organism lives.
Cm – The ghost shrimp (*Callianassa*), which lives in a complex vertical to horizontal catacomb-like network of burrows lined with mud-rich fecal pellets.
Om – The soda straw worm (*Onuphis microcephala*), which lives in a vertical mucoid tube covered with sand grains.
Dc – The periscope worm (*Diopatra cuprea*), which lives in a vertical chitinous tube with a periscope-like entrance. (Modified after Howard and Dorjes, 1972.)

TIDAL CHANNELS

On a global scale, there are four basic styles of channel morphology that occur in natural river and tidal

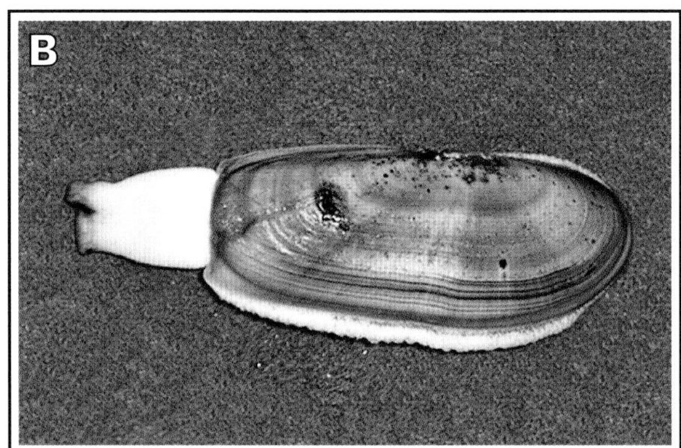

FIGURE 81. Infauna on tidal flats. (A) The ghost shrimp (*Neotrypaea californiensis)* Whidbey Island, WA by Dave Cowles. (B) The razor clam (*Silqua patula*) Westport, WA, courtesy of Wikimedia Commons.

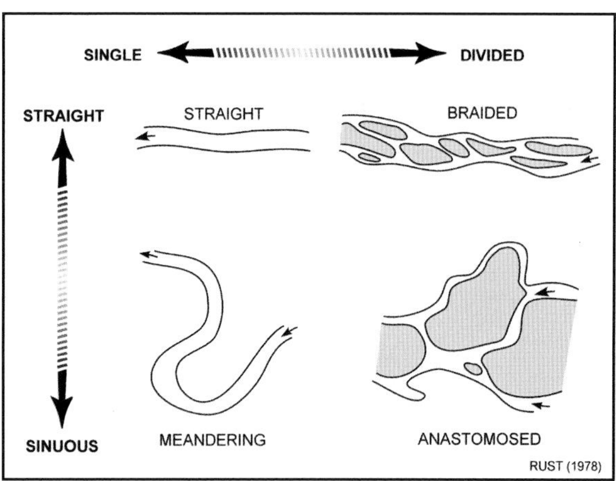

FIGURE 82. Channel types, classified as to whether they are straight or sinuous and whether they are single channels or a complex divided channel (after Rust, 1978).

systems, as illustrated in Figure 82. **Straight channels** are rare in general, so be suspicious if you see one on the coast, because it has most likely been dug for navigation, drainage, or other purposes. As a rule, **braided channels** occur under conditions of: 1) Relatively steep channel slope; 2) High bedload content (sedimentary material on channel bottom; e.g., gravel and coarse sand, that moves along the bottom by rolling and bouncing), and 3) Flashy discharge (high flow during periods of exceptionally high water and low flow during the rest of the time). **Meandering channels** tend to have: 1) Flatter slopes than braided channels; 2) More sediments suspended in the water column than moving along the bottom; and 3) More steady discharge. Meandering channels are by far the most common channel type in the coastal water bodies of the Pacific Northwest Coast. **Anastomosing channels** typically occur on extremely flat slopes in waters with a very high ratio of suspended to bedload sediment.

9 CONTINENTAL SHELF AND SUBMARINE CANYONS

Though not a place you are apt to visit during your explorations on the Pacific Northwest Coast, unless you own or hire an ocean-going vessel, the continental shelf represents the seaward end member of the coastal zone continuum. The continental shelf is defined as a wide and gently sloping submerged surface that connects the beach zone (mean sea level) with an abrupt increase in slope, or **shelf break**, where the shelf edge joins a steeply descending planar surface called the **continental slope** (Figure 83).

Another of the major contrasts between the leading-edge and trailing-edge coasts of North America is the width of the continental shelf. Off the coast of the southeastern U.S., the shelf is quite wide, extending an average of about 60 miles off the shore, with the shelf break located at depths of around 600 feet. In contrast, according to Landry and Hickey (1989), the width of the continental shelf off Washington and Oregon varies between 15 and 36 miles, with the shelf break located at depths of 495 to 660 feet. In places, the continental shelf

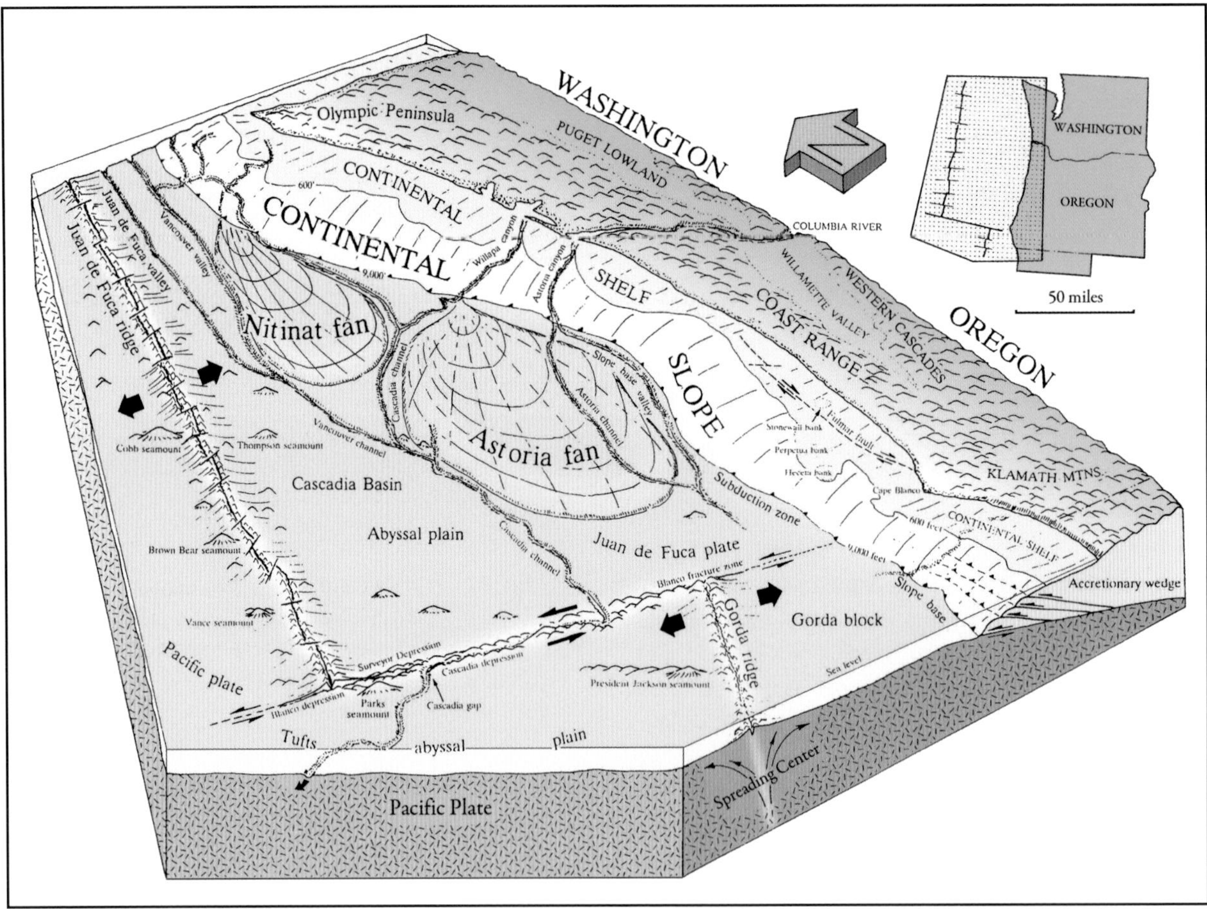

FIGURE 83. Physiography and structure of the continental slope and abyssal plain off Oregon and Washington. Modified after Orr and Orr (2000).

is carved into by submarine canyons starting in water depths of 430 feet, reducing the shelf width to 9-18 miles.

Therefore, the continental shelf off Oregon and Washington is generally quite narrow. The shelf off the U.S. East Coast is wide because it is underlain by relatively horizontal sedimentary rock layers that range in age back to the late Triassic or early Jurassic (to the past 200 million years or so), the time when the present ocean began to take shape as a result of the opening of the mid-Atlantic rift zone. The reason the shelf off Oregon and Washington is much narrower is because this coastal area is a location where long-term subduction of the seafloor along the leading edge of the North American Plate has been taking place for millions of years.

Of considerable relevance, no doubt, as far as the coastal zone is concerned, is the role of numerous submarine canyons located along the edge of the continental shelf (see Figure 83). Submarine canyons are steep, V-shaped valleys eroded into the continental slope. They typically head near the shelf break (contact between continental shelf and continental slope) and end near the bottom of the continental slope. There has been much speculation in the scientific literature regarding their origin. Many of them are located off major rivers, such as the Mississippi and the Ganges/Brahmaputra, where the largest submarine canyon on Earth, the "Swatch of No Ground," feeds the largest submarine fan in the World, the Bengal Deep-Sea Fan. One idea is that the rivers played a role in carving the canyons during the lowstands of the Pleistocene epoch, when the river mouths were near the edge of the shelf. According to that reasonable explanation, the sediment dumped at the river mouth was then carried deeper by **turbidity currents** (that are illustrated in Figure 84). The fact that some canyons are not located off major rivers (at the present time) implies that other processes may be involved, such as seafloor slumping and other forms of mass wasting. Along the Cascadia Subduction Zone, turbidites are triggered by large earthquakes (Goldfinger et al., 2012). In any event, there is considerable evidence that, at the present time, turbidity currents help to keep the canyons eroding and many submarine deposits (fan-like lobes in some cases) are deposited by such density currents. The coastal scientists who originally conceived the idea of littoral cells for the California coast (Inman and Chamberlain, 1960; Bowen and Inman, 1966) cited turbidity currents and other mechanisms, such as sand slumps or debris flows carrying sand into submarine canyons, as a principle way sand is lost from those cells.

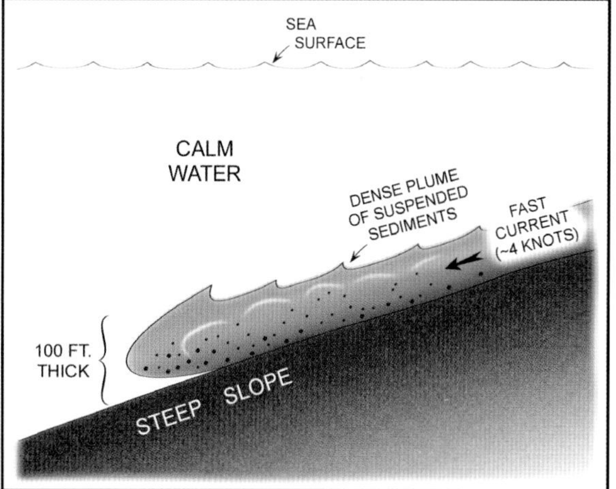

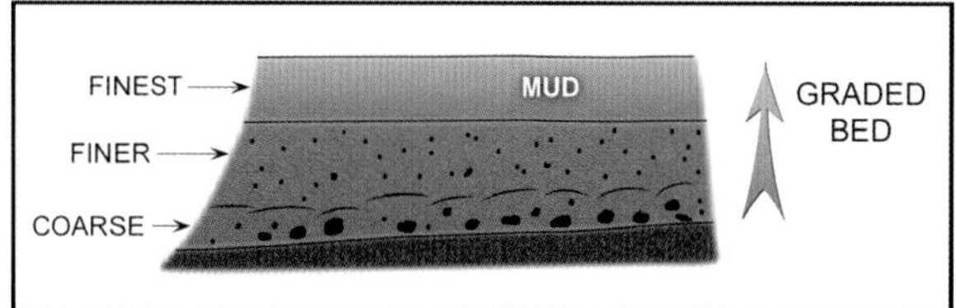

FIGURE 84. Turbidity currents and their deposits (turbidites). (A) General model. These sedimentary plumes come in a wide variety of sizes. (B) General pattern of fining upward (grading) of a turbidity-current deposit. (C) Graded bed in a turbidity-current deposit in the Carmelo Formation at the Point Lobos State Natural Reserve, CA.

The diagram in Figure 83 illustrates the structure of the continental slope and abyssal plain off Oregon and Washington. Dominating the scene are two massive submarine fans, the Nitinat Fan to the north and the

Astoria Fan off the Columbia River mouth. Orr and Orr (2000) state that *cutting completely across the continental shelf and slope, submarine canyons and channels are important avenues for dispersing sediments by turbidity currents that originate in shallow water. Of these, Astoria Canyon, a major submarine feature off the Oregon Coast, heads 10 miles west of the mouth of the Columbia River and extends across the Astoria submarine fan. … That fan extends from 6,000 feet along the continental slope to the abyssal plain below 9,000 feet and covers more than 3,500 square miles.*

10 SHORE EROSION AND RETREAT

INTRODUCTION

Shore erosion is a common phenomenon on many coastal areas. In fact, almost no coastal area of the World that has been developed by man is free of problems caused by such erosion. For the Pacific Northwest Coast, a distinction should be made between shore erosion and **shore retreat**. Beaches and foredunes erode during storms but can build back as part of the beach cycle; however, once a cliff erodies, it will never return to that position relative to the high-water line. Therefore, we discuss shore erosion processes along the Pacific Northwest Coast for: 1) Beaches and foredunes; and 2) Sea cliff retreat and landslides.

EROSION OF BEACHES AND FOREDUNES

The U.S. Geological Survey has conducted regional surveys as part of their National Assessment of Shoreline Change. Ruggiero et al. (2013) provided the results for the beaches along the Pacific Northwest Coast, which can be summarized as follows:

- The average rate of long-term and short-term shoreline change along beaches was about 3 feet per year;
- About 36% of the shoreline was determined to be eroding, more in Oregon than Washington because of the influence of the Columbia River where beaches north of the jetties have built out; and
- In Oregon, many coastal areas are either less accretional, changed from accretional to erosional, or more erosional when comparing the long- and short-term rate, possibly because of recent sea-level rise and an increase in storm wave heights.

It is the interaction of **sediment supply** and **water-level changes** that controls beach erosion. Along some coasts, fluctuations in sediment supply prevail, and on others, rapid changes in water level are most important. The most serious erosion problems typically occur either where man, or some natural process, interferes with sediment supply, or where some phenomenon causes an unusually rapid rise in water level (e.g., abrupt lake level rise in Lake Michigan or around sinking abandoned delta lobes on the Mississippi Delta). Erosion along the beaches of the Pacific Northwest Coast is a function of both sediment supply (because there are limited modern sand sources within the littoral cells) and water-level changes (short-term during storms and long-term related to sea-level rise).

To understand which natural sources are most important in providing sand to the shore at any given site, a **sediment budget** would have to be calculated in which the sediment additions (credits) and losses (debits) are determined and equated to the net gain or loss over a selected segment or littoral cell. Detailed sediment budgets for Oregon and Washington littoral cells are generally lacking, except for those associated with the Columbia River (Ruggiero et al., 2005).

There are a number of natural processes that could produce a deficit of sand at a particular beach along the Pacific Northwest Coast, including:

1) Transport of sand offshore during storms beyond the depth from which the sand is normally returned to the beach during calm periods. The sand can be carried far offshore onto the continental shelf or lost to submarine canyons, as shown in Figure 61; and

2) Wind transport of the sand landward to coastal foredune systems and out of the normal zone of the beach cycle (Figures 50 and 61).

3) Trapping of sediment in estuaries that are carried to the coast by rivers and transported by tidal currents. This process can be enhanced where earthquakes lowered the land elevation, essentially increasing the volume of the "sink" in the estuary, such as Komar and Styllas (2006) found for Tillamook Bay.

4) Transport of sand during tsunamis.

Short-term water-level changes, both seasonally and during storms (and especially during *El Niños*, as shown in Figure 31), can cause higher rates of beach erosion by increasing the elevation of waves that then break and surge higher onto the beach. During very large storms, the waves can inundate the beach and erode the foredunes. The eroded sand is carried offshore and deposited in nearshore bars, where it is eventually transported back on the beach during lower-wave conditions (Figure 51B). In the long term, beach erosion will likely increase because of two factors: 1) Increased sea levels; and 2) More frequent and larger waves during storms.

Man-induced changes that could produce a deficit of sand at a particular beach, include:

1) Dams on rivers. Sediment contribution of any river discharging on the coast is of importance to the sediment budget of that coast. If a dam is built on the river or its tributaries to create a reservoir for storage of water, the current velocities in the river are reduced to such an extent that practically all sandy sediment carried by the river settles in the reservoir. The water discharged from the dam contains very little sand; therefore, the sand supply formerly delivered to the coast by that river is cut off. The Elwha River in the Strait of Juan de Fuca is a classic example, which is discussed in detail in Section III.

2) Sand mining. This is not a practice that is normally accepted on U.S beaches. The extraction of sand from sand bodies in the nearshore zone, such as ebb-tidal deltas, for beach nourishment projects is a type of sand mining. Dredging of those shoals may have two side effects: a) Alteration of depth contours that may change wave-refraction patterns and cause a focusing of wave energy at different locations on the shore; and b) Elimination of a natural breakwater that acts to reduce wave energy arriving at the shore.

3) Construction of shore perpendicular structures. Jetties are long parallel structures built at a river mouth, a tidal inlet, or even an artificially dug harbor, to stabilize the channel, prevent its shoaling by littoral drift, and protect its entrance from waves (see example in Figure 85A). Jetties are also designed to direct or confine the flow to help the channel's self-scouring capacity. Usually, jetties extend through the entire nearshore zone to beyond the breaker zone to prevent sedimentation in the main channel. Therefore, the jetties can interfere with the seasonal transport of sediments within the littoral cell. Ten inlets in Oregon and three inlets in Washington have jetties at their entrances. Komar et al. (1976) and Komar (1983) note that, along the Oregon Coast, jetties behave as minature headlands, dividing the shore into sub-cells with a littoral cell. Sand accumulates both north and south of the jetties, with seasonal reversals but zero net change.

4) Construction of seawalls and revetments. Seawalls are vertical, hard structures very commonly built to protect man-made structures. Thousands of examples could be cited to demonstrate the effectiveness of such structures in saving property, at least in the short term. However, erosion of the sand on the beach itself is usually accelerated in front of seawalls because of wave reflection from their hard, vertical faces. According to Silvester (1977), when waves are obliquely reflected from such seawalls, energy is applied *"doubly"* to the sediment bed and *"hence expedites the transmission of material down coast."* That process is illustrated in Figure 86. Over the past few decades, there has been a hue and cry by a number of concerned scientists against building seawalls along open ocean beaches, to the extent that some states now ban new seawalls. Revetments serve the same purpose as sea walls, but they are typically made of materials such as boulder-sized chunks of rock, called riprap. Riprap revetments do not reflect waves as severely as sea walls. As shown in Table 2, exposed, solid man-made structures (mostly seawalls and revetments) are present along 16 miles of the outer coast of Oregon and Washington.

SEA CLIFF RETREAT AND LANDSLIDES

Sea cliff erosion rates vary both along the coast and over time. As shown in Figure 25, the southern part of the Oregon Coast and near the Oregon-Washington border are rising faster than the current rate of sea-level rise. Therefore, cliff erosion is not prevelant in these areas. However, the north-central part of the coast is rising slower than the current rate of sea-level rise. Therefore, there is more cliff erosion in this area because of the relatively higher water levels.

On sea cliffs fronted by beaches, the beaches act as a buffer to erosion, except during storms and particularly during periods of high water levels as discussed above. These higher water levels can reach the toe of the cliff and waves can erode the cliff face. Cliffs composed of sedimentary rock or Pleistocene terrace sand are more readily eroded by waves, compared to those composed

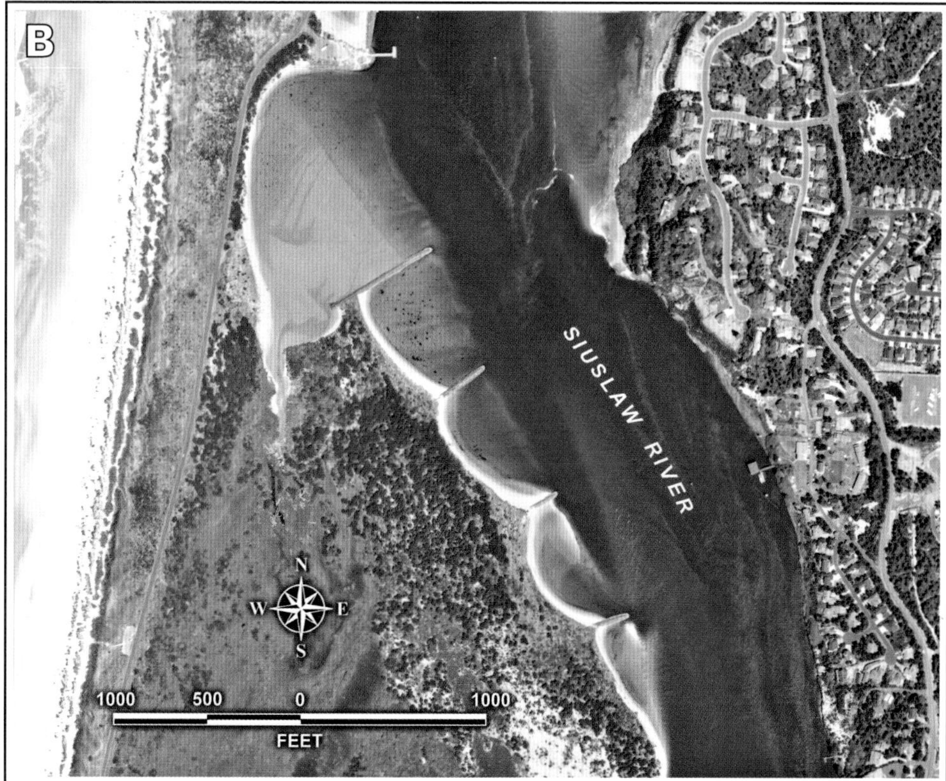

FIGURE 85. (A) The jetties at the mouth of the Siuslaw River near Florence, OR. NAIP 2011 imagery, courtesy of Aerial Photography Field Office, USDA. (B) Groins on the southwest shore of the Siuslaw River near Florence, OR. NAIP 2009 imagery, courtesy of Aerial Photography Field Office, USDA.

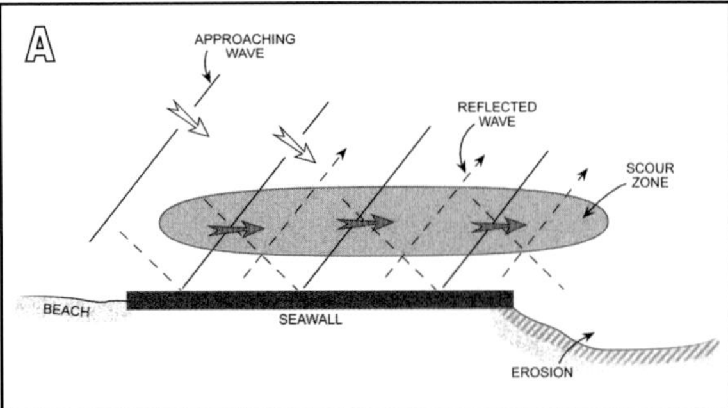

FIGURE 86. Seawalls and beach erosion. (A) Illustration of erosion in front of and on the downdrift side of a seawall as a result of waves reflecting from the seawall (highly modified after Silvester, 1977). The approaching wave crests meet the reflecting wave crests at approximately right angles, generating a flow parallel with the shore in the downdrift direction. This current scours a channel a few yards seaward of the wall. The beach downdrift of the seawall will also erode due to loss of sand into the scour zone. (B) Waves reflecting off a seawall at Hampton Beach, New Hampshire. Solid arrow indicates approaching wave and dashed arrow indicates reflected wave moving away from the seawall at a 90-degree angle to the approaching wave. Photograph by A. D. Hartwell taken circa 1969.

of more resistant basalts. Most of the headlands, sea stacks, and other rocky promontories are composed of basalt and erode very slowly, indicating their ability to withstand wave attack. However, cliff composition is only one of the many factors identified by Komar (1998a; 2004) that affect sea cliff erosion rates. The other factors include:

- Orientation of the sediment layers, joints, and faults in the rock cliff, which can allow preferential erosion along weaker layers.
- Inclination of the rock layers, where those that slope seaward are more prone to have massive landslides.
- Height and slope of the cliff face.
- Whether or not the eroded cliff material accumulates as talus that can provide some protection. Resistant rocks can break off as individual blocks, and terrace sands and gravels can accumulate at the base of the cliff. In contrast, fine-grained materials tend to be washed away.
- Presence of drift logs, which can act as buffers. However, Komar (1998a) notes that the logs are often floated away during storms.
- Rain wash and groundwater seepage, which can trigger landslides and wash sediment away. On the Oregon Coast, landslides are more frequent during months with the highest precipitation.
- Vegetation cover, which can slow the effects of rain wash and groundwater seepage.
- Human factors, including carving graffiti on the cliff face, digging tunnels, and placing riprap structures.

It should be noted that sea cliff erosion is a significant source of modern beach sediment, so a little erosion helps maintain the beach. However, on shorelines where sea cliff erosion is an important sand source, placement of seawalls and revetments to slow cliff erosion and protect property can reduce the amount of sand contributed to the beach, which could affect the sand budget (Shih and Komar, 1994).

Sea cliff erosion is augmented considerably when a landslide moves a mass of debris from the coastal cliff into the intertidal zone. Landslides are an example of the process called mass wasting. Harden (2004), in addition to illustrating the general characteristics of landslides with the diagram in Figure 87A, made the following points about those features:

1) In addition to feeding the longshore sediment transport system, landslides have dammed rivers, destroyed hundreds of homes, and killed several people *in the last 30 years*;

2) A slow movement of material, called creep, occurs *on virtually all hill slopes*;

3) When mass movement occurs abruptly, *a visible landslide scar is left in the area where debris was detached, which are typically bowl-shaped, with a steep slope at the head of the slide area* (Figure 87);

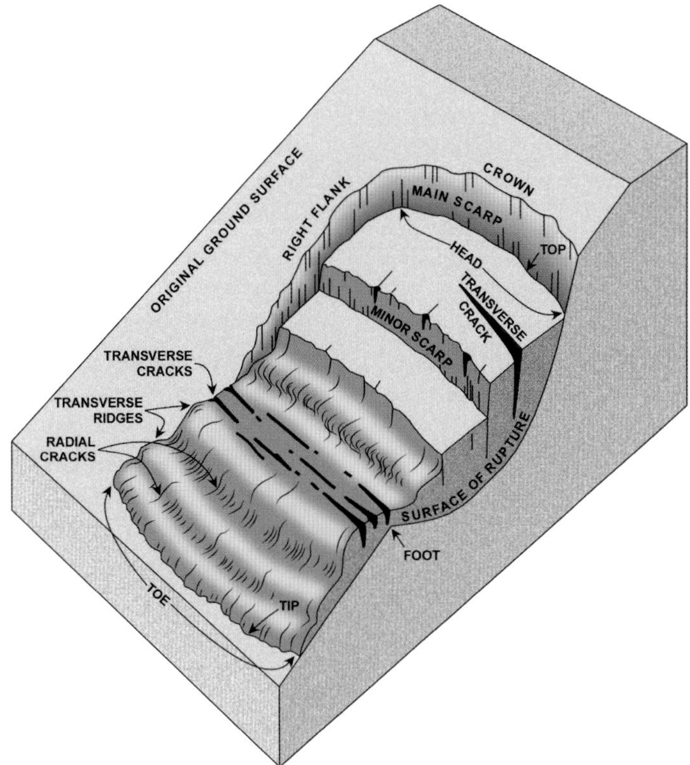

FIGURE 87. Landslides. (A) The features of a typical landslide (modified after Harden, 2004; original source: Highway Research Board, 1958). (B) Upper parts of the Oso Landslide that occurred in northwest Washington on 22 March 2014. Photo taken on 27 March 2014 by Jonathan Godt, USGS.

4) The lower end of the slide debris is a *lumpy and disrupted mass*; and

5) Debris flows, that are created *when rock and soil become completely saturated, sometimes becoming so mobile that they can obtain speeds of hundreds of kilometers per hour*.

Needless to say, landslides are exceedingly dangerous and houses should never be built on slopes prone to such slippage. A classic example of what happens when building on unstable cliffs is the Jump Off Joe landslide in Newport, Oregon in 1941-42 (Komar, 1998a).

Because of their location on the tectonically active western side of the North American Plate, several conditions enhance the probability of landslides along the Washington and Oregon Coasts, such as abundant steep slopes, uplifted naturally weak materials, including unconsolidated sediments or relatively young sandstones and mudstones, and the common occurrence of extremely sheared and faulted rocks. Other notable favorable conditions for the generation of landslides include frequent earthquakes, wet winter storms, and big waves, especially during the *El Niño* storms that under cut the rock cliffs back of the beach. Just as for beaches, sea cliff erosion will likely increase in the future because of two factors: 1) Increased sea levels; and 2) More frequent and larger waves during storms.

METHODS COMMONLY USED TO PREVENT COASTAL EROSION

Engineers have attempted to curtail coastal erosion for centuries. Their success rate has been variable, depending upon the vagaries of the sediment supply, changes in water levels, and storm-wave conditions. For purposes of discussion, we divide the techniques used to prevent such erosion into two classes, "hard" engineering solutions and "soft" engineering solutions.

In the past, engineers have usually dealt with shore erosion by building resistant, permanent features that reflect or dissipate incoming waves. Some of these "hard" solutions include:

1) Seawalls, revetments, and bulkheads. In places where fixed property, such as highways and large hotels, are threatened by erosion, these three types of features are commonly built. Seawalls, massive concrete structures designed to hold the line against storm-wave erosion, are a double-edged sword, so to speak because, while they tend to succeed in keeping the road or hotel in place, at least until a major storm hits, the beach sand itself is usually sacrificed, as is illustrated in Figure 68. A bulkhead is made of pilings, composed of a wide range of materials, driven into the ground. Revetments are constructed by armoring the slope or face of a dune or bluff with one or more layers of rock, concrete, or asphalt;

2) Groins. These features (Figure 85B) are commonly made of rubble stone, but they may consist of wood, sand bags, gabions (rocks or gravel in wire mesh), or other materials. They work best where waves approach the coast obliquely. Whereas they work well if installed properly in the right place, groins have not proved to be an effective solution to beach erosion in many localities, especially where the dominant wave crests approach parallel to the beach; and

3) Offshore breakwaters. These features, which are usually composed of riprap or heavy concrete blocks of miscellaneous shapes, are built offshore, detached from the beach in segments several tens of feet long oriented parallel to the beach. They reduce wave energy on their landward sides, which causes sand moving along shore in the longshore sediment transport system to accumulate in their lee. Such an obstacle to the movement of the sand along shore will commonly cause erosion on its downdrift side. Thus, many engineers recommend placing nourished sand in the shelter of these structures so that sand can continue moving along the shore.

There are many workers in the area of coastal erosion, particularly coastal geologists, who prefer to use solutions to coastal erosion that do not involve hard structures, called "soft" solutions. Two of the lines of reasoning used to support that position are that: 1) Hard structures, such as seawalls, accelerate sand loss, and 2) Once in place, hard structures are difficult to remove, making it virtually impossible to correct a mistake.

Some examples of these types of "soft" solutions include:

1) Setback lines. The "softest" of the soft solutions is the construction a setback line landward of the high-tide line seaward of which building is prohibited. They work best and are easiest to implement in areas that have not been developed as yet. The criteria used to establish such lines include analysis of historical changes in the position of the high-tide line, preservation of the line of foredunes, defining areas of flooding and storm wave

uprush, and so forth. Where such setback lines have been judiciously applied, they usually work. Our group designed setback lines for a development on Kiawah Island, South Carolina in 1974, and no serious beach erosion has occurred there since that time except in one place where the setback was not adhered to (Hayes and Michel, 2008). Komar et al. (1999) proposed a model to develop setback distances on the Oregon coast based on predicted foredune erosion during 50- and 100-year storms, though such decisions are made at the local level, both in Oregon and Washington;

2) Sand bypassing. As noted earlier, jetties constructed to stabilize navigational channels usually block the flow of sand along the shore, with severe erosion commonly developing downdrift of the jetties. This is one of the more common major causes of beach erosion problems around the World. That type of erosion problem may be at least partly solved by installing mechanical sand bypassing systems, such as land-based dredging plants, pumping systems, and so on, that move the sand from the updrift to the downdrift sides of the jetties. Such systems have been established in many areas;

3) Relocating a tidal inlet. Migrating tidal inlets erode the shore as they move down the coast. Relocating the inlet back up the coast in the direction from which it came would relieve the down shore area of erosion, at least until the inlet migrates back. To our knowledge, one of the major examples, and maybe the only one up to that time, of this process was carried out by our group in March 1983 at Captain Sams Inlet on the South Carolina Coast, when the inlet was moved to a new dredged channel approximately 4,000 feet up the coast (to the northeast). This stopped the erosion of the inlet into a development to the south, as well as provided a huge volume of sand to an eroding beach to the south when the abandoned ebb-tidal delta of the original inlet was driven ashore by wave action. At an erosional area along a golf course that had been fortified by sand bags and riprap, the sand beach has built out 1,000 feet as a result of that project (Hayes and Michel, 2008). Compared to establishing hard structures or jetties, that was a relatively inexpensive process. The downside is that it has to be repeated about every 14 years or so, but it would take many repeats to reach the costs of other types of protection; and

4) Beach nourishment. This is the most commonly used of the "soft" solutions for sand beaches. Numerous projects of that type have been carried out around the World, some of which involve moving millions of cubic yards of sand. Sources for such sand include dredging sand deposits on the continental shelf, dredging sand bodies associated with tidal inlets, moving sand by various mechanisms from adjacent beaches that have an abundance of sand, and hauled or pumped in from land-based sources. The major objection to beach nourishment schemes is that they are not permanent. Watching millions of dollars worth of sand wash away within a few years does not appeal to either public servants or even casual observers. Therefore, such projects require a careful analysis of monetary costs and benefits balanced against the aesthetic, economic, and recreational value of maintaining a sand beach in place.

Historically, mostly small beach nourishment projects have been carried out in Oregon and Washington. However, in summer 2010, 366,000 cubic yards of dredged sand from the mouth of the Columbia River were placed on Benson Beach, Washington. Most of the nourishment material was transported into the nearshore bars during moderate wave conditions 16 days later (U.S. Geological Survey, 2012).

Komar et al. (2003) and Komar (2007) have proposed "Design with Nature" approaches, which include artificial cobble berms that mimic natural berms and beach nourishment to replace sand lost by human causes, such as dams and jetties.

CONCLUDING REMARKS

And there are times when the people and the hills and the Earth, all, everything except the stars, are one, and the love of them all is strong like a sadness.

John Steinbeck – TO A GOD UNKNOWN

In co-author Hayes' book entitled *Black Tides*, published by The University of Texas Press in 1999, that deals with our experiences responding to most of the major oil spills that have occurred since 1974, reference was made in the last chapter to a short article entitled "Ensoulment of Nature" by Gregory A. Cajete. That paper was published in the book *Native Heritage* in 1995. Cajete stated in the article that the importance the American Indians put on connecting with their place of origin (i.e., the environment) is not just a romantic notion out of step with the times, but rather *the quintessential ecological mandate of our times*. He stated further that the *native peoples experienced nature as a part of themselves and themselves as a part of nature*. Under that concept, **ensoulment** means that the human participates with the Earth as if it were a living being. Hayes concluded this subject with the following comment:

If we all adopted this frame of mind, I'm sure it would be one more step in the honorable direction toward making the Earth a better place to be, which is something all of us tree-huggers and right-thinking engineers want to do, correct?

[NOTE: Hayes' love of the concept of ensoulment took a hard blow recently when he read Yuval Harari's 2014 book, *Sapiens: A Brief History of Humankind*. In that book, Harari pointed out that 45,000 years ago, the *Homo sapiens* settled in Australia with the result of the complete extinction of the Australian megafauna. Then 16,000 years ago, *Homo sapiens* settled in North America with the result of the complete extinction of the North American megafauna. All for something to eat, we guess.

In July 2015, the authors of this book paid a visit to a Stone Age archaeological site at Céide Fields in County Mayo on the west coast of Ireland. When the *Homo sapiens* first arrived there about 5,500 years ago, the hillsides were covered with a massive forest of large trees. Then they proceeded to eliminate the forests in order to establish farm fields. That massive area is now covered with grass in sheep fields. No more of the giant trees. Thus, an extensive ecosystem was completely wiped out by the new visitors. Ensoulment indeed.]

Anyway, getting back to the ensoulment concept, Hayes concluded that final chapter in *Black Tides* with a quote from the Bureau of American Ethnology Collection (published in Brown, 1970), which follows:

The old men
say
the Earth only
endures.
You spoke
truly.
You are right.
The Earth only endures.

Of course, all of the animals and plants that lived on the Earth have not fared so well at times, particularly after the *Homo sapiens* became dominant. Another example of those hard times being when the Earth was struck by a large asteroid or comet, such as the one proposed to have caused the Cretaceous/Tertiary extinction event about 65 million years ago.

[NOTE: One hypothesis for that extinction event was given as follows by NOVA Evolution (2012): *For months, dense clouds of dust blocked the Sun's rays, darkening and chilling Earth to deadly levels for most plants, and in turn, many animals. Then, when the dust finally settled, greenhouse gases created by the impact caused temperatures to skyrocket above pre-impact levels.*

In just a few years, according to this hypothesis, *these frigid and sweltering climatic extremes caused the extinction of not just the dinosaurs, but of up to 70 percent of all plants and animals living at that time*. This is one hypothesis, there are several others. However, there is excellent evidence that a large asteroid did hit the Earth just off the coast of the Yucatan Peninsula around that time.]

But, sure enough, the Earth itself has endured, so far.

SECTION II
Coastal Ecology Overview

11 INTRODUCTION TO COASTAL ECOLOGY

The material presented in this section is based primarily on a detailed synthesis of the coastal environments and biological resources of the Oregon and Washington Coasts conducted by our group (RPI) and published as part of the Environmental Sensitivity Index (ESI) database in 2014 (NOAA/RPI, 2014). The part of the project related to the types of habitats present along the coast was discussed in some detail in Section I (the coastal types are listed in Table 2). The project also included compilation of data on key biological and socio-economic resources for the purpose of oil and chemical spill planning and response; however, these materials are also very useful for other environmental and natural resource planning purposes, as well as providing key basic information for this naturalist's guide.

The biological information presented in the atlas was collected, compiled, and reviewed with the assistance of biologists and resource managers from the following agencies:

- Bird Research Northwest
- Cascadia Research Collective
- Crescent Coastal Research
- NOAA Northwest Fisheries Science Center
- NOAA Southwest Fisheries Science Center
- Northwest Forest Plan Interagency Regional Monitoring Program
- Olympic Coast National Marine Sanctuary
- Olympic National Park
- Oregon Biodiversity Information Center, Portland State University
- Oregon Department of Fish and Wildlife
- Oregon State University
- Portland Audubon
- Seattle Audubon Society
- University of Washington
- U.S. Fish and Wildlife Service
- U.S. Fish and Wildlife Service Division of Migratory Bird Management
- U.S. Geological Survey
- Washington Department of Fish and Wildlife
- Washington Department of Natural Resources
- Washington Pacific Coast Treaty Tribes: Hoh Tribe, Makah Tribe, Quileute Tribe, Quinault Tribe, and Shoalwater Bay Tribe
- The Xerces Society for Invertebrate Conservation

Figure 88 is an example of the ESI biological resource maps that are generated from these data. Animal and plant species are represented on the ESI maps as polygons, points, or lines. Species are organized into groups and subgroups based on their behavior, morphology, taxonomic classification, and spill vulnerability and sensitivity. The icons below are used to represent this grouping on the maps:

MARINE MAMMAL
Dolphin/Porpoise
Pinniped
Sea Otter
Whale

TERRESTRIAL MAMMAL
Small Mammal
Ungulate

BIRD
Alcid/Pelagic Bird
Diving Bird
Gull/tern
Passerine Bird
Raptor
Shorebird
Wading Bird
Waterfowl

FISH
Fish

REPTILE
Turtle

INVERTEBRATE
Barnacle
Bivalve
Cephalopod
Crab
Gastropod
Insect
Shrimp

BENTHIC
Algae
Coral
Kelp
SAV

HABITAT
Plant
FAV

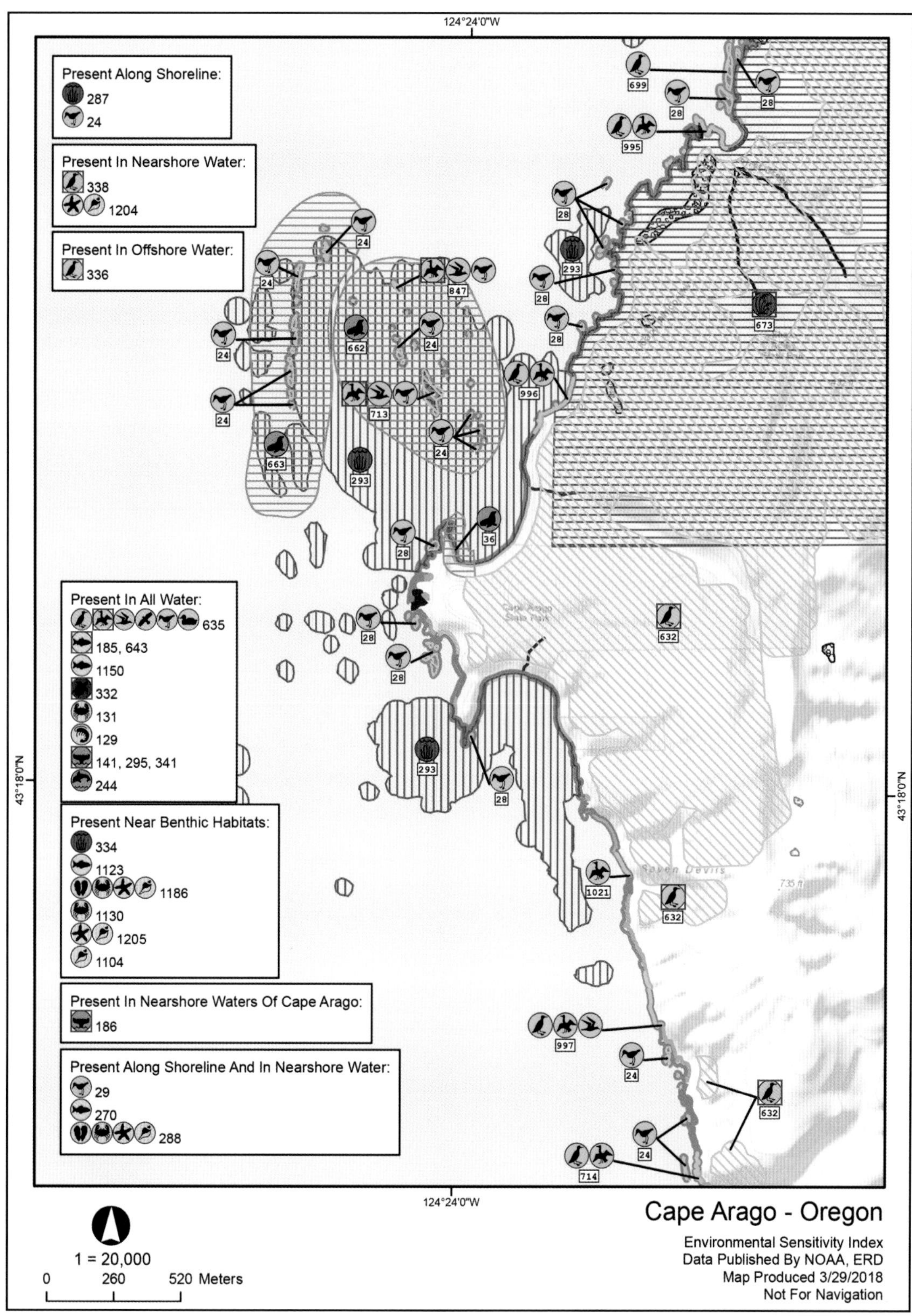

FIGURE 88. An example of the ESI biological resource maps that are generated from these data. Animal and plant species are represented on the ESI maps as polygons, points, or lines. Species are organized into groups and subgroups based on their behavior, morphology, taxonomic classification, and spill vulnerability and sensitivity.

Associated with each icon is a "Resources at Risk" or RAR number that refers to a table for each map that lists the species name, status as threatened or endangered under the Endangered Species Act, the concentration (e.g., number of birds in a nesting colony, or general concentration such as low, medium, or high), the month it is present, and "life history" information such as months for nesting, pupping, spawning, etc.

For species that are found throughout general geographical areas or habitat types on certain maps, displaying the polygons for these species would cover large areas or would obscure the shoreline and biological features, making the maps very difficult to read, a real challenge for the biologically rich coasts of Oregon and Washington. In these cases, the RAR number is shown in a box on the lower left of the map that lists the location of the species associated with the RAR number. Figure 88 shows all of these mapping conventions.

12 KEY BIOLOGICAL RESOURCES OF OREGON AND WASHINGTON

In the following sections, we provide an overview of the major groups of animals and habitats that are present along the outer coasts of Oregon and Washington and the Strait of Juan de Fuca. Site-specific details for some areas are provided in Section III.

MARINE MAMMALS

Introduction

Marine mammals are abundant along the Oregon and Washington Coasts, and can often be sighted from shore. Species that can be seen along the Oregon and Washington coast include whales, dolphins, porpoises, pinnipeds (seals and sea lions), and sea otters.

Whales

Sighting whales from the coast or on a whale-watching boat trip is one of the greatest thrills you can experience. You most often first see the spout, the moist air being exhaled from the blowhole. In fact, expert whale watchers can identify the whale by the size and shape of the spout. You can even hear the sounds of spouting, if the whales are close. Along the coast, you will most likely see gray, humpback, blue, and minke whales. In some areas of the Strait of Juan de Fuca and Puget Sound, killer whales are commonly sought and seen. Also, the chance of seeing each species varies by season. So, if seeing a whale is an important objective of your visit to the coast, you will need to plan your visit during the peak seasons. A whale-watching boat (or kayak in calm waters) trip is a great way to really make sure you see whales. There are boat charters operating out of Brookings, Charleston, Depoe Bay, Garibaldi, Newport, Rockaway Beach, and South Beach (Oregon) and Anacortes, Bellingham, Everett, La Conner, Port Angeles, Seattle, and San Juan Islands (Washington), among others. In Oregon, there are 29 whale-watching viewing parks, and the *Whale Watching Spoken Here* program places volunteers at these sites during the peak gray whale migration periods to help visitors see and identify whales. The Whale Watching Center in Depoe Bay, Oregon is open year-round and is a great place to learn more about whales. Listed below are eleven species of whales that can be found off of the Oregon and Washington Coasts that were included in the ESI atlas.

Baird's beaked whale (*Berardius bairdii*): They are present March-October in deep offshore waters, but have been rarely sighted.

Blue whale (*Balaenoptera musculus*): Federally listed as endangered. They are present June-November, feeding mostly on krill (shrimp-like planktonic crustaceans). They are occasionally seen but usually no closer than 10 miles offshore.

Fin whale (*Balaenoptera physalus*): Federally listed as endangered. They are present May-November, but are rarely seen because they occur in deep offshore waters.

Gray whale (*Eschrichtius robustus*): They are present year-round and are by far the most commonly observed whale. The entire population (about 19,000) migrates from northern feeding areas December-January with a late December peak; they migrate from southern breeding and calving areas March-June with a late March peak. Figure 89 shows the gray whale migration routes, from their summer feeding area in the Bering and Chukchi Seas to their winter breeding and calving area from San Diego to Baja, Mexico. On their way north late in the season, the females and calves tend to hug the shore to avoid killer whales, which are known predators of the calves. There is also a resident population that can be seen year-round.

Humpback whale (*Megaptera novaeangliae*): Federally listed as endangered. There are three populations; the California/Oregon/Washington stock (with about 1,900 individuals) winters in Central America and Mexico and spends the summer in the region. Sightings of humpbacks appear to be increasing.

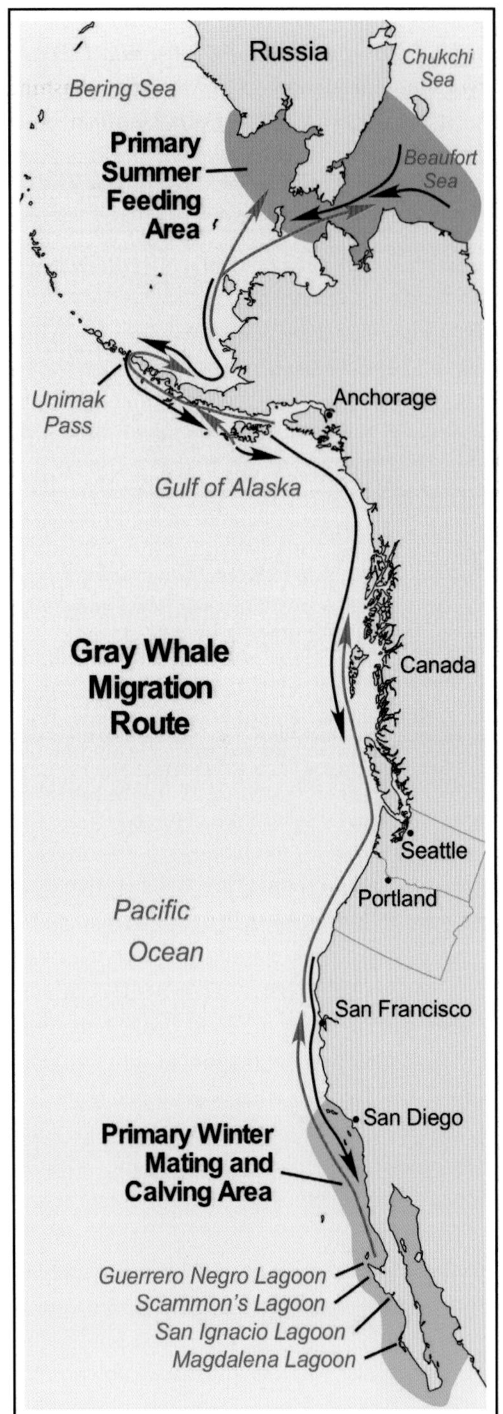

FIGURE 89. The migration routes of gray whales from their summer feeding areas in the Bering Sea to their winter mating and calving areas in southern California and Mexico. They migrate north from February to May, with a peak in March, and south from December to February, with a peak in January. Modified illustration from MTYcounty.com (2002).

Killer whale (*Orcinus orca*): They are present year-round and have been categorized into 'populations' that rarely interact and do not interbreed. Four distinct populations are recognized and rarely interact and do not interbreed: southern residents, northern residents, transients, and offshores. Southern Resident and Transient killer whales are the only populations that regularly enter the coastal waters of Oregon and Washington, whereas offshore whales mainly inhabit open ocean off the outer coast. Offshore killer whales are most commonly seen in spring as they prey on gray whale calves migrating north. The Southern Resident killer whales (Federally listed as endangered) are most frequently sighted in Puget Sound, where they spend several months in summer and fall to feed on Chinook salmon. That population is composed of three family groups that have been named J, K, and L pods. As of December 2017 there were only 76 individuals left, and the population is not growing. Their small population size and social structure puts them at high risk of further decline. Nearly all of the Strait of Juan de Fuca and Puget Sound has been designated as critical habitat for the Southern Resident killer whales (except Hood Canal).

Minke whale (*Balaenoptera acutorostrata*): They are present May-November. Individual whales can be seen in the Strait of Juan de Fuca and more rarely offshore.

Sei whale (*Balaenoptera borealis*): Federally listed as endangered. They are present June- September. They occur in deep offshore waters, thus have been rarely sighted.

Short-finned pilot whale (*Globicephala macrorhynchus*): They are present year-round but generally stay offshore and have been rarely sighted.

Sperm whale (*Physeter macrocephalus*): Federally listed as endangered. They are present year-round. They are found far offshore in very deep water, thus have been rarely sighted.

Dolphins and Porpoises

There are eight species of dolphins and porpoises in the waters off Oregon and Washington: bottlenose dolphin, northern right-whale dolphin, Pacific white-sided dolphin, Risso's dolphin, short-beaked common dolphin, striped dolphin, Dall's porpoise, and harbor porpoise. Oregon is the northern extent of many species' ranges. Large pods are most often seen during the summer by offshore fishers and during whale watching cruises. Harbor porpoises are present year-round, and individuals can be seen alone in shallower water of harbors and bays. Pods of Dall's porpoises are regularly seen in the Strait of Juan de Fuca.

Pinniped Haulouts and Rookeries

Pinnipeds are a group of carnivorous marine mammals with fin-like limbs. Five species of pinniped are found along the Oregon and Washington Coast.

California sea lion (*Zalophus californianus*): They mostly winter in Oregon and Washington and return to offshore islands from the Channel Islands to Mexico to breed. From September to May, you can see California sea lions in many bays and estuaries, and hauled out on offshore rocks, often in the same locations as Steller sea lions. They are frequently seen throughout Washington, on logbooms, marina docks, jetties, and navigation buoys, as well as offshore rocks and islands. The Oregon/Washington stock population is estimated to be 25,000 individuals. California sea lions are charismatic and intelligent, and are often used for animal shows in zoos, circuses, and aquariums. This species feeds heavily on endangered and threatened stocks of salmon and steelhead in the Columbia River system, particularly where fish congregate below the Bonneville Dam and Willamette Falls. Federal and state agencies have identified sea lion predation as a key factor in the continued decline of salmon runs, and efforts to haze and/or relocate the sea lions have not been successful (individuals relocated to southern California have traveled thousands of miles to return to their feeding sites). As a result, a controversial program to cull animals that are observed feeding on salmon in these locations has been implemented.

Northern elephant seal (*Mirounga angustirostris*): Seen occasionally hauled out on sand beaches, mostly as sub-adults that come ashore to molt. Their name comes from the distinctive elephant-like proboscis found on adult male seals, which is used to make noises that assert dominance during the breeding season. Although these animals look like they are in trouble, you should not approach them; most of the time, they will survive on their own.

Northern fur seal (*Callorhinus ursinus*): Present year-round but not commonly seen because they spend most of their time offshore.

Pacific harbor seal (*Phoca vitulina*): The most common pinniped in the region. They are often seen in the surf zone "watching" you on the shore. They prefer nearshore waters and haul out on sandy tidal flats in bays and estuaries. There are hundreds of haulout and pupping sites throughout Oregon and Washington waters, where up to 100-500 individuals can be seen. The latest population count was conducted during the 1999 pupping season, with 5,735 animals in Oregon and the Columbia River and 10,430 animals in Washington (NMFS, 2014).

Steller sea lion (*Eumetopias jubatus*): Federally listed as threatened in the Aleutian Islands, but were delisted in 2013 elsewhere. The largest breeding sites in U.S. waters south of Alaska are in Oregon, at Orford Reef (Port Orford) and Rogue Reef (Gold Beach), producing 1,500 pups annually. During the breeding season (late May to July), Oregon regulations prohibit boaters from approached within 500 feet of the rookery. Steller sea lions are also found year-round in smaller numbers at Three Arch Rocks, Sea Lion Caves, and Cape Arago State Park. The only rookery in Washington waters was recently established at the Carroll Island and Sea Lion Rock complex, with more than 100 pups born there in 2015. They can be found during fall and winter hauled out along the outer Washington coast from the Columbia River to Cape Flattery. The non-pup population estimate as of 2015 is 5,634 in Oregon and 1,407 in Washington (NMFS, 2016).

Sea Otters

The northern sea otter (*Enhydra lutris kenyoni*) is the smallest marine mammal in North America. They spend all of their time in the water. Unlike other marine mammals that have blubber to keep them warm in water, sea otters rely on their thick fur to survive. Therefore, they are at great risk from oil spills, which can foul their fur and quickly result in hypothermia and death. Sea otters eat 25% of their body weight each day because of the energy needed to keep warm in cold water. They almost went extinct Worldwide due to the fur trade, and they were extirpated (locally extinct) in Washington and Oregon by the 1920s. Sea otters rebounded in Alaska once commercial harvesting was outlawed, though the population along southwest Alaska was federally listed as threatened in 2005. In 1969/1970, 59 sea otters were relocated from Alaska to Washington, and the 2016 population is about 1,800 animals distributed primarily between Pillar Point in the Strait of Juan de Fuca to just south of Destruction Island on the outer coast, though most animals are south of La Push (Jeffries et al., 2017). A great place to view sea otters is along the trail to Cape Flattery. There are no resident sea otters in Oregon, though individuals are occasionally seen along the Oregon coast.

MARINE AND COASTAL BIRDS

Nesting Seabirds

The rocky coasts and waters along the Oregon and Washington Coasts teem with seabirds. In Oregon, based on the latest compilation in 2006 (representing field survey data collected from 1979 to 2006), there are an estimated 1.29 million nesting seabirds representing 15 species at 393 colonies (Naughton et al. 2007). A similar effort is underway in Washington; however, data were not available as of the publication of this book. The number of seabird colonies and estimated number of breeding birds by species in Oregon are listed in Table 3. It should be noted that many colonies support multiple species.

TABLE 3. Seabird colonies in Oregon based on data compilation in 2006 (Naughton et al., 2007).

Species	Number of Nesting Colonies	Estimated Number of Breeding Birds
Fork-tailed Storm-Petrel	13	100s
Leach's Storm-Petrel	24	482,000
Double-crested Cormorant	36	30,400
Brant's Cormorant	120	21,200
Pelagic Cormorant	236	10,100
Black Oystercatcher	214	470
Western/Glacous-winged Gull	253	32,300
Ring-billed Gull	1	400
Caspian Tern	3	19,000
Common Murre	89	685,000
Pigeon Guillemot	289	4,500
Cassin's Auklet	14	400
Rhinoceros Auklet	21	500
Tufted Puffin	59	4,600
Total Estimates Numbers	393	1,290,000

Common murres represent 53% of the nesting seabirds in Oregon, followed by Leach's storm petrels at 37%. The density of birds present at a site varies by species. For example, black oystercatchers are solitary nesters that defend breeding territories, whereas colonies of common murres can have tens of thousands breeding birds, with the largest at the Shag Rock colony at Three Arch Rocks, hosting 120,500 breeding birds. By looking at the detailed data, about 20% of the Leach's storm petrel breeding population was at the Goat Island colony. The East Sand Island colony in the Columbia River estuary hosted all of the breeding caspian terns and over 80% of the double-crested cormorants in Oregon. Nearly 65% of the rhinoceros auklet breeding population was on Goat Island in 1988.

Other seabirds that you might see in nearshore waters include the marbled murrelet (Federally listed as threatened in California, Oregon, and Washington), which nests in old growth forests but feeds in nearshore waters. As of 2015, there are an estimated 7,500 marbled murrelets in Washington (with over 75% in Puget Sound and the Strait of Juan de Fuca) and 11,000 in Oregon, out of a total population of between 263,000-841,000 individuals (U.S. Fish & Wildlife Service, 2016). The Oregon population appears to be stable; however, the Washington population has been decling at 4.4% per year since 2001.

Waterfowl

The ESI maps for Oregon and Washington include 26 species of waterfowl (ducks, geese, swans). Many species overwinter in the coastal waters and bays, whereas others are summer breeders or year-round residents. The lower Columbia River estuary is a very important wintering area for waterfowl. Nearly every estuary along the coast supports large numbers of wintering waterfowl.

Shorebirds

Estuaries along the Oregon and Washington coasts are extremely important migratory stopover sites for shorebirds as they travel to and from summer breeding areas in the north and wintering areas in the south. These species take advantage of the abundant prey present in estuarine tidal flats to fuel up before continuing along their journey. Grays Harbor Estuary is recognized as a site of hemispheric importance in the Western Hemisphere Shorebird Reserve Network, regularly hosting tens and sometimes hundreds of thousands of birds per day during peak migration periods. Many of these species have smaller numbers of winter residents. Migrant and wintering shorebirds can be found in different habitats based on their species preferences: on rocky shores you will see oystercatchers and turnstones; on sand beaches and tidal flats, you will see sandpipers, sanderlings, dunlins, and black-bellied plovers.

Shorebirds that breed along the coast include the western snowy plover, which is the only shorebird along the West Coast that is federally listed as threatened. Nesting activity extends from 15 March to 15 September on about 25 beaches in Oregon and five in Washington. These beaches have signs that indicate the dry sand beach is closed during that period, except along designated trails. These small shorebirds breed on coastal beaches from southern Washington to southern Baja, Mexico. They nest just above the high-tide level in sparsely vegetated areas, so the nests are susceptible to damage during storms and disturbance by people and particularly dogs. The chicks are precocial, meaning that they leave the nest within hours after hatching to search for food. The chicks also are not able to fly for about a month after hatching, making them vulnerable to predators and disturbance. In 2006, their nesting populations in the three western states were estimated to be: California = 1,723; Oregon = 178; and Washington = 70 (USFWS, 2007). The recovery goal is for maintaining an average of 3,000 breeding adults for 10 years, with 250 in Oregon/Washington and 2,750 in California. There is a concerted effort among federal and state agencies, local communities, and an extensive volunteer network to protect the nests during incubation (cages are placed over the nests as soon as they are found), control predators (birds, foxes, raccoons, coyotes, feral cats), and enforce restrictions on beach access during nesting season. These efforts have really paid off in some areas. Nesting season counts for 2015 show that there were 1,920 in California, 276 in Oregon, and 64 in Washington, for a total of 2,260 adults (Audubon California, 2017). To account for birds missed during the surveys, a multiplier of 1.3 is applied, so the total estimated number of breeding adults is 2,938, very close to the inital goal. These conservation efforts are needed to meet the second part of the recovery plan, that is, to maintain these numbers for 10 years.

FISH AND SHELLFISH

Anadromous Fish

The most iconic fishes of the Pacific Northwest are salmon (Chinook, chum, coho, and sockeye) and steelhead trout—anadromous fish that live in salt water for 2-5 years but migrate to freshwater habitats to spawn. Almost every fast-flowing river or stream in Oregon and Washington is used by these fish for spawning and as habitat for newly hatched fish. The adult fish return to their natal stream (where they were born) to spawn. Over time, some of these runs of fish became reproductively isolated and genetically distinct from other runs or populations of the same species. The construction of dams for hydropower and other reasons, degradation in spawning and nursery habitat (both stream habitat complexity and water quality), and overfishing led to dramatic reductions in these fish populations. As a result, the federal government has listed 28 distinct salmon and steelhead populations (referred to as 'evolutionarily significant units' or ESUs) as either threatened or endangered under the Endangered Species Act, such as the endangered Upper Columbia River Spring-run Chinook salmon and the threatened Oregon Coast coho salmon.

This designation is a big deal—for landowners, industry, fishers, and anyone who wants to take an action that may affect these listed species and their habitat. Lots of efforts are underway to remove dams, improve fish passages around dams, directly improve stream habitat (adding wood, complexity, vegetation along the stream, connecting side channels, etc.), indirectly improve stream habitat through watershed restoration, improve water quality, and reintroduce hatchery-reared native fish into its historical range, among others.

You can still fish both commercially and recreationally for these fish; however, there are many regulations, and they frequently change. So, if you plan to fish for salmon or steelhead, you need to check with the state agency that manages those resources before you cast a line.

Surf smelt is another fish species that is popular with recreational fishers. They spawn in the intertidal zone on gravel beaches; there is a summer spawn and a fall/winter spawn with timing that varies by location. Families go out with dip nets or A-frame nets to catch these fish as they come ashore to spawn. From what we understand, they are quite tasty. They are also a very important "forage fish" that eats plankton, then is eaten by a lot of other fish, marine mammals, birds, and humans. Thus, surf smelt (and a similar species, sand lance) are key links in the marine food web, transferring energy from plankton to larger marine animals (and humans).

Shellfish

Clamming and crabbing are very popular for both residents and visitors to the Oregon and Washington Coasts. People seek gaper, butter, cockle, littleneck, softshell, and purple varnish (a recently introduced and very common species) clams in bays. Razor clams are found along more exposed beaches; be sure to head out at the lowest part of the tide. People seek Dungeness, red rock, and Pacfic rock crabs. Though you can catch crabs in most every estuary, in Oregon, Coos, Yaquina, and Tillamook normally provide the best year-round opportunities. Seems like everywhere in the Strait of Juan de Fuca is great for crabbing. If you don't have your own gear, you can rent what you need in many coastal towns. Again, be sure that you have your license and understand the current regulations.

INTERTIDAL BIOLOGICAL COMMUNITIES

Rocky Intertidal Communities

The abundance of rocky shores in Oregon and Washington (see Table 2) provides many opportunities to explore their biological communities, which are among the most studied and diverse of all intertidal habitats. There can be hundreds of species of plants and animals present! The 2016 book by Robert Steelquist—*The Northwest Coastal Explorer: Your Guide to the Places, Plants, and Animals of the Pacific Coast*—is a great guide. Rocky intertidal communites have very distinct distributions by tidal elevation (Figure 90), which are based on species' ability to survive direct wave action, duration of exposure during low tide, solar radiation, predation, competition, and associations with other species that provide habitat, such as algae, mussel beds, and kelp. A short summary of the four zones follows.

The **splash** (or spray) zone occurs above all but the highest tides and thus is almost always exposed to air and has relatively few species. You will most often find lichens, small algae, snails, and some crabs. The **high intertidal zone** is exposed to air for several hours, so the animals in this zone usually have protective shells (such as barnacles) or can breathe air and move to lower levels when the tide is out (such as crabs). This zone is where you will first see plants such as rockweed and larger algae. Many animals live under the algae, protected from long exposure to solar radiation and heat, such as snails, limpets, and chitons, which also feed on the algae. Barnacles can be very dense in places, covering the available substrate. The **mid intertidal zone** is exposed to air for a few hours once or twice a day, and the diversity of seaweeds and animals increases substantially. Dominant species include sea anemones and dense beds of mussels and goose barnacles. Sea stars move into this zone to feed on these animals (and other sea stars!), and they can be seen on vertical surfaces, in crevices, and under rocks at low tide. The **low intertidal zone** is exposed only during the lowest of tides and has the greatest diversity. You will see more sea stars, crabs, sea urchins, sea anemones, sponges, tunicates, fish, and coralline algae here. The presence of surf grass (a flowering marine plant) is a good indicator of the mean lower low water level, so you need to be out at the lowest of tides in the tidal cycle to see it. However, it is well worth planning your visit at these lowest of tides; it is very exciting to see the bright green surf grass (only at the more exposed locations), as well as the many animals in this zone. Just be sure to keep a close watch on the water levels and have a ready retreat plan as the waters rise. Tidal pools also provide a way to see these animals and plants of this lowest intertidal zone. And please, don't trample them!

Where rocky shores are adjacent to sand beaches, you will notice that the attached biota can be reduced due to "sand-blasting" of the hard rock surface during high wave energy events. Sand can also cover part of the rocky platform, which can smother and kill the attached biota.

Sand Beach Communities

The abundant sand beaches along the outer coast have a biological community that is adapted to living in an environment that dries out at low tide and is exposed to strong wave action that can quickly erode the beach. Though there are many very small species

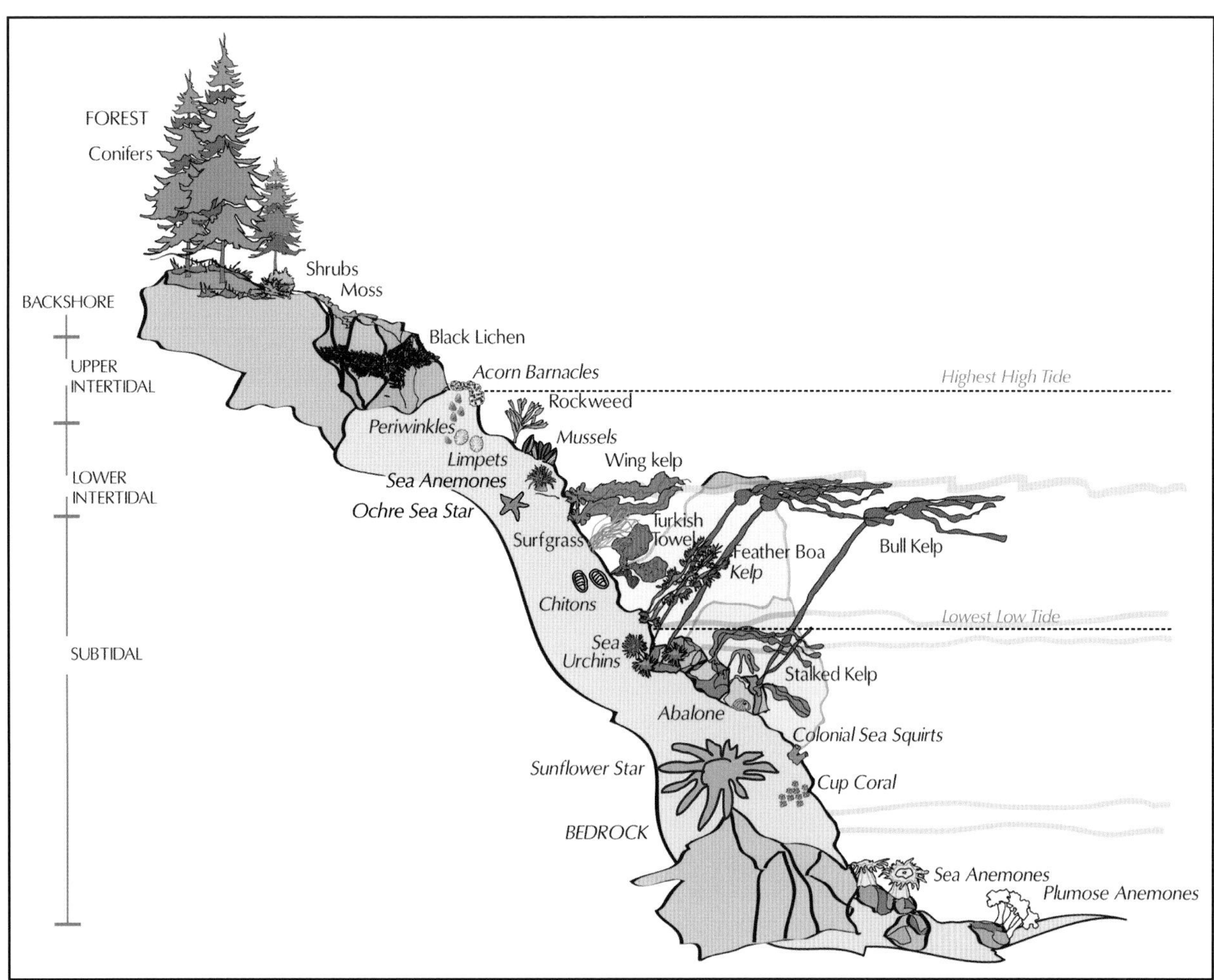

FIGURE 90. The intertidal zones and dominant species on the rocky shores of Oregon and Washington. Species are distributed within the intertidal zone based on their ability to withstand exposure to air and sun during low tide. Modified illustration from the *Coastal Shore Stewardship: A Guide for Planners, Builders, and Developers*, courtesy of the Stewardship Centre for British Columbia.

that live between the grains (called meiofauna), the animals that you can see are relatively few in number, with most beaches having less than 20 species. Sand beach communities are distributed by tidal zone. The **supratidal zone** is above the normal high tide, and the beach **wrack** that accumulates there provides shelter and food for kelp flies, beetles, amphipods, and isopods. Wrack is usually composed of kelp and eelgrass, and it is a very important contributor of nutrients and organic matter to a habitat that can have very little of these key contributions to the food web. Some of the animals live in the sand under the wrack during the day to escape the heat, dessication, and predation, coming out at night to feed. Pick up any pile of wrack and you will see lots of them. In the swash zone, where the waves wash up and down the beach during the tides, species such as clams, amphipods, mole crabs, and tube-building worms have adapted as rapid or deep burrowers, to prevent being exposed by waves or to reduce the risk of being eaten by birds. One of the dominant species is the mole crab, which is an important food source for birds and fish; it feeds by extending a feathery antenna into the overlying water to filter out plankton as the wave recedes, then scrapes the food into its mouth. They can be hard to find because they are patchy and highly active, moving up and down the beach with the tides. At the very lowest tidal levels, razor and littleneck clams can be found and, on some beaches, surf smelt.

FEDERALLY MANAGED LANDS

The Oregon Islands National Wildlife Refuge (NWR) was established in 1935 and includes all the rocks, reefs, and islands (at last count, there were 1,853 of them!) along the Oregon coast above the mean high water line except Chiefs Island and Three Arch Rocks NWR. The rocks and islands are designated as National Wilderness Areas, with the exception of Tillamook Rock, meaning that you can only view them from land or by boat (staying 200 yards offshore).

Other coastal NWRs in Oregon include Cape Meares (mostly vertical cliffs and old-growth forests), Bandon Marsh (large tidal salt marsh near the mouth of the Coquille River), Nestucca Bay (wetlands that are important wintering habitat for geese), and Siletz Bay (formerly diked agricultural land that has been restored into productive estuarine habitat).

Coastal NWRs in Washington include Flattery Rocks, Quillayete Needles, and Copalis, which were established in 1907 and include all of the about 800 offshore rocks, reefs, and islands stretching from Cape Flattery to just south of Copalis Head, with the exception of those features that are parts of designated Native American reservations. You can't actually visit these NWRs because they have been designed as wilderness areas; however, there are many trails and locations from which these fantastic areas and the numerous nesting birds and marine mammals can be viewed.

In southern Washington, the Willapa NWR consists of extensive wetlands, dunes, and grasslands that support over 200 bird species, including over 30 species of waterfowl, and 30 species of shorebirds. Grays Harbor NWR consists of 1,500 acres of mudflats and wetlands that are a major staging area for migrating shorebirds and wintering area for thousands of dunlin. In the Strait of Juan de Fuca, Dungeness NWR is a 5-mile long sand spit building out into Dungeness Bay that offers great bird and marine mammal viewing.

The Olympic Coast National Marine Sanctuary consists of 3,188 square miles of the marine waters from Cape Flattery to the Copalis River and extends 25 to 50 miles offshore. All marine sanctuaries only include the water; there is no land within the sanctuary. However, much of the land adjacent to the marine sanctuary (from Shi Shi Beach to just north of the Queets River) is part of the Olympic National Park, with the exception of areas designated as Native American reservations. For example, all of the land around Cape Flattery is part of the Makah Indian Reservation. There are over 70 miles of rugged coastline in the Olympic National Park, making it a fantastic area to visit.

Other federal and state properties are discussed in Section III.

SECTION III
Major Compartments

13 INTRODUCTION

In this section, we discuss and identify opportunities for you to learn about and enjoy the diverse ecosystems that exist within the four major geomorphological Compartments located along the outer coast of Oregon and Washington and the Strait of Juan de Fuca. All of the major Compartments are subdivided into Subcompartments (called Zones) in order to focus more sharply on individual features and locations that illustrate unusually compelling aspects of the major Compartment itself. For each Zone, we provide a summary of the regional geological framework, general description, general ecology, and key places to visit. The four primary Compartments are labeled on Figure 1 and listed below:

Compartment 1: SOUTHWEST COAST (California Border to Florence Jetties)

Compartment 2: CENTRAL UPLAND (Florence Jetties to Tillamook Head)

Compartment 3: CENTRAL ESTUARIES (Tillamook Head to Point Grenville)

Compartment 4: OLYMPIC PENINSULA (Point Grenville to East End of Dungeness Spit)

14 COMPARTMENT 1 - SOUTHWEST COAST (California Border to Florence Jetties)

INTRODUCTION

The coastal habitats present in this Compartment are listed in Table 4. There are a variety of habitats along the shore, with five of them occurring along greater than 10% of its length. Of those, salt- and brackish-water marsh is the longest, occurring along 25.9% of the length of the shore (229 miles). Next comes exposed tidal flats at 22%, sheltered tidal flats at 20.3%, freshwater marshes at 15.4%, and fine- to medium-grained sand beaches at

TABLE 4. Coastal habitats present in Compartment 1 (Southwest Coast).

COMPARTMENT 1 - SOUTHWEST COAST (California Border to Florence Jetties)			
ESI No.	**Habitat Definition**	**LENGTH (MILES)**	**% ABSOLUTE LENGTH**
1A	Exposed Rocky Shores	54.28	6.13%
1B	Exposed, Solid Man-Made Structures	4.17	0.47%
2A	Exposed Wave-Cut Platforms in Bedrock	70.68	7.99%
2B	Exposed Scarps and Steep Slopes in Clay	0.17	0.02%
3A	Fine- to Medium Grained Sand Beaches	118.59	13.40%
3B	Scarps and Steep Slopes in Sand	2.71	0.31%
4	Coarse-Grained Sand Beaches or Bars	11.94	1.35%
5	Mixed Sand and Gravel Beaches or Bars	37.36	4.22%
6A	Gravel Beaches or Bars	38.07	4.30%
6B	Riprap	35.13	3.97%
6D	Boulder Rubble	12.25	1.38%
7	Exposed Tidal Flats	194.94	22.03%
8A	Sheltered Rocky Shores and Sheltered Scarps	49.09	5.55%
8B	Sheltered, Solid Man-Made Structures	4.94	0.56%
8C	Sheltered Riprap	25.81	2.92%
8F	Vegetated, Steeply-Sloping Bluffs	0.20	0.02%
9A	Sheltered Tidal Flats	180.05	20.34%
9B	Vegetated Low Banks	48.09	5.43%
10A	Salt- and Brackish-Water Marshes	228.98	25.87%
10B	Freshwater Marshes	135.88	15.35%
10C	Swamps	37.51	4.24%
10D	Scrub and Shrub Wetlands	40.72	4.60%
	Total ESI Length (Miles)	1,331.54	150.45%
	Absolute Shoreline Length (Miles)	885.01	100.00%

13.4%. The abundance of sheltered habitats reflects the fact that 14 estuaries (including the smaller ones) with their associated sheltered habitats are present along the shore of this Compartment.

For purposes of discussion, we have further subdivided the Compartment into three Zones (as mapped in Figure 91):

1) California Border to Cape Blanco
2) Cape Blanco to Cape Arago
3) Cape Argo to Florence Jetties

ZONE 1. CALIFORNIA BORDER TO CAPE BLANCO

Introduction

The dominant geomorphological features of this 63-mile-long Zone, shown in Figure 92, are:

1) A number of prominent rocky headlands (e.g., Cape Ferello, Cape Sebastion, Port Orford Heads, and Cape Blanco);

2) Ten river outlets, two of which, the mouths of the Rogue and Chetco Rivers, have jetties. Three other modest-sized rivers emptying along this Zone include the Winchuck, Pistol and Elk Rivers;

3) Abundant rocky erosional remnants (sea stacks and sea arches);

4) A variety of beach types, ranging from boulder beaches to long, commonly cuspate, medium-grained sand beaches; and

5) A series of raised marine terraces centered around Port Orford and Cape Blanco.

Geologic Framework

A segment of the Klamath Mountains is located on the landward side of this Zone (Figure 93). Alt and Hyndman (1978) discussed the origin of those mountains, beginning with the time when North America separated from Europe and Africa 200 million years ago and started moving to *about 1500 miles from where it had been*. They also observed that it seems safe to assume that at least that amount of the *Pacific Ocean floor has disappeared beneath the western edge of the continent to be consumed back into the interior of the Earth*.

Then, with regard to the origin of the Klamath Mountains per se, as Pacific Ocean crust began its long slide beneath the continent, it *telescoped together the sedimentary rocks of the old coastal plain and continental shelf* (formerly located west of the original coast), *rudely jamming them against the edge of the continent to make the Blue, Wallowa, and Klamath Mountains.* Those ranges all consist of old former coastal and shelf sediments, scrambled together in a conglomeration virtually impossible to decipher.

Alt and Hyndman (1978) reported further that when the collision between the North American continent and the Pacific seafloor began about 200 million years ago, *the Blue, Wallowa, Klamath and Sierra Nevada of California were all part of one continuous coastal range forming in the same way and at the same time. Then sometime between 150 and 100 million years ago a very strange thing happened to the Klamaths – they broke away from the rest of the chain and moved westward about 60 miles to become an island standing offshore from the coast.* The location of that broken-off piece of the Klamaths, that now abuts this Zone as part of the Rogue River-Siskiyou National Forest, is shown in Figure 93.

With regard to the composition of the rocks in this broken-off piece of the Klamath Mountains, they are complexly tilted and folded, being composed of Jurassic greenstone and mudstone, Cretaceous sandstone, and several other rock types, including famous deposits of serpentinite.

This Zone has the most continuously rocky shores in Oregon, as well as the highest abundance of sea stacks we have ever seen. And there is a reason for both of these features—the geologic history, of course. First, this part of the Oregon Coast has been rising higher in the last 100,000 years than sections to the north (see Figure 25 and Kelsey et al., 1994), so waves are scouring vertical rock cliffs rather than building beaches. Second, there are few large rivers to bring sediments to the shore. Third, in a story that will be re-told many times in this book, the types of rocks exposed at the shore also explain the abundance of sea stacks. Essentially, most of the sea stacks in this Zone are composed of more resistant rocks, mostly basalt and thick sandstones that have not been fractured, compared to fractured and finer-grained mudstones and siltstones, which are more easily eroded (Lund, 1975). As the cliffs retreat, these more resistant rocks get left behind as sea stacks and islands. This process continues today; thus, some of the cliffs that you stand on to view the coast will someday become sea stacks.

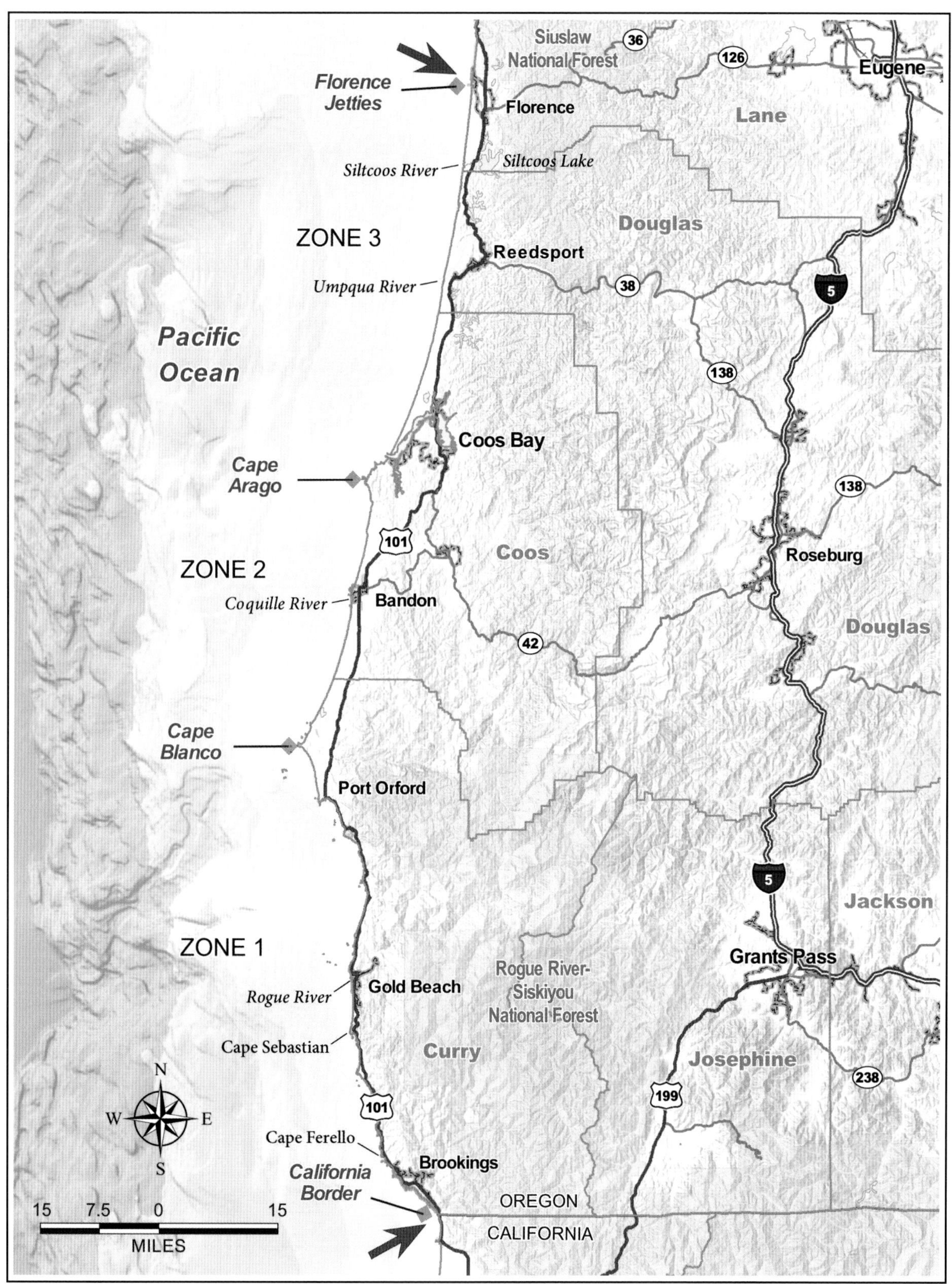

FIGURE 91. Map of Compartment 1.

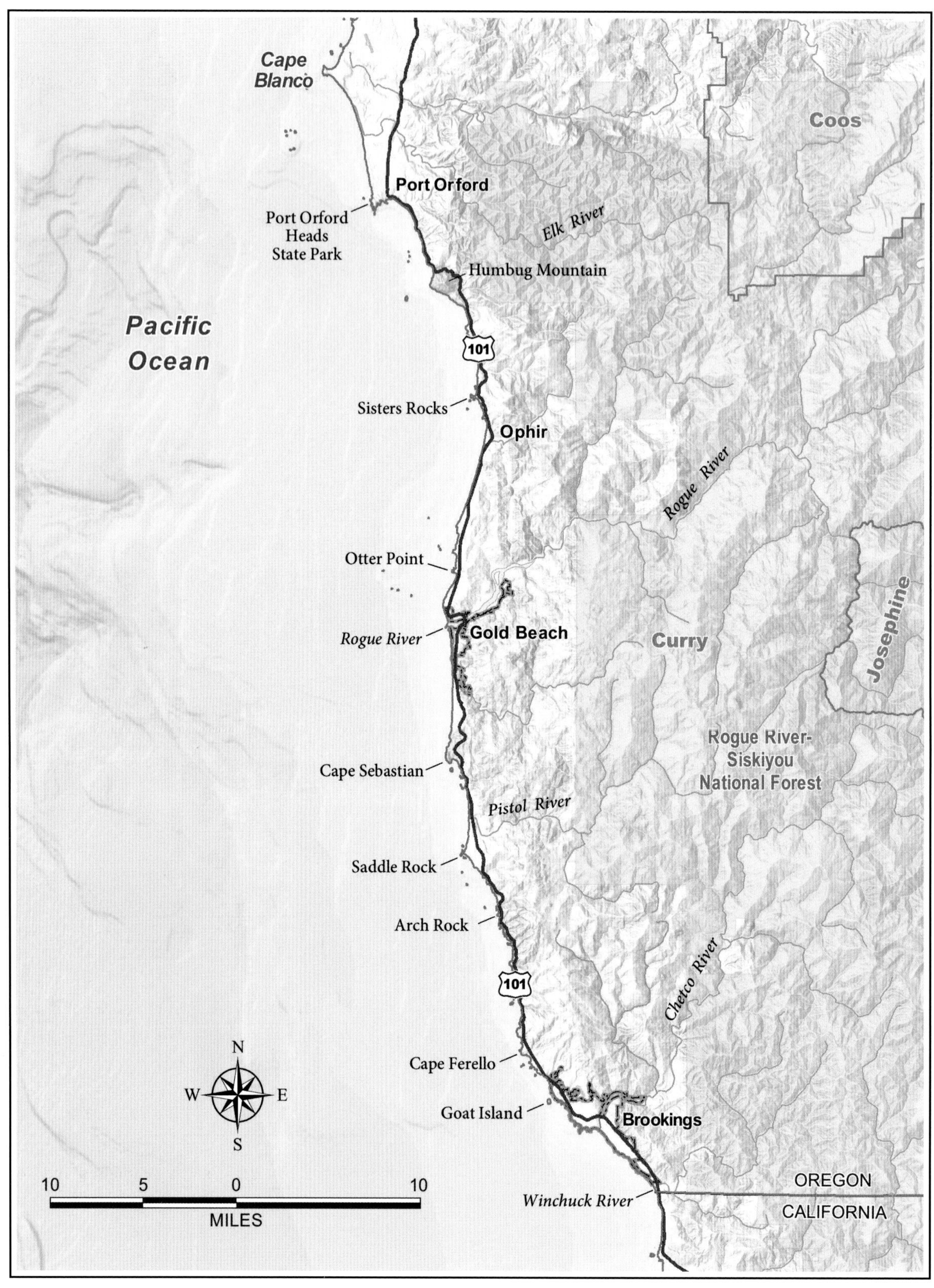

FIGURE 92. More detailed map of Zone 1 of Compartment 1.

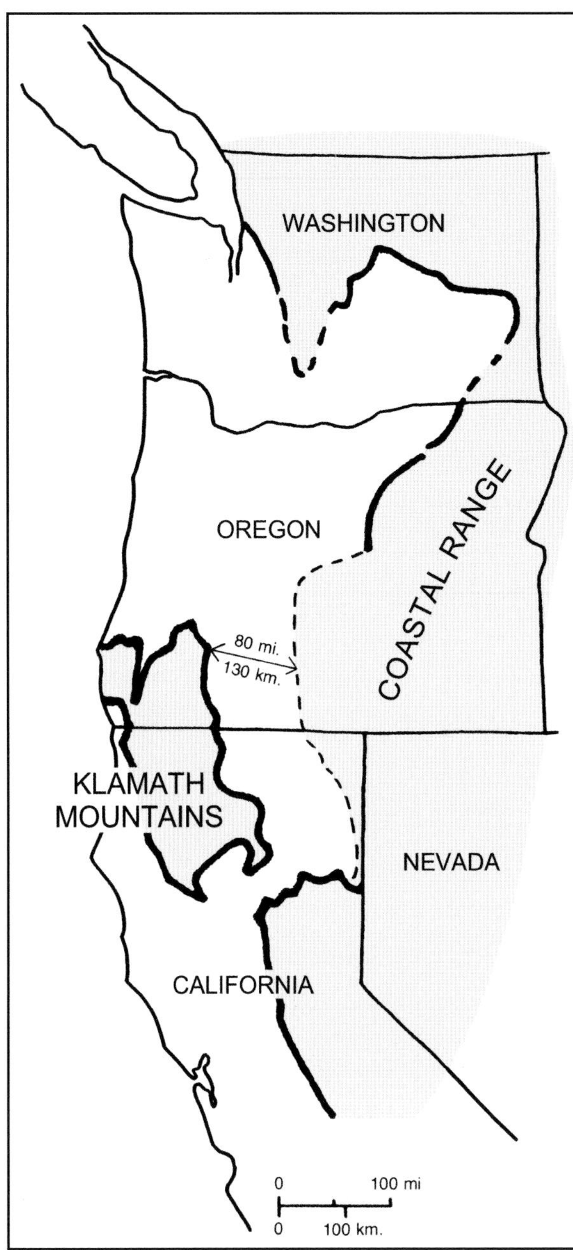

FIGURE 93. Location of the fragment of the Klamath Mountains that separated from the rest of the coastal range about 100 million years ago. Modified after Alt and Hyndman (1978).

General Description

Between the mouths of the Winchuck and Chetco Rivers, a flat surface up to 1 mile wide separates the upland area from the beach. That is a **raised marine terrace** that was cut ~80,000 years ago and since then has been uplifted about 100 feet (Alt and Hyndman, 1978). Measurements of the uplift since the terraces were formed is one of the methods used to compare uplift rates along the coast. McVay Rock and Hasting Rock were large sea stacks on that old wave-cut terrace; however, McVay Rock has been mostly quarried for stone, so there is not much left to see (Lund, 1975).

For the next 2 miles or so north of the McVay Rock area, the nearshore zone is crowded with sea stacks. The beach to the south of the Chetco River mouth has excellently developed cusps on a sand and gravel beach, at least part of the time.

The shore north of the Chetco River mouth is composed of a series or rocky headlands with an abundance of sea stacks located just offshore of them. A little ways north of there is Harris Beach State Park, which claims to contain the "largest offshore island in Oregon" - Goat Island.

Boy oh boy, this is sea stack land!!!

Between Goat Island and the next major headland to the north, Cape Ferello, the shore remains much the same. About half way along is a rocky zone called Rainbow Rock that is composed of folded thin beds of colorful chert, a very hard sedimentary rock composed almost entirely of silica that was described as follows by Alt and Hyndman (1978):

This kind of bedded chert forms far out on the ocean floor in deep water, beyond the reach of sand and muds washed in from the continent. The only sediments that settle on those remote reaches of the seafloor are the microscopic skeletons of one-celled animals called radiolaria and plants called diatoms, both of which are made of silica. Other kinds of shells dissolve in those deep, cold waters. The tiny shells accumulate very slowly to form an ooze with a consistency about like toothpaste which eventually hardens by recrystallization into chert. So what we see at Rainbow Rock are deep-sea oozes that almost certainly accumulated many hundreds of miles offshore and finally made it into the Coast Range where they are mixed with sandstones that were deposited much closer to shore.

From Cape Ferello to Saddle Rock, "sea stack land" continues with a few good-sized headlands and some scattered short sand beaches. Arch Rock, shown in Figure 94, is a spectacular example of a sea stack. You would be hard pressed to see more sea stacks and sea arches anywhere in the abundance they occur at along that stretch of coast.

FIGURE 94. Arch Rock, located 5 miles south of the mouth of Pistol River, OR. Source: public domain.

The shore character changes north of Saddle Rock, with long sand spits bordering both sides of the mouth of the Pistol River (Figure 95), reflecting the bimodal wind pattern shown in Figure 73. The long sand beach south of the river mouth is backed by some noteworthy vegetated dunes (again reflecting the WNW winds during the dry season), the first significant patch of coastal dunes north of the California border.

North of the sand spit, there are three rocky headlands with their accompanying sea stacks, culminating with the prominent Cape Sebastion that has overlooks offering spectacular long-range views along the coast. Cape Sabastian and Hunter Island are composed of a massive, gray sandstone that is obviously more resistant to erosion compared to adjacent rocks that have thin beds of sandstone and mudstone (Lund, 1975). There are two modest-sized rocky protuberances north of Cape Sebastion. From there, the shore is composed of sand beach, built with sediments transported to the shore by Hunter Creek and Rogue River.

When co-author Hayes was a young teenager checking out books from the Asheville, NC library, one of his favorites was *Tales of Freshwater Fishing* by Zane Grey, published in 1928, a book with several stories about fishing the Rogue River, where the famous author had a fishing camp. For years, Hayes had always wanted to go there, and we finally did, birding this time and not fishing.

This is one of the better known rivers in Southern Oregon, with healthy numbers of wild and hatchery steelhead, cutthroat trout and rainbow trout. Chinook salmon run in both spring and fall, bringing lots of people from all over the state and country to fish for them. The river is about 215 miles long from its headwaters near Crater Lake National Park to its mouth.

FIGURE 95. Mouth of Pistol River, OR. Note field of vegetated dunes south of the river mouth. NAIP 2014 imagery, courtesy of Aerial Photography Field Office, USDA.

Alt and Hyndman (1978) pointed out the following interesting facts about the geology in the vicinity of Gold Beach. *Gold Beach is within a big belt of serpentinite spectacularly exposed in several road cuts both north and south of town. The rock is unmistakable - green and sheared into minute slivers but containing solid chunks which have rounded outlines and brightly polished surfaces. These chunks appear to have slipped through the sheared groundmass as it moved as though they were so many watermelon seeds. Some have such fascinating shapes and colors that they qualify as natural sculpture.*

The serpentinite and accompanying basalt were originally formed on an ocean floor about 150 million years old.

A continuous, 3-mile stretch of sand and gravel beach extends from the north jetty at the mouth of the Rogue River to Otter Point, a modest-sized rocky headland surrounded by sea stacks. You should notice something different about Otter Point—the top is flat! It is a remnant of a wave-cut terrace created about 80,000 years ago. Continuing north from Otter Point there are three more rocky protuberances within the next several miles that are, as usual, surrounded by numerous sea stacks. Nesika Beach is backed by prominent cliffs of Pleistocene marine terrace deposits of sand and gravel overlying sheared sedimentary (mudstone and sandstone) rocks that are currently being eroded at about 2 feet per year, which is some of the highest rates measured in Oregon (Ruggiero et al., 2013). Those sheared rocks and the overlying sediments can't stand up to the pounding of storm waves, especially during *El Niño* storms.

From Otter Point to Nesika Beach, the beach is mostly sand. However, as you get closer to Cedar Creek, the amount of gravel increases, likely due to the historical transport of gravel by the creek. The sand and gravel beach at the mouth of Cedar Creek at Ophir is very wide and a great place to view the sea stacks to the north. A fairly prominent rocky headland called Sisters Rocks marks the north end of that rocky area.

The coast takes a sharp turn to the northwest to the high rock cliff at Humbug Mountain, probably the most prominent headland along this section of the Oregon Coast. Alt and Hyndman (1978) discussed the geology of Humbug Mountain, saying that *it is a huge mass of gravelly conglomerate deposited during Cretaceous time, a little more than 100 million years ago.*

Inspection of Figure 48A will show you that there are five raised marine terraces between Humbug Mountain and Cape Blanco. The terrace just to the north of Humbug Mountain, Poverty Ridge Terrace, has been raised 800 feet. Those terraces decrease in elevation in a northerly direction, with the next one, Indian Creek Terrace having been raised 600 feet. It is followed by Silver Butte Terrace raised 180 feet, Pioneer Terrace raised 160 feet, and, finally, Cape Blanco Terrace, raised 150 feet.

The shore from Humbug Mountain to Port Orford is mostly sand beach with the usual compliment of relatively abundant sea stacks offshore. Interestingly, that sand beach has a distinctly crenulate-bay shape at its northwest end that indicates alongshore sediment transport toward the southeast.

The massive rock cliffs of Port Orford Heads, the location of Port Orford Heads State Park, were discussed by Alt and Hyndman (1978), noting that they have *actually been scraped off the seafloor onto the edge of the continent as the oceanic crust slid out from under them into an ocean trench. So these rocks are severely deformed and recrystallized enough to be very hard; they are no longer soft mudstone and sandstones. They began as sediments laid on the seafloor back in Jurassic and Cretaceous time, between 200 and 100 million years ago in round numbers, and then jammed into the Coast Range during that same time interval.* Those rocks occur in the wave-cut cliffs for miles along the coast to the north of Port Orford. The vast majority of them are dirty sandstones now hardened by recrystallization into very solid rock, *which is usually rather dark and massive looking.*

[NOTE: A "dirty" sandstone is one that contains an abundance of muddy sediment in between the sand grains, which means that the sand was deposited rather abruptly without much reworking of the sand by waves and/or currents.]

Continuing north to the mouth of the Elk River, the shore is composed of sand beach that provides a partial border for Garrison Lake. Garrison Lake is a typical "sand dune lake" in that it was a coastal valley that was inundated as sea level rose and a sand spit and dune system built across its mouth (contributed to by sand transported to the coast by the Elk River?), allowing a freshwater lake to form behind the dunes. On older maps, the lake is shown as Garrison Lagoon, reflecting its history of the sand spit breaching during large storm events.

The beach between the Port Orford headland to the south and Cape Blanco to the north (the westernmost point in Oregon) is a classic littoral cell, with the headlands as natural barriers to sediment transport either north or south. The Elk River spit diverts the river up to 12.6 miles to the north indicating a dominant northerly transport direction, though the river can temporarily cut through the beach and dunes during high river flow, as occurred during the 1996 flood. Check out Figure 48A to see a sketch of the five raised marine terraces behind Cape Blanco. When you are in this area, try to see if you can locate them.

According to a study of shoreline change over the long-term period from the 1880s to 2002 and the short-term period from the 1960s to 2002 by the U.S. Geological Survey (Ruggiero et al., 2013), most beaches in this Zone are experiencing short-term erosion rates of less than 3 feet per year, though often the long-term net change is close to zero. Some sections undergo short-term erosion due to the migration of river mouths, such as at Hunter Creek (see Figure 69B where the creek is very close to the structures at its northern extent), and during *El Niño* years, when a change in winter wind/wave patterns results in erosion "hot spots" as shown in Figure 41. Hunter Creek (just south of Gold Beach) is a great example of migrating creek mouths that you can see yourself by using the historical imagery option on Google Earth (it looks like a clock in the tool bar). For this location, there is imagery dating from May 1994 to present, and you can see how much the creek mouth has migrated over time both to the north and south. Google Earth is pretty amazing in that way. Remember that this part of the Oregon Coast is rising faster than the current rate of sea-level rise, thus the sea cliffs and beaches are pretty stable.

General Ecology

This Zone has 28 seabird nesting colonies, with two colonies that have over 100,000 nesting birds: 1) Goat Island, with recent counts of over 100,000 Leach's storm petrels, but also cormorants, murres, rhinoceros and Cassin's auklets, tufted puffins, and pigeon guillemots; and 2) A group of offshore islands at Crook Point that supports the second largest concentration of nesting seabirds in Oregon, with recent counts of over 187,000 Leach's storm petrels. Black oystercatchers can be seen on most rocky shorelines. You can see thousands of seabirds at many observation points along the coast during migration and a long list of species at any time. The Oregon Coast Birding Trail has great regional maps that list sites to stop at and what species you will see at different times of the year.

Marine mammals along this coast include migrating gray whales in spring and fall, humpback whales in summer, killer whales from fall to spring, and harbor porpoise, which are highly abundant year-round. There are two large seal/sea lion rookeries: 1) Rogue Reef north of the Rogue River, with >500 Steller sea lions (pupping June-July) and 100-500 Pacfic harbor seals (pupping May-June); and 2) Orford Reef south of Cape Blanco with >500 Steller seal lions). Pacific harbor seals haul out on rocky shores and sand beaches/bars throughout the area.

Just about every stream connected to the ocean has runs of cutthroat and steelhead trout, coho salmon, Chinook salmon (spring and fall runs), and Pacific lamprey.

Places to Visit

Highway 101 is a very scenic drive all the way from the border to Port Orford. There are three state recreational areas and parks to stop at to walk along the beach, watch for birds and whales, go surf fishing and clamming, or just enjoy the great views: Crissey Field State Recreation Area, McVay Rock State Recreation Area, and Chetco Point Park.

Harris Beach State Park is where you can see, in the distance, Goat Island (the largest island along the Oregon coast) with its very large seabird nesting colony, including hundreds of tufted puffin, rhinoceros auklet, and Cassin's auklet. There are smaller colonies on the rocky islands just off the beach. Sea stacks dot the ocean just off shore.

And sometimes, starting to turn over a big rock in the Great Tide Pool – a rock under which he knew there would be a community of frantic animals – he would drop the rock back in place and stand, hands on hips, looking off to sea, where the round clouds piled up white with pink and black edges. And he would be thinking, What am I thinking? What do I want? Where do I want to go? There would be wonder in him, and a little impatience, as though he stood outside and looked in on himself through a glass shell. And he would be conscious of a tone within himself, or several tones, as though he heard music distantly.

John Steinbeck – SWEET THURSDAY

The Samuel H. Boardman State Scenic Corridor offers views of many sea stacks including Arch Rock and Natural Bridges, as well 27 miles of the Oregon Coast Trail. The Pistol River State Scenic Viewpoint is located at the Pistol River (see Figure 95) with its dunes and dynamic river outlet. Look for ponded water behind the spits where there can be lots of shorebirds and waterfowl in season.

Just offshore of Crook Point is Mack Reef, a group of small islands that is home to the second largest concentration of nesting seabirds in Oregon. The headland, which is part of the Oregon Island NWR, is closed to all public entry, but you can walk along the adjacent publically accessible beaches where the views of the headland and sea stacks are spectacular.

You should stop at the scenic overlook near the mouth of Meyers Creek, which provides a stunning view of the Meyers Creek Rocks. Be sure to note the tombolo that connects them to the shoreline.

The prominent Cape Sebastion is the centerpiece of the Cape Sebastion State Scenic Corridor. If you hike the moderately difficult 1.5 mile trail to the Cape, you will have a great view of Hunters Island, which hosts 12 species of nesting seabirds, including over 30,000 Leach's storm petrels. You might even see marbled murrelets in summer. Oregon State Parks (2014a) gives the following description of Cape Sebastion State Park:

The most striking feature of this park is the panoramic view, both parking lots are over 200 feet above sea level. At the south parking vista, you can see up to 43 miles to the north with Humbug Mountain filling the view. Looking south, you can see nearly 50 miles toward Crescent City, California and Point St. George Lighthouse. .... A deep forest of Sitka spruce covers most of the park and a 1.5 mile walking trail takes you out to the lower levels of the cape. If you're lucky, the trail will give you a chance to enjoy awe-inspiring views of the gray whales on their bi-annual migrations.

"I've been to the ridge and looked down," Thomas said. "It's wild over there, redwoods taller than anything you ever saw, and thick undergrowth, and you can see a thousand miles out on the ocean. I saw a little ship going by, half-way up the ocean."

John Steinbeck – TO A GOD UNKNOWN

At Gold Beach you may want to check out the serpentenite rocks described earlier. Otter Point and the Otter Point State Recreation Area offer overlooks with great views of the many sea stacks and eroding rock cliffs. About 2 miles offshore is Rogue Reef consisting of four islands and numerous rocky islets awash at high water. There can be over 5,000 common murres nesting on the islands. Hundreds of harbor seals haul out on the southernmost island, Pyramid Rock, so you should see lots of them feeding close to shore. Steller seal lions are abundant and have a rookery with up to 550 animals on Needle Rock. The nearshore rocks host smaller numbers of nesting pelagic cormorants, pigeon guillemots, gulls, and black oystercatchers. The rocky islands off Hubbard Mound also host large numbers of nesting seabirds (one island has over 20,000 common murres) and seal/sea lion haulouts, but there is no public access except by boat.

The Arizona Beach State Recreation Site, which has a pullover site by the beach, is located 4 miles north of Ophir. It was described as follows by Oregon State Parks (2014b):

Arizona Beach is a two-third mile stretch of sand bookended by two rocky headlands. The headlands shelter the beach from prevailing winds, creating temperatures warm enough to give the site its name.... Mussel and Myrtle creeks flow through the park east of U.S. 101. The wetlands attract elk and a variety of waterfowl.

Humbug Mountain State Park is a place you do not want to miss. There you will have access to the beach and a steep hiking trail with great views. Marbled murrelets nest in the adjacent forests so you may see them in nearshore waters in summer. Island Rock is about 1.4 miles offshore and a major seabird nesting colony with 12 species, including up to 20,000 common murres and 300 tufted puffins.

The rocky shores at Port Orford Heads State Park have eight seabird nesting colonies with small numbers of pigeon guillemots, cormorants, and gulls, and multiple harbor seal haulouts. Harbor porpoises are highly abundant. Gray whales feed in this area from June to November. According to Oregon State Parks (2014c):

The Port Orford Lifeboat Station was constructed in 1934 by the Coast Guard to provide lifesaving service to the southern portion of the Oregon Coast until 1970. A museum, operated by the Cape Blanco Heritage Society, is now housed in the station. ... The Park has excellent hiking trails on the headlands affording spectacular views up and down the Pacific Coast.

Cape Blanco is noted for its beautiful lighthouse that is open for tours from April to October. It is the southernmost lighthouse in Oregon, and Cape Blanco is the westernmost point in the state. Just north of the

Cape is a group of small islands called Gull Rock that supports seven species of nesting seabirds, including over 40,000 common murres, and numerous harbor seal haul outs. Marbled murrelets nest in adjacent forests and can be seen in nearshore waters in summer. Thousands of seabirds fly by during migration.

ZONE 2. CAPE BLANCO TO CAPE ARAGO

Introduction

The dominant geomorphological features of this 36-mile long Zone are:

1) Long sand beaches;

2) A number of relatively prominent rocky headlands (e.g., Cape Blanco, Tower Rock, Table Rock Area, and Cape Arago);

3) Eleven tidal inlets, one of which, the mouth of the Coquille River, has jetties; and

4) A series of four raised marine terraces in the vicinity of the town of Bandon (see Figure 48B).

Geologic Framework

In contrast to Zone 1, marine terraces of weakly consolidated Quaternary sands and mud cover most of the coast for 2-4 miles from the shoreline. However, the underlying bedrock has been exposed in coastal cliffs as the shoreline retreats. The bedrock is composed mostly of sedimentary rocks of various ages and degree of faulting and metamorphism. Headlands, islands, and offshore sea stacks occur where the rocks are more resistant to erosion by the waves: that is, where the sandstones are thickest and the metamorphic rocks are hardened but not sheared (Lund, 1973). Because some of the sedimentary rocks are tilted to the east by up to 30 degrees, there can be wide wave-cut platforms with abundant tidal pools.

General Description

This Zone can be characterized as long sand beaches with small sections of rocky headlands and sea stacks at the southern, central, and northern portions. The abundance of sand and a lowland behind the shore allows for the deposition of wind-blown sand forming extensive dunes, particularly behind New River Spit and south and north of Bandon. The net sand transport direction is north, as indicated by the nearly 10-mile-long New River Spit and the accumulation of sand on the south side of the jetties on the Coquille River. The New River Spit has a pattern of being breached during storms on the south end, and the north end, so the dunes never build up as they do along the more stable central portion (Komar, Diaz-Mendez, and Marra, 2001). The shoreline north of the jetties has experienced short-term erosion rates of 6.5 feet per year, whereas most of the rest of the shoreline is stable (Ruggiero et al., 2013).

The Coquille River estuary is where researchers (Witter et al., 2003) were able to determine the dates that different layers of tidal flat mud and upland soils were deposited, which showed that this area experienced twelve earthquakes that caused the land to sink 4 to 10 feet each time in the last 6,700 years. The youngest buried soil layer represents the *Orphan Tsunami of 1700* discussed earlier. They found evidence that the tsunami traveled over 6 miles up the estuary. This sudden sinking led to all kinds of shoreline change, including the erosion of the rocky shores south of the Coquille River and formation of the cluster of sea stacks that you see there now. However, due to uplift along this part of the coast, that zone of rocks, including Face Rock, Table Rock, Cat and Kittens Rock, are now close to shore.

As can be seen in Figure 48B, the uplift of the Bandon area has created a series of raised marine terraces, namely Metcalf Terrace at 500 feet, Seven Devils Terrace at 300 feet, Pioneer Terrace at 150 feet, and Whisky Run Terrace at 60 feet. Alt and Hyndman (1978) discussed as follows the experience of driving on one of those terraces, specifically along U.S. Highway 101 between Port Orford and about 6 miles north of Bandon:

The road follows the surface of a marine terrace, that was leveled by wave action when it was submerged offshore and then raised above sea level. The old shoreline is near the base of the hills east of the highway. Big, craggy rocks rising out of the fields and nestled here and there in the woods are old sea stacks that once stood proudly in the surf. Except for the old sea stacks, there is no bedrock along the highway because the terrace surface is covered by sand and mud laid down along the beach and offshore while the area was underwater.

According to Alt and Hyndman (1978), *the sea cliffs around Cape Arago just west of Coos Bay are probably the best place to see good outcrops of the Coaledo Formation especially when the tide is out.* They described that formation as *a thick series of sand and muds deposited near shore, possibly in a bay, during Eocene time. This formation contains several substantial coal seams which supported large mines many years ago.*

General Ecology

There are 23 seabird nesting colonies in this Zone, mostly clustered on the rocky headlands and islands south of Bandon and at Cape Arago. The colonies are dominated by common murres; the largest single common murre colony in Oregon is on Cat and Kittens Rock, with over 38,000 birds; North Coquille Point Rock comes in second place, with over 10,000 birds. There are thousands of nesting cormorants and gulls, and a few tuffed puffins nesting on Elephant Rock and Face Rock. Pairs of black oystercatchers can be found on most rocky areas. Marbled murrelets nest in adjacent forests and can be seen in nearshore waters. Thousands of wintering shorebirds are present October to May. The flooded pastures and small wetlands along the Coquille River valley near Coquille are important wintering areas for tens of thousands of northern pintail ducks, and thousands of other ducks and geese. Western snowy plovers (federal and state listed as threatened) nest on the New River Spit and in the Bandon State Natural Area. There are restrictions regarding beach access and whether dogs are allowed at all or need to be on leash during nesting season from 15 March to 15 September.

Marine mammals along this coast include migrating gray whales, humpback whales in summer, killer whales from fall to spring, and harbor porpoise are highly abundant year-round. Harbor seals haul out on rocky shores and sand beaches/bars throughout the area. Even the smaller streams support runs of coho and Chinook salmon, steelhead, cuttrout trout, and lamprey.

Places to Visit

The first place you should head to is the beach south of the south jetty at the mouth of the Coquille River to see the spectacular assemblage of sea stacks and rocky protuberances. There are three relatively small parks in the area that provide adequate parking, among other facilities. You can also get great views from the Face Rock State Scenic Viewpoint.

The Bandon Marsh National Wildlife Refuge has two parcels at the mouth of the Coquille River estuary that are great places to see large numbers of many different birds, including large numbers of migrating shorebirds with peak numbers from late April to early May and again in early September. There can be tens of thousands of western sandpipers, thousands of least sandpipers, dunlins, and short-billed dowitchers, and hundreds of sanderlings, semipalmated plovers, black-bellied plovers, and black turnstones.

The spacious Bullards Beach State Park is another good place to visit, with highlights such as a mile-long pathway to the beach that crosses grassy fields, lowland forests, and sand dunes, views of the Coquille River, and a historic lighthouse.

Doc walked on the beach beyond the lighthouse. The waves splashed white beside him and sometimes basted his ankles. The sandpipers ran ahead of him as though on little wheels. The golden afternoon moved on toward China, and on the horizon's edge a lumber schooner balanced.

John Steinbeck – SWEET THURSDAY

The next place we recommend that you visit is the Seven Devils State Recreation Site. It has a good parking area right by the beach, is in a relatively isolated area so you can have a lot of the beach to yourself, and it is a great place to look for agates (best after storms).

Cape Arago State Park is at the north end of this Zone. The unique shape of the Cape reflects its complex geology. Thick layers of hard sandstones are more resistant to erosion by waves, and faults have weakened the rocks, forming linear slots in the rock platform. As you walk along the North Cove beach, look at the cliffs to see many different sedimentary structures.

At Cape Arago, you will have the unique opportunity to see all four of the common pinnipeds in Oregon: Steller sea lions, California sea lions, elephant seals, and harbor seals. Simpson Reef, Shell Island, and the adjacent reefs are the largest haul outs for marine mammals on the Oregon Coast. You can spend hours here in awe at the views, waves, and so many marine mammals and birds. Sunset Bay is a great location to watch the sun set.

ZONE 3. CAPE ARAGO TO FLORENCE JETTIES

Introduction

The dominant geomorphological features of this

51-mile-long Zone are:

1) Several good-sized coastal dune fields;

2) Long and straight sand beaches, some with rhythmic topography (defined in glossary);

3) A large estuarine complex at Coos Bay; and

4) Three major river outlets – Umpqua, Siuslaw, and Coos Rivers – and eight inlets, three of which, the mouths of the Umpqua, Siuslaw, and Coos Rivers, have jetties.

Geologic Framework

Except for the headland of Cape Arago, the surface geology map of this Zone shows the coastal region is covered by marine terraces composed of Quaternary sediments. From Cape Arago to Coos Bay, the sedimentary rocks are highly tilted to the east and fractured. According to Lund (1973), *the edges of these beds are exposed to wave attack along a southwest trending coast. Erosion, directed along soft sedimentary layers and fractures, has shaped a shore that is distinctly different from that of any other part of the Oregon coast.* The rest of the shoreline is backed by a wide coastal plain of marine terrace deposits with extensive dune fields.

General Description

Northeast of Cape Arago there are 3.4 miles of highly tilted bedrock exposures, such as at Shell Island, Simpson Reef, Rock Reef, Gregory Point, and Yoakam Point, to name a few. A geological cross-section across the Coos Bay area that illustrates the highly folded nature of the rocks in this area is given in Figure 96 and explains why the rocks are so tilted in that area.

Coos Bay, illustrated in Figure 97, is by far the largest of the eleven larger estuarine systems on the Oregon Coast. The following rivers and streams enter the headwaters of the estuary (from south to north): Winchester Creek, John B. Creek, Elliot Creek, Davis Creek, Noble Creek, Wilson Creek, Morgan Creek, Daniels Creek, Coos River, Millicoma River, Kentuck Creek, and Palouse Creek. There are also several features called sloughs that border the main estuary.

In Section 1, we defined estuaries as flooded, lowstand river valleys subject to tides and recipients of significant fresh water influx. With a mean tidal range at the entrance of 6.7 feet and freshwater influx from the numerous sources listed above, two of these criteria have clearly been met for the Coos Bay Estuary. When sea level was down ~400 feet during the Wisconsin Glaciation 20,000 years ago, the valley now occupied by Coos Bay was obviously carved by the Coos River and accompanying streams, satisfying the third criteria.

The stretch of the beach north of the north jetty at Coos Bay was one where our research team spent some time because a freighter called the M/V *New Carissa* ran aground there during a storm on 3 February 1999 (see Figure 98). The vessel had no cargo but contained 400,000 gallons of heavy fuel oil. As described in the U.S. Coast Guard report (Hall, 1999), because of concerns that the hull would fail during efforts to pull it off the

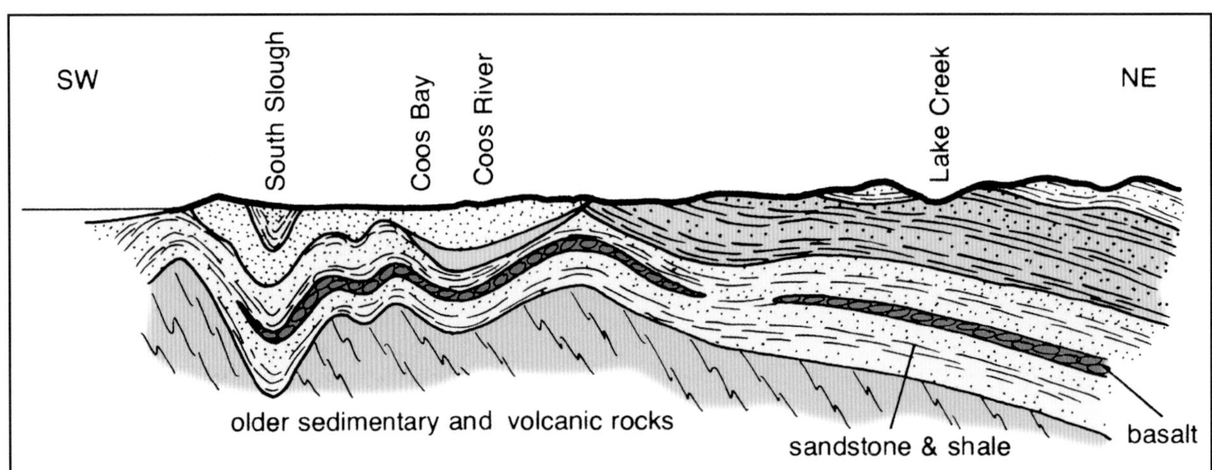

FIGURE 96. Geological cross-section in the Coos Bay area illustrating the folded nature of the rocks in that area. Modified after Alt and Hyndman (1978).

FIGURE 97. Coos Bay estuary, OR. NAIP 2011 imagery, courtesy of Aerial Photography Field Office, USDA.

FIGURE 98. Photograph of the grounded M/V *New Carissa* taken 4 February 1999. The ship was intentionally ignited to burn the fuel oil and later broke in two. Source: public domain.

beach during the severe storm conditions, which could release the oil, the decision was made to attempt to burn the oil onboard. It took two attempts, but after using 400 pounds of explosives to rupture the fuel tanks, on the 8th day after the grounding, the oil was ignited and burned for 33 hours, consuming about half of the oil onboard. During the burn, the vessel broke in two and continued to release oil. While preparing to tow the bow section to sea and scuttle it, attempts to remove the remaining viscous oil failed. On 1 March, 26 days after the grounding, the bow section was pulled off the beach and towed offshore. However, because of >65-knot winds and 30-foot seas, the towline broke 50 miles offshore and the bow section went aground again (!) at Waldport on 3 March. It was refloated and sunk at sea 280 miles offshore in 11,000 feet of water on 11 March. Plastic explosives were used to open up the hull to help it sink; however, it took a final hit from a torpedo to sink it before nightfall. Back on the beach at Coos Bay, much of the oil remaining in the stern section was removed. The stern section remained aground for over nine years, finally being dismantled and removed from the beach in 2008. This event shows how complicated vessel groundings can be.

According to Ruggiero et al. (2013), the shoreline north of the north jetty built out by over 3,000 feet after the jetty was constructed in the late nineteenth century. However, since the 1960s, this beach has been erosional, whereas most of the other beaches in the Zone have been slightly depositional.

The land to the east of the beach is part of the Siuslaw National Forest as well as the Oregon Dunes National Recreation Area. Those dunes were discussed briefly in Section 1. According to Oregon's Adventure Coast (2014), this is the largest expanse of coastal dunes in North America. It extends for 40 miles along the coast with sand dunes that reach as high as 600 feet. The large barren dune field near Lakeside and south of the Umpqua River is illustrated in Figure 99; there is a similar, large barren dune field near Dunes City.

The mouth of the Siuslaw River has a heavy armor of two confining jetties, illustrated in Figure 85A. And speaking of armoring, there is a series of four groins about two miles inside the jetties on the southwest shore of the Siuslaw River (Figure 85B).

General Ecology

There are over 20 seabird nesting colonies in this Zone. Five double-crested cormorant colonies with hundreds of birds each are located inside the three estuaries. When you cross the Umpqua River north of Reedsport, you can see one of these colonies in a patch of dead trees covered in white guano on Bolon Island just west of the road. The colonies in Coos Bay are located in a variety of habitats, with about a hundred pigeon guillemots nesting at the Sitka Dock, pigeon guillemots nesting on the Charleston Bridge over South Slough, two cormorant colonies near the McCullogh Bridge where U.S. 101 crosses the Coos River, and pelagic cormorants and pigeon guillemots nesting at Coos Head. The rocky headland and islands between Point Arago and Gregory Point support 14 more traditional colonies with all three cormorants, gulls, common murres, pigeon guillemots, and black oystercatchers. Chiefs Island, right off Gregory Point, also hosts rhinocerous auklets and tuffed puffins.

All of the sand beaches in this Zone are important for western snowy plovers nesting, supporting 63% of all the nests in Oregon in 2015 (Audubon California, 2017). Federal, state, and other agencies have been implementing conservation measures to help this threatened species recover. Measures include removing invasive beachgrass and leveling steep dunes formed by the vegetation (plovers nest in unvegetated areas between the high-tide and the dunes), placing shell material in really good nesting areas, and predator control through placement of cages over some of the nests as soon as they are found by volunteers (there is not enough funding to place cages over all nests). When volunteers monitor

FIGURE 99. Coastal dune field near Lakeside, OR and just to the south of Winchester Bay. Arrow points to the mouth of Tenmile Creek. NAIP 2011 imagery, courtesy of Aerial Photography Field Office, USDA.

nests, they document, to the best of their ability, the causes of nest failures (usually the eggs are eaten), but most of the time, the cause is "unknown." Researchers have been placing cameras at nests to better understand why nests fail (Gaines et al., 2016). The results were surprising. Of 130 nests with cameras, 70 failed, and of these, 74% of the failures were due to predation. And guess what species was the most common nest predator? Northern harriers! This was a surprise to everyone; these raptors fly from the nest and leave very little evidence of their brief raid. Crows, coyotes, and foxes were also feeding on the eggs, as expected.

North of Point Arago is Simpson Reef, which is the northernmost pupping site for northern elephant seals, as well as being the largest marine mammal haul out on the Oregon Coast. Nearby Shell Island is a major haulout for Steller and California sea lions. These are "must-see" sites on your trip along the southern Oregon Coast. As is the case for the entire Oregon Coast, you may see migrating gray whales, humpback whales in summer, killer whales from fall to spring, and harbor porpoise year-round. Pacific harbor seals haul out on rocky shores and sand beaches/bars throughout the area.

The Umpqua River is one of the most productive systems for salmon in Oregon, particularly for coho salmon. It also has some of the highest returns in the region for fall Chinook salmon and high numbers of spawning nests (called redds) for winter steelhead. The Coos Bay estuary is also important spawning and rearing habitat for coho and Chinook salmon. Both of these systems have extensive estuarine habitat, which is important because juvenile salmon spend 1-2 years in the estuary before they undergo smoltification, which allows them to live in salt water.

Places to Visit

The peninsula just to the north of Cape Arago has great rocky shores and wave-cut platforms. The variations in the tilt of the sedimentary rocks relative to the shoreline orientation create very interesting features on the wave-cut platforms, where the more resistant layers are a little higher than the adjacent weaker layers. You can explore these features at two parks: Shore Acres State Park and Sunset Bay State Park.

In Coos Bay, South Slough Reserve has the distinction of being designated in 1974 as the first unit in the National Estuarine Research Reserve System (NERRS), a network of estuary habitats protected and managed for the purposes of long-term research, education, and coastal stewardship. There is a visitor center and many foot and water trails. It is the only such site in Oregon.

The Oregon Dunes National Recreation Area is a really special place to visit (Figure 99). The dunes extend along nearly 50 miles of the shoreline and as far as 2-3 miles inland. Some are nearly 600 feet high. Geotimes (2004) mentioned that *if you are interested in visiting, you should consider planning a trip while the dunes are largely free of vegetation. August is the peak tourist season, but Oregon's mild winters make the dunes enjoyable year-round.... In the winter, visitors can find rare occurrences of yardangs — elongate, wind-sculpted ridges — among the dunes.* Jessie Honeyman Memorial State Park is one of the best places to experience the dunes.

William Tugman State Park is located on the side of Eel Lake and, despite its close proximity to U.S. 101, this Park is relatively unknown. Supposedly the fishing is great, and there are trails around the south end of the lake, which is forested and undeveloped thus providing opportunties for wildlife viewing.

In the Umpqua River estuary, you will see hundreds of harbor seals hauled out on the tidal flats at low tide and thousands of western sandpipers on their spring and fall migrations. The Bolon Island Tideways State Scenic Corridor is a great place to visit because of its history: Native Americans used the site, then it was a busy sawmill and dock. The hiking trail half way around the island provides views of the large double-crested cormorant nesting colony, as well as the Umpqua River and wildlife. You might even see bald eagles. The Umpqua River supports runs of threatened stocks of coho and Chinook salmon and summer and winter runs of steelhead. Dams and habitat degradation has severely reduced these fish populations (thus their designation as threatened), and hatcheries capture adults to raise young fish for release to augment the recreational fishery. Restoration of the watershed has been a key goal to restore these fish stocks.

There are only two beach access roads in this Zone: South Jetty and Siltcoos River. The Siltcoos River drains the adjacent freshwater Siltcoos Lake (the biggest lake on the Oregon Coast; refer to Figures 91 and 101), and there is a 3-mile canoe trail through undeveloped forests, dunes, and then the wide estuary. coho salmon have a fall run up the river to spawn in the gravel bars in the tributaries to the lake.

15 COMPARTMENT 2 - CENTRAL UPLAND (Florence Jetties to Tillamook Head)

INTRODUCTION

The coastal habitats present in Compartment 2 are listed in Table 5. This Compartment has a variety of habitats along the shore, with five of them occurring along greater than 10% of its length. Of these five habitats, salt- and brackish-water marsh is the longest, occurring along 37.9% of the length of the shore (203 miles) (those tidal channel shorelines add up fast!). Next comes exposed tidal flats at 30.8%, sheltered tidal flats

TABLE 5. Coastal habitats in Compartment 2 (Florence Jetties to Tillamook Head) (NOAA/RPI, 2015).

COMPARTMENT 2 – CENTRAL UPLAND (Florence Jetties to Tillamook Head)			
ESI VALUE	**ESI DEFINITION**	**LENGTH (MILES)**	**% ABSOLUTE LENGTH**
1A	Exposed Rocky Shore	67.18	12.58%
1B	Exposed, Solid Man-Made Structures	2.31	0.43%
2A	Exposed Wave-Cut Platforms in Bedrock	38.10	7.13%
2B	Exposed Scarps and Steep Slopes in Clay	0.55	0.10%
3A	Fine- to Medium Grained Sand Beaches	101.41	18.99%
3B	Scarps and Steep Slopes in Sand	3.53	0.66%
4	Coarse-Grained Sand Beaches or Bars	0.00	0.00%
5	Mixed Sand and Gravel Beaches or Bars	12.88	2.41%
6A	Gravel Beaches or Bars	47.55	8.91%
6B	Riprap	33.84	6.34%
6D	Boulder Rubble	9.80	1.84%
7	Exposed Tidal Flats	164.52	30.81%
8A	Sheltered Rocky Shores and Sheltered Scarps	11.76	2.20%
8B	Sheltered, Solid Man-Made Structures	1.32	0.25%
8C	Sheltered Riprap	25.50	4.77%
8F	Vegetated, Steeply-Sloping Bluffs	0.31	0.06%
9A	Sheltered Tidal Flats	135.08	25.30%
9B	Vegetated Low Banks	4.19	0.78%
10A	Salt- and Brackish-Water Marshes	202.52	37.93%
10B	Freshwater Marshes	31.89	5.97%
10C	Swamps	11.23	2.10%
10D	Scrub and Shrub Wetlands	15.72	2.94%
	Total ESI Length (Miles)	921.19	172.52%
	Absolute Shoreline Length (Miles)	533.97	100.00%

at 25.3%, fine- to medium-grained sand beaches at 19%, and exposed rocky shores at 12.6%. These numbers are greater than 100% because some parts of the shore are composed of more than one habitat type. The abundance of sheltered habitats reflects the fact that 11 estuaries (including some smaller ones) with their associated sheltered habitats are present along the shore of this Compartment.

For purposes of discussion, we have further subdivided this Compartment into three Zones shown in Figure 100:

1) Florence Jetties to Yaquina Head
2) Yaquina Head to Cape Lookout
3) Cape Lookout to Tillamook Head

ZONE 1. FLORENCE JETTIES TO YAQUINA HEAD

Introduction

The dominant geomorphological features along this 46-mile-long Zone, illustrated in Figure 101, are:

1) Sand beaches and dune fields on both sides of the Siuslaw River;
2) Rocky shores at Heceta Head and south of Yachats;
3) Mostly coastal plain and sand beaches from Yachats north;
4) Two prominent estuaries - Alsea Bay and Yaquina Bay; and
5) Thirteen inlets, only one of which, the mouth of the Yaquina River, has jetties. Two other modest sized rivers empty along this Zone, the Yachats and Alsea Rivers.

Geologic Framework

Two types of rocks dominate this Zone. First, basaltic rocks of Eocene age (50 million years old) extend along the shoreline and many miles inland from Cox Rock to Yachats, creating this rugged rocky coast. Yaquina Head is another basaltic headland. In contrast, the rest of the Zone is underlain by much softer sandstones and mudstones of Miocene age (20 million old) that have been eroded at different sea levels and now form a low coastal plain covered by marine terrace deposits.

General Description

A sand beach and inland dune system extends for 6.3 miles from the north jetty at the Siuslaw River to a rocky headland called Cox Rock (Figure 101). A sand spit has diverted the outlet of Sutton Creek to the south, likely caused by the buildup of high dunes between the creek and the shoreline by strong northwest winds. The outlet migrates between the vegetated bluff at Heceta Beach on the south and the vegetated dunes to the north. In Figure 74, you can see the main outlet in the center and an older outlet to the south that has been closed off.

Bedrock headlands of basalt with pocket sand and gravel beaches extend between Cox Rock and Heceta Head. The sand beaches usually have a high-tide berm composed of coarse gravel, as shown in Figure 102. Aptly named Sea Lion Point forms rugged cliffs with a large sea cave system. These rocky headlands and platforms occur where masses of hard rocks, composed of basalt that flowed out on the seafloor millions of years ago, are more resistant to erosion compared to the softer, sedimentary rocks on either side. Some of the basalt is fractured, allowing waves to erode the weaker areas, forming caves, sea stacks, and long ridges that extend across the shoreline. The prominent rocky headland called Heceta Head (Figure 103) is a good example, with a beautiful lighthouse on its southern end that some say is "The most photographed lighthouse in the whole World!"

From Heceta Head to Searose Beach, the shoreline consists of sand beaches, gravel beaches, and scattered rocky platforms and sea stacks. According to Alt and Hyndman (1978), golf-sized rounded agates eroded out of the basalts can be found on these beaches after major storms.

From Searose Beach to Yachats, those resistant basaltic rocks crop out along the shoreline, creating a predominantly rocky coast with some sand beaches with high-tide berms composed of coarse-grained gravel. There are some outstanding spots to view blow holes and other features of eroding bedrock exposures along the shore. Those features have exotic names like Cook's Chasm, Spouting Horn, Thor's Well, Double Trouble, and Devils Churn (a deep channel eroded along a fracture in the rocks). Cape Perpetua itself is composed of resistant basalt agglomerate cut by dikes (Alt and Hyndman, 1978). The cove at Yachats was likely formed by the valley cut by the Yachats River during lower sea levels.

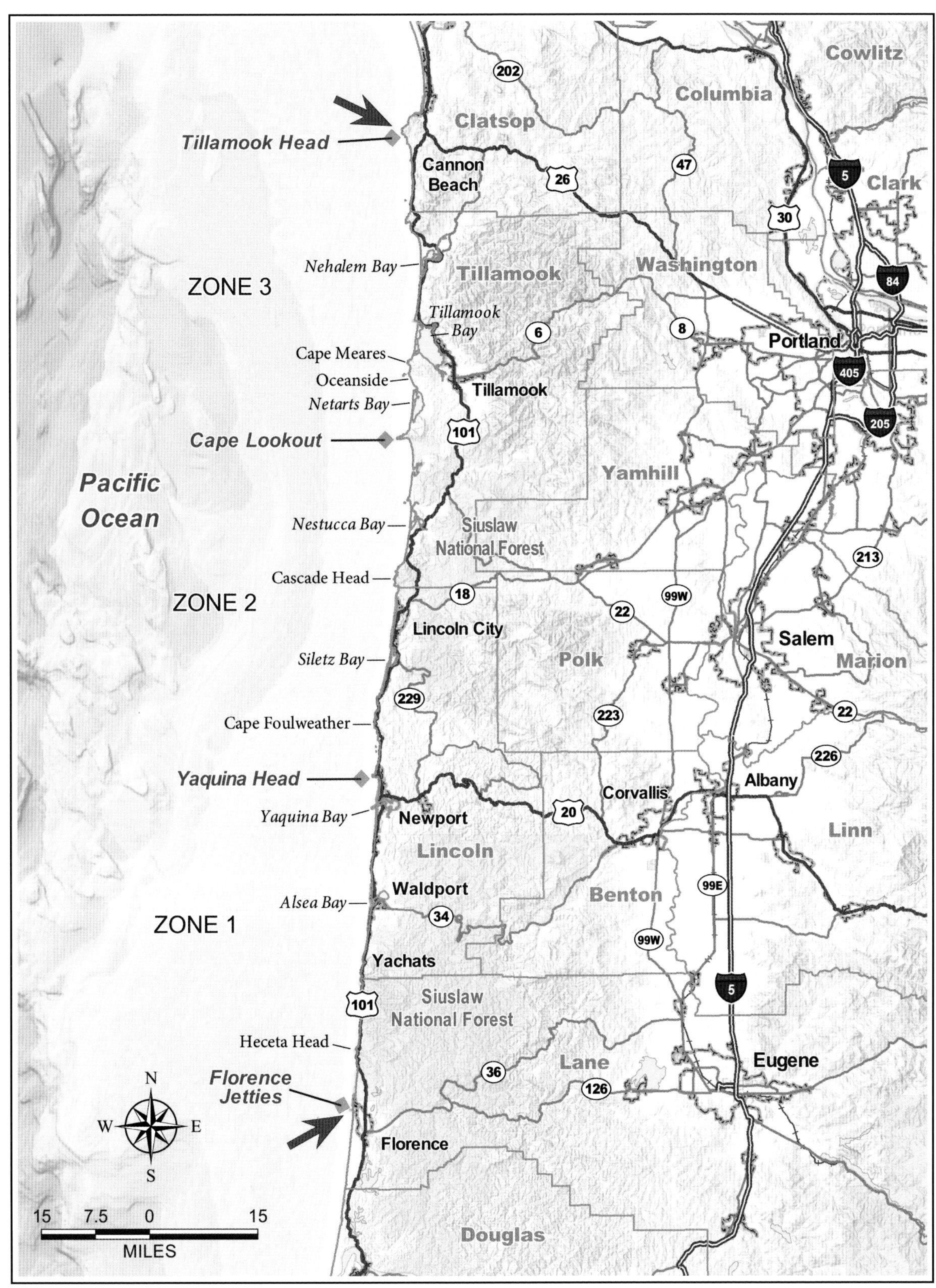

FIGURE 100. Map of Compartment 2.

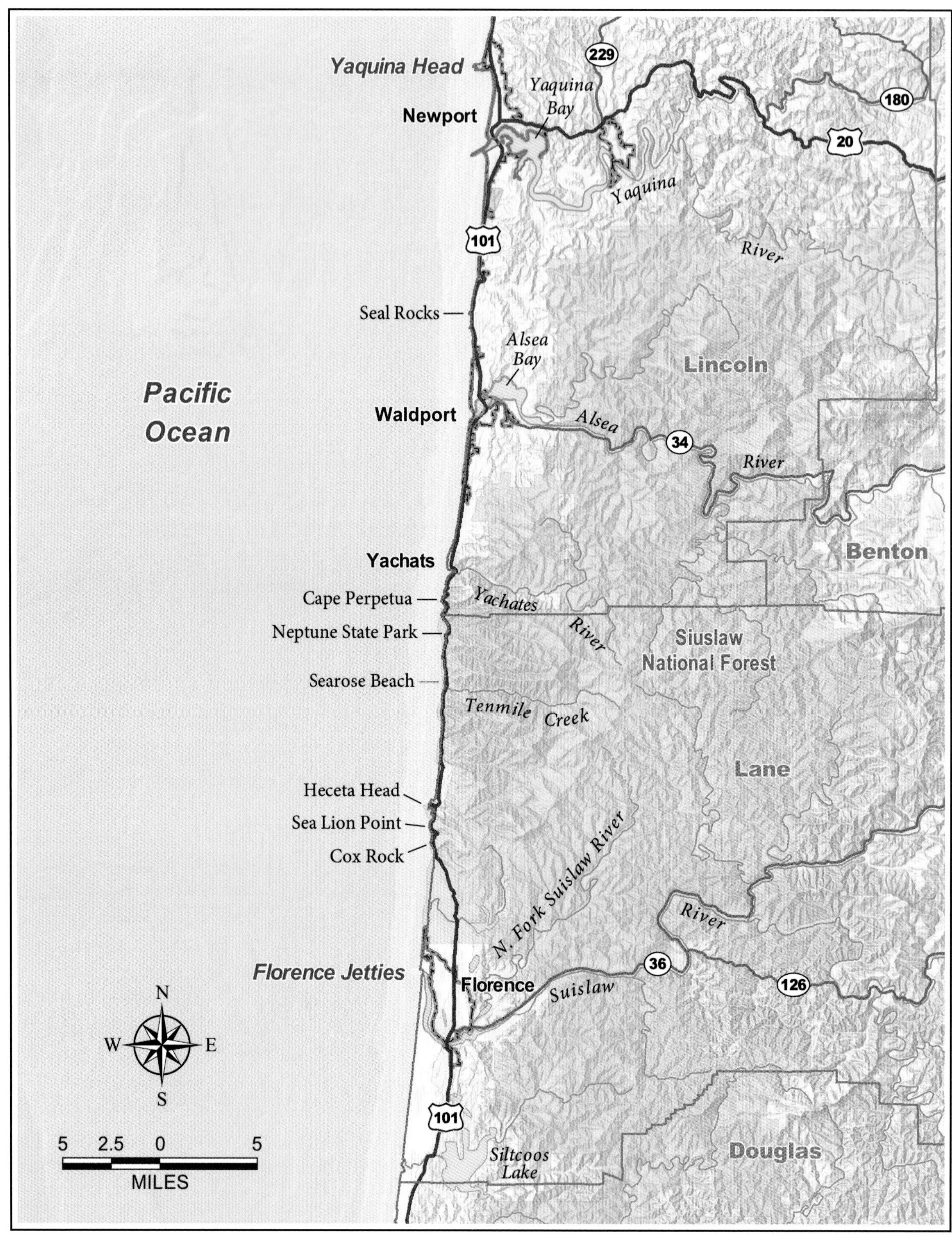

FIGURE 101. More detailed map of Zone 1 of Compartment 2.

FIGURE 102. Oblique view of the Cape Cove sand beach located one half mile south of Heceta Head, OR. Arrow A points to the sand beach, arrow B points to the high-tide berm composed of coarse-grained gravel, and arrow C points to Highway 101. Photo from June 2011, courtesy of Oregon ShoreZone, CC-BY-SA.

A raised marine terrace backs the shoreline between Yachats and Newport. From the north end of the bedrock shore in Yachats, a sand beach backed by low dunes in front of coastal bluffs stretches for 7.2 miles to the bridge over the Alsea River (shown in Figure 104) at Waldport. The estuary landward of the river mouth has a large flood-tidal delta complex, with abundant marsh systems. The channel at the mouth of the Alsea River switches from south to north, depending on the intensity of *El Niño* storms, resulting in erosion of the north spit. You can see the switching and subsequent erosion using the historical imagery slider (from 1994 to present) on Google Earth. Usually, the channel is pointed to the west; when strong SW waves build the offshore bar to the north, the channel is forced to migrate north, eroding the northern spit. The 1997-98 *El Niño* caused severe erosion. However, eventually the sand returns.

The shoreline between Waldport and Newport is mostly sand beach, except for a line of rock in the intertidal zone and sea stacks just offshore at Seal Rock (Figure 105). The intertidal rocks are tilted sedimentary rocks; the sea stacks are masses of hard basaltic rock. The basalt at Seal Rocks has a different origin than the basalt further to the south, which flowed onto the seafloor as indicated by pillow basalt features. It is hypothesized that the lava that formed the basalt at Seal Rock was part of the massive lava flows of the Columbia River Basalt Group that erupted on land and flowed 300 miles to the coast (Beeson, 1979). At Seal Rock, the basalt invaded layers of soft, wet sediments and found fissures where it was able to form a mass and cool 15 million years ago. The sediment/basalt layers were buried, consolidated, then raised up. The softer sedimentary rocks around the basalt have been eroded (you can see them as tilted sedimentary strata in the intertidal zone and cliffs). However, the hard basalt is eroding slower,

FIGURE 103. Heceta Head Lighthouse, OR. Public domain photo taken 18 April 2006, courtesy of Mary Beth Seibert.

creating the sea stacks. Elephant Rock is the largest one; in Figure 106A, you can see the rocky overhang at the contact between the basalt and the underlying sedimentary rocks, where the waves are eroding the softer sedimentary rocks faster than the overlying basalt. When the overhang reaches a certain size, slabs of the basalt slough off the rocks.

The Yaquina River enters the ocean at the busy port of Newport. Yaquina Bay is a small, drowned river valley that has not filled with much sediment since the last sea-level rise. Jetties built in 1896 confine the river mouth, with sand accumulation north and south of the jetties, backed by dunes. These jetties are somewhat unique in that they are oriented to the southwest rather than straight to the west. Komar et al. (1976) used historical maps to document the accumulation of sand north and south of the jetties since 1860, showing that sand built up more on the south side due to the oblique orientation. However, over time, an equilibrium shoreline was established that has not changed much. The striking protuberance of Yaquina Head, with its prominent lighthouse (pictured in Figure 107), extends 0.8 miles out into the Ocean.

Komar (1998a) has a whole chapter in his book on the Pacific Northwest Coast called The Jump-Off Joe Fiasco, which *was the most divisive land-use battle ever fought on the Oregon coast.* Just north of Nye Beach, the Jump-Off Joe sea arch was a popular tourist attraction. In spite of a 1942 landslide that destroyed a dozen homes, a new condominium was built on the bluff in 1982. However, cliff erosion caused the foundation to fail, even before the project was complete.

General Ecology

There are 45 seabird nesting colonies in this Zone, with 11 colonies at Yaquina Head where over 70,000 common murres nest. Western snowy plovers have reduced nesting on the beaches in this zone, compared to beaches to the south. Possible factors could include: more development along the sand spits where plovers prefer to nest, or beaches backed by steep cliffs with little ideal nesting habitat. The old-growth coastal forests from Florence to Waldport support nesting marbled murrelets. The Alsea and Yaquina estuaries are important wintering areas for waterfowl and shorebirds.

Migrating gray whales can be seen close to shore at the Florence jetties and rocky headlands. Stonewall

FIGURE 104. The mouth of the Alsea River, OR. Note the large flood-tidal delta and the wetlands up the channel. NAIP 2014 imagery, courtesy of Aerial Photography Field Office, USDA.

FIGURE 105. Crenulate bay near the town of Seal Rock, OR. Arrow points to Elephant Rock (see photographs in Figure 106). NAIP 2014 imagery, courtesy of Aerial Photography Field Office, USDA.

FIGURE 106. Elephant Rock, OR, which is composed of columnar basalt (Alt and Hyndman, 1978). (A) Oblique aerial view looks east. Vertical view is in Figure 105. Arrow points to basalt overhang. Photo from June 2011, courtesy of Oregon ShoreZone, CC-BY-SA. (B) Ground photo of Elephant Rock showing columnar basalt, taken September 2018.

and Heceta Bank are considered to be biologically important feeding areas for humpback whales. Steller and California sea lions haul out on rocky shores, and also isolated beaches.

Streams in this Zone support salmon and trout runs, though habitat degradation, dams, and other factors have reduced their populations, as all along the Pacific Coast. The Alsea River was once the most productive coho salmon stream in Oregon, and conservation efforts are trying to bring it back to its former status. It has a healthy fall Chinook salmon run, as well as coho salmon, and steelhead. The Yaquina River basin is the southern extent of chum salmon range in Oregon.

Places to Visit

Darlingtonia State Natural Site is a very popular stop along U.S. 101, with its 18-acre botanical park that is dedicated to preservation of a single species—the cobra lily. It is a carnivorous plant that lures insects into its hollow tube; when they fall into liquid on the bottom, they are digested into food for the plant.

Heceta Head Lighthouse State Scenic Viewpoint is a place you certainly would not want to miss. There are 7 miles of trails, the historic lighthouse and assistant keeper's house, close viewing of the seabird nesting colony, and one of the best places to see migrating gray whales close to shore.

There are four pull offs in the Neptune State Scenic Viewpoint that provide excellent views of the coast where you can see mirating whales, sea lions, brown pelicans in summer, and marbled murrelets year round. There are stairs to the beach where you can explore caves and tidepools. At Cape Perpetua, you can visit exotic features of the bedrock along the shore at the Devils Churn, Thor's Well, and spectacular blow holes. Definitely a spot worth the visit. Strawberry Hill, a site south of Neptune State Park, offers excellent tidepooling and rocky intertidal exploration.

There is what we interpret to be a 350-yard-wide raised marine terrace populated with a number of houses to the south of the Yachats River. Hard to see from the ground, though. There are multiple places to stop, view, explore, rest, and camp between Yachats and Waldport.

Seal Rock State Wayside is where you see the unique Elephant Rock (Figure 106, with its nesting colony of cormorants and gulls) and other very interesting rock features (columnar basalt, concretions, and linear ridges) and tidepools with particularly abundant rocky fauna. Several pairs of black oystercatchers nest there, and rocky-associated shorebirds are seasonally abundant. Maxine Centala (2013) published a free, downloadable book on the geology of Seal Rock written for the lay person.

There are state park beaches on both sides of the Yaquina jetties at Newport. South Beach State Park south of the jetties has wide, vegetated low dunes, reflecting its higher amount of sand accumulation since construction of the jetties and its little change over time. At Yaquina Bay State Park north of the jetties, the back beach dunes are unvegetated and often covered by low, active dunes,

FIGURE 107. Oblique aerial view of Yaquina Head, OR. Public domain photo taken 18 April 2006, courtesy of Mary Beth Seibert.

indicating that periodic erosion has occurred. No sand is transported to the north around the jetties, and the orientation of the jetties allows storm waves from the southwest to erode the north beach. The Yaquina Bay Lighthouse was built on a bluff on the north side of the channel in 1871, before the construction of the jettes and the build-out of the shoreline, so it is set back from the shore. It has been restored as a working lighthouse and is open to the public. Both of these state parks offer good digging for razor clams. Agate Beach State Recreation Site is another good beach for clamming.

Attractions in Newport include marine life exhibits at the Hatfield Marine Science Center, the Oregon Coast Aquarium, and the Newport Research Center for the NOAA Northwest Fisheries Science Center. Newport is also the homeport for the ships operated by the NOAA Marine Operations Center for the Pacific. The large white NOAA ships spend a lot of time at sea collecting data on marine mammals, coral reefs, fisheries, and seafloor bathymetry.

ZONE 2. YAQUINA HEAD TO CAPE LOOKOUT

Introduction

The dominant geomorphological features of this 46-mile-long Zone, illustrated in Figure 108, are:

1) Several rocky headlands and some good-sized rocky zones;

2) Many miles of developed areas with Route 101 and houses close to the beach;

3) Sand spits south of the mouth of Siletz Bay and north of the mouth of Nestucca Bay;

4) Eight miles of sand beach and a zone of wind-blown sand dunes south of Cape Lookout; and

4) A couple of smallish rivers - Salmon and Nestucca Rivers - and 11 inlets, none of which have jetties.

Geological Framework

As described by Alt and Hyndman (1978), the dark, basaltic rocks that form the rocky headlands at

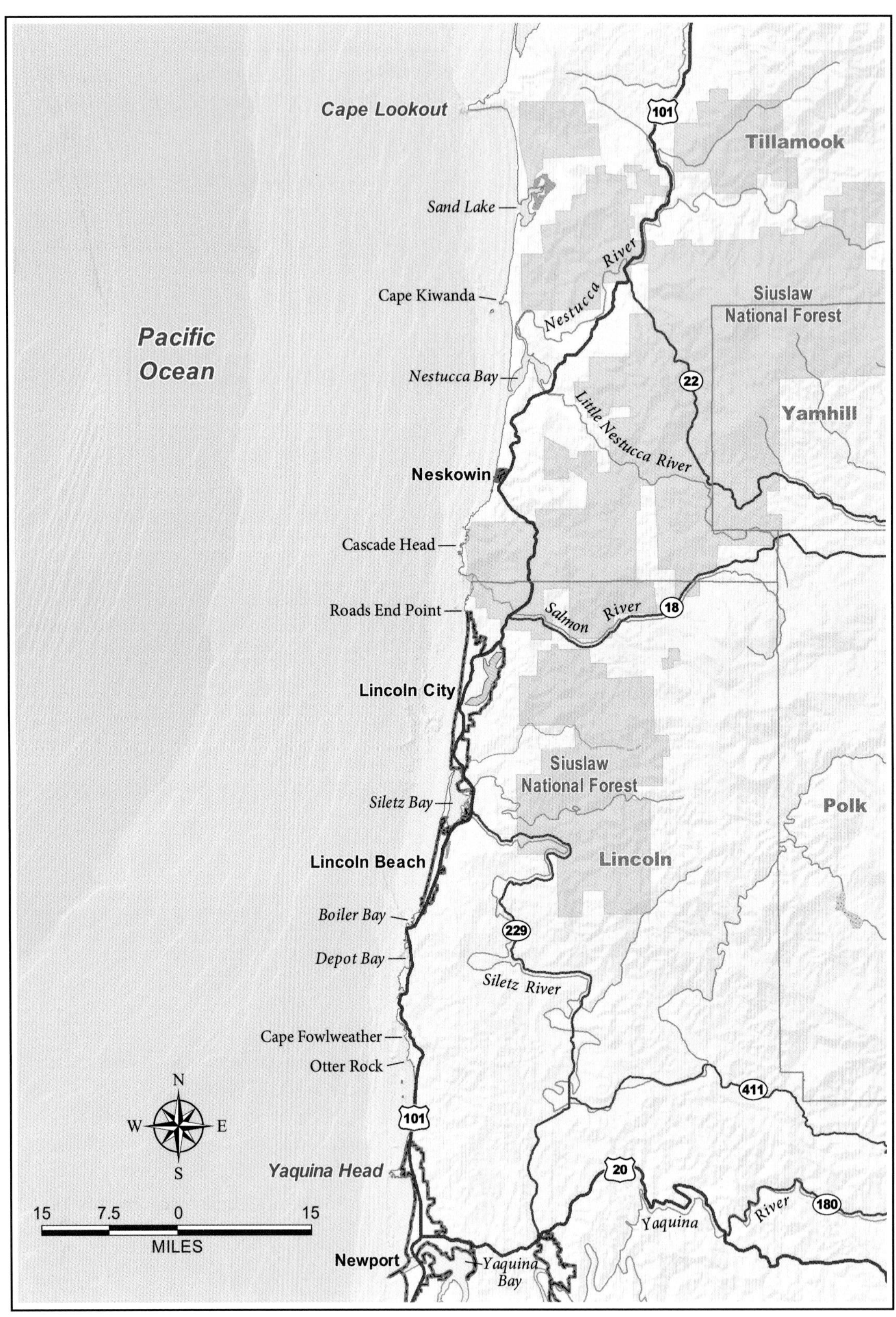

FIGURE 108. More detailed map of Zone 2 of Compartment 2.

Yaquina Head, Cape Foulweather, and Cape Lookout are composed of different type of basalt than previously discussed. These rocks formed during the eruption of a volcano near Cape Foulweather about 15 million years ago. Some of the basalt was ejected into the air creating beds of volcanic ash, some flowed on land down the side of the volcano, and some flowed into water. Thus, these basaltic rocks vary widely in look and strength. They said Cape Foulweather is obviously a wreck of a volcanic island that stood a short distance offshore during Miocene time and has since raised above sea level as the Coast Range rose.

Alt and Hyndman (1978) also noted that bedrock exposed in the sea cliffs behind the beaches between Lincoln Beach and Oretown is sandstone, mudstone and basalt formed on the seafloor during Eocene time, 50 million or so years ago. The softer sedimentary rocks crop out along the smooth stretches of coastline, whereas the harder basalts are exposed in the rocky coast between Neskowin and Cascade Head.

General Description

The bedrock headland called Yaquina Head (shown in Figure 107) is an outstanding natural area home to seabirds and harbor seals. It also has a beautiful lighthouse. A sand beach extends north to Otter Rock, with the main highway being located a short distance behind the beach. According to Ruggiero et al. (2013), this stretch of beach is one of the most sand-starved littoral cells in Oregon. With no transport of sand around the headlands to the north and south, the only current source of sand is from erosion of the bluffs behind the beach; however, these bluffs are in mudstone, which doesn't create much sand. Erosion rates between the 1960s to 2002 averaged 3.6 feet per year.

Otter Rock and Finger Rock to its north have that unique flat surface indicative of a raised marine terrace carved by waves when sea level was higher and/or the land was lower. The famous Devil's Punchbowl was formed when two sea caves met and the roof collapsed. Between Otter Rock and Depoe Bay, the shoreline consists of rocky basalt cliffs with pocket gravel beaches. We were curious how Cape Foulweather (shown in Figure 109) got its name, since it didn't seem to be any "fouler" than the other rocky points along the Oregon Coast. Turns out, it was the first promontory on the Pacifc Northwest Coast named by Captain James Cook during his third sail around the World. On 7 March 1778, he wrote in his journal (Cook, 1842):

The land appeared to be of moderate height, diversified with hill and Valley and almost everywhere covered with wood. There was nothing remarkable about it except one hill…At the northern extreme the land formed a point which I called Cape Foulweather from the very bad weather we soon after met with.

Siletz Bay, shown in the vertical image in Figure 110, is a classic example of a valley carved by the Siletz River during the last low sea level; when sea level rose to its current level about 5,000 years ago, erosion of the cliffs to the south provided the sand to build Salishan Spit across its mouth. Siletz Bay has nearly filled in with sand and mud. Note the rhythmic topography on the outer beach and the well-developed flood-tidal delta just inside the inlet.

As you well know now, the rocky shores of Roads End Point and between Cascade Head and Neskowin are composed of more resistant basaltic rocks. However, note in Figure 111 that Roads End Point has some erosional scallops. They are forming because the resistant basaltic rocks have been almost eroded away, exposing the softer sedimentary rocks (the basalt looks darker on the photograph, compared to the lighter sedimentary rocks). Eventually, the basalt "wall" will break up, creating sea stacks offshore and rapidly eroding cliffs behind the shore. This process is also occurring along the southern end of Cascade Head.

The sand beach in front of the town of Neskowin is an erosional hotspot that has led the beachfront homeowners to construct a riprap seawall along the entire length of the town. However, with rising sea level and increasing wave heights, the seawall has failed during recent major storms. Ruggiero et al. (2013) noted that erosion and flooding along the beaches of Neskowin are being perceived as a harbinger of the possible effects of future climate change for other communities along the Oregon coast.

A vertical image of Nestucca Bay is shown in Figure 112. Note the north to south orientation of the sand spit that deflects the mouth of the Nestucca River to the south and the variety of intertidal sand shoals in the lower portion of that small estuary.

A mostly natural sand beach extends for 4.2 miles to the next tidal inlet, the small estuary at Sand Lake that has a highly meandering outlet channel (see Figure 67B). Sand Lake and Salmon River are the only two estuaries on the Oregon Coast that have been designated as "natural," meaning that they lack maintained jetties or channels

FIGURE 109. Cape Foulweather, OR. Public domain photo taken on 13 October 2008, courtesy of Marcea Palmer.

and have little commercial or residential development. The main part of the sand dunes at Sand Lake is a single large parabolic dune (see diagram for this dune type in Figure 72) that is 3 miles long and nearly 1 mile wide. The western and eastern edges are forested, with active dunes in the middle and at the southern end.

The north end of Zone 2 of Compartment 2 is the spear-shaped Cape Lookout that projects straight offshore for 1.7 miles. Alt and Hyndman (1987) pointed out that the Cape is the last remnants of a Miocene volcanic island.

General Ecology

There are 13 seabird nesting colonies in this Zone, mostly occupied by gulls, cormorants, pigeon guillemots, and black oystercatchers. It is interesting to see the changing use of some of the offshore rocks. Gull Rock, offshore of Otter Rock headland, hosted thousands of common murres in the 1960s and 1970s, then tens of thousands in the late 1980s until 1997. Surveys in 2001-2003 found no common murres (Naughton et al., 2007). What caused them to abandon this site? Rob Suryan, a seabird ecologist at Oregon State University, speculated that predation of chicks by eagles and egg-robbing by gulls forced the murres to move to Yaquina Head, where that colony grew as the one at Gull Rock declined.

With all the great looking and undeveloped sand beaches along this Zone, no snowy plover nesting was reported in 2015 for the monitored beaches at Nehalem Spit, Bayocean Spit, Netarts Spit, Land Lake Spit, and Nestucca Spit (Audubon California, 2017). However, thousands of migrating and wintering shorebirds and ducks feed on the tidal flats and shallow waters in the estuaries in Siletz Bay, Salmon River, and Nestucca Bay. Marbled murrelets nest in old-growth forests and may be seen in nearshore waters in summer.

Good places to see migrating gray, humpback, and killer whales include Boiler Bay State Scenic Viewpoint, Rocky Creek State Scenic Viewpoint, Cape Foulweather, and of course the Whale Watching Center in Depoe Bay. Whale Cove, just south of Depoe Bay, is the oldest

FIGURE 110. Entrance to Siletz Bay, OR. Arrow points to flood-tidal delta. Note that the Salishan Spit is building in a northerly direction. NAIP 2011 imagery, courtesy of Aerial Photography Field Office, USDA.

FIGURE 111. Area in the vicinity of the mouth of the Salmon River, OR. Arrow points to Roads End Point. NAIP 2011 imagery, courtesy of Aerial Photography Field Office, USDA.

marine reserve (meaning that all marine life is protected, so you have to enjoy it from nearby viewpoints or by boat) in Oregon, and in 2014, it became part of the Oregon Islands National Wildlife Refuge. Harbor seals rest and pup in the Cove, and can be seen throughout the area.

Coho and chum salmon, steelhead, and coastal cutthroat trout occur in many streams, and Chinook salmon have fall runs in the larger streams. The Salmon River fall Chinook salmon stocked by the hatchery at

FIGURE 112. Nestucca Bay, OR. Note the rhythmic topography along the beach north of the entrance to the Bay. NAIP 2014 imagery, courtesy of Aerial Photography Field Office, USDA.

Otis are monitored as an indicator stock in the health of ocean fisheries under the Pacific Salmon Treaty. What this means is that coded-wire tags are implanted in about 200,000 young fish (smolts), then efforts are made to capture all returning fish (they spawn at 4-5 years old) to determine the percent survival in the Ocean. For the brood years 1983-1994, 1.6% of the Chinook smolts returned as adults (Oregon Department of Fish and Wildlife, 2016). This survival rate is pretty good for hatchery-reared fish.

The Siletz River basin is unique in that it contains viable runs of seven species of anadromous fish (spring and fall Chinook salmon, coho salmon, chum salmon, summer and winter steelhead, and sea-run cutthroat trout). Also, it is the only Coast Range basin in Oregon with a native run of summer steelhead (Wilson, 2008).

Places to Visit

The Yaquina Head Outstanding Natural Area (Figure 107) is a must-see location, with its lighthouse, interpretive center, abundant seabirds, harbor seals, whales, tidal pools. and wild flowers in season. It is ranked as a Globally Important Bird Area. There are viewing decks where you can observe the seabird nesting colonies very closely; the murre colony on Colony Rock is one of the largest on the Oregon Coast. Beverly Beach State Park is a great place to view the headlands to the north and south.

You can explore both geology and tide-pool fauna at Devil's Punchbowl State Natural Area at Otter Rock, with a large rock formation shaped like a punchbowl as well as wide, wave-cut rock platforms and steep cliffs in the rock.

You should definitely take Otter Crest Loop between Otter Crest State Park and the Rocky Creek State Scenic Viewpoint, which provides expansive views at 500 feet above the Ocean, sea cliffs, seabirds, and whales. There is a parking lot just south of Whale Cove where you can look down on the birds flying by. Depoe Bay is home to the Whale Watching Center, a good place to stop by for the latest information on where to best see whales during your travels along the coast. Just to the north, Boiler Bay State Scenic Viewpoint is considered to be one of the best sites in Oregon to see ocean-going birds (such as shearwaters, jaegers, albatrosses, grebes, pelicans, loons, oystercatchers, and murrelets). It is called Boiler Bay because there is a ship's boiler from a 1910 explosion exposed at low-tide. If you are up for a scamper down the rocky ledge, Boiler Bay offers an excellent rocky intertidal exploration opportunity, with clear zonation of intertidal species along the rocky benches.

The Siletz Bay National Wildlife Refuge covers much of the eastern part of the Bay, being one of the few "places of refuge" for ducks that feed in offshore waters during storms along this coast, in addition to wintering and migrating shorebirds and waterfowl, diving ducks crowd in as storms move through. Same for Nestucca Bay National Wildlife Refuge. However, in addition to extensive salt marshes and tidal flats, habitats in this refuge include pastures and grassland (which you can see in Figure 112), thus thousands of geese (Canada, white-fronted, dusky Canada, and Aleutian cackling) winter there. The Nestucca Bay area supports 10 percent of the World population of dusky Canada geese, and 100 percent of a very unique subpopulation of Aleutian Canada geese.

Cascade Head is a 270-acre preserve managed by The Nature Conservancy. The moderately difficult trails provide stunning high views of the Salmon River estuary to the south and the open ocean to the west. It hosts amazing meadows of native grasses and flowers, including 99 percent of the World's population of the Case Head catchfly. The Oregon silverspot butterfly, a federally listed species, is known to occur in only four other locations in the World.

Cape Kiwanda is yet another must-see site. It is unusual in that this headland is not made of basalt like many of the others along the Oregon Coast. Instead, it is composed of sandstone, which normally would be eroded away by wave attack. So, why is it still there? According to Lund (1974), this point of sandstone owes its survival in small part to the basalt dike on its south side but more importantly to Haystack Rock, a basalt sea stack four-tenths of a mile to the southwest. At one time, the promontory extended to Haystack Rock, which defended the sandstone from severe winter wave attacks from the southwest. Erosion on the flanks of the promontory finally separated the basalt from the sandstone, isolating it as a sea stack. Haystack Rock still gives some protection to the Cape by receiving part of the assault of the storms from the southwest, but the Cape is being visibly eroded, principally by undercutting along the sea cliffs and by rock fall. Waves have sculptured the layers of sandstone into cliffs and caves.

Cape Lookout, underlain by layers of resistant basalt from a 15-million-year-old volcano, is a narrow peninsula that juts out 1.7 miles into the Ocean (Figure 113), making the tip a great place for whale watching

FIGURE 113. Oblique aerial view of Cape Lookout, OR taken on 11 October 2008, courtesy of Cape Lookout State Park.

and views up and down the coast. It is quite popular, so be prepared for crowds on holidays and great weather days.

ZONE 3. CAPE LOOKOUT TO TILLAMOOK HEAD

Introduction

The dominant geomorphological features of this 63-mile-long Zone, illustrated in Figure 100, are:

1) Counting Cape Lookout, there are four major rocky headland areas;

2) Three fairly good-sized estuaries - Netarts, Tillamook, and Nehalem Bay, each of which is partially closed off by sand spits;

3) Many miles of sand beach; and

4) Tweleve tidal inlets, two of which have jetties - the entrance to Tillamook Bay and the entrance to Nehalem Bay.

Geological Framework

The general sketch map of the geology of this Zone in Figure 114 shows there are two large areas of basalt present, a band of 20-million-year old Miocene basalt near the coast, and a belt of Eocene seafloor basalt deposited 50-60 million years ago further to the east. Starting at the north end of the Zone, Alt and Hyndman (1978) noted that Tillimook Head exists because there is a large intrusion of basalt sandwiched between layers of mudstone which just happens to be exposed right at sea level where it can withstand the pounding surf. Had the layer of basalt been either higher or lower, Tillamook Head would not be there. Figure 115 shows a geological section across U.S. 101 at Tillamook Head, showing the basalt that forms this headland.

Alt and Hyndman (1978) pointed out further that between Tillamook Head and Arch Cape the bedrock is almost entirely soft mudstones in which the waves have shaped a long, smooth stretch of coast. Smaller

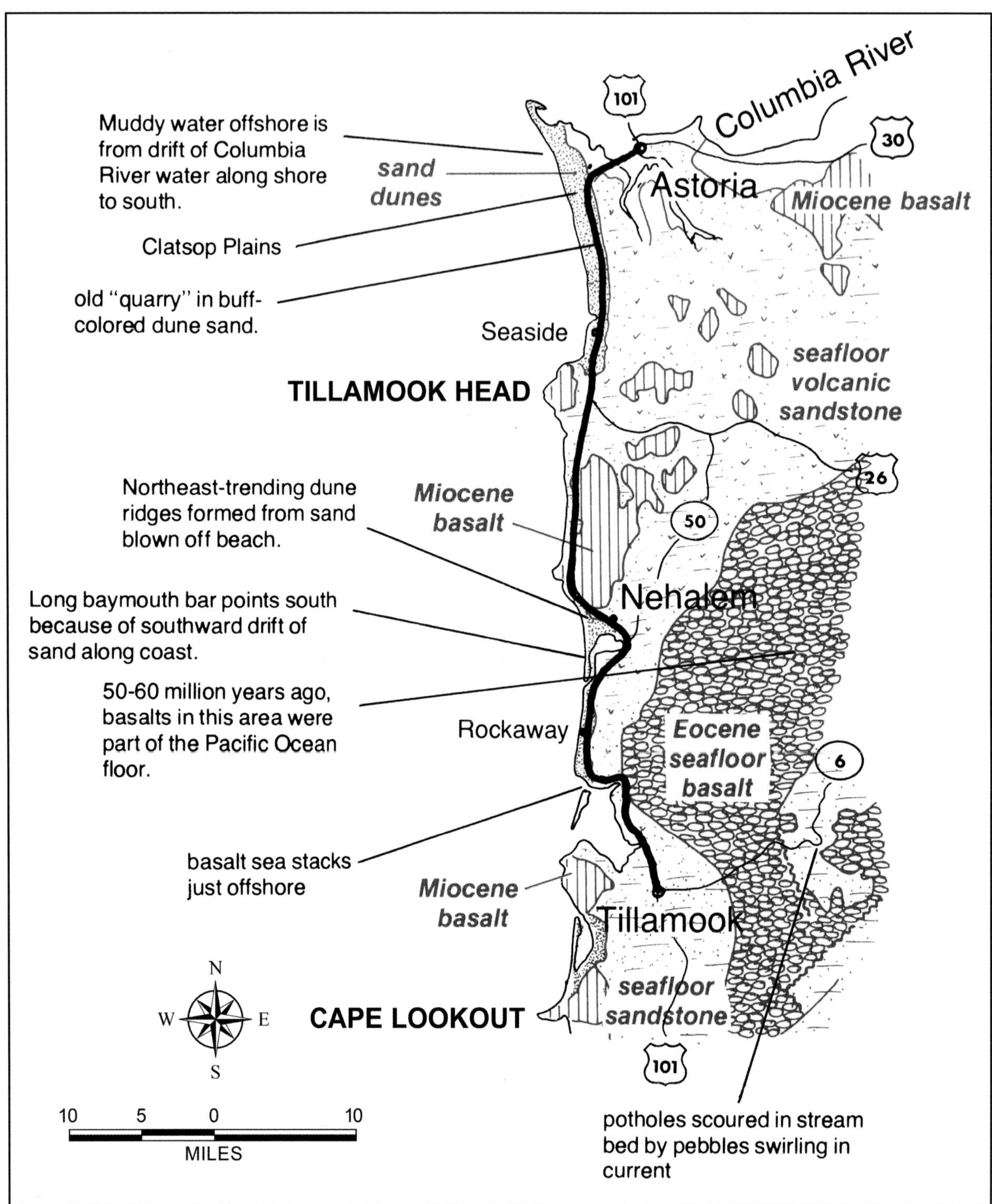

FIGURE 114. Sketch map showing the general geology of Zone 3 (Compartment 2) – Cape Lookout to Tillamook Head (and beyond). Modified from Alt and Hyndman (1978).

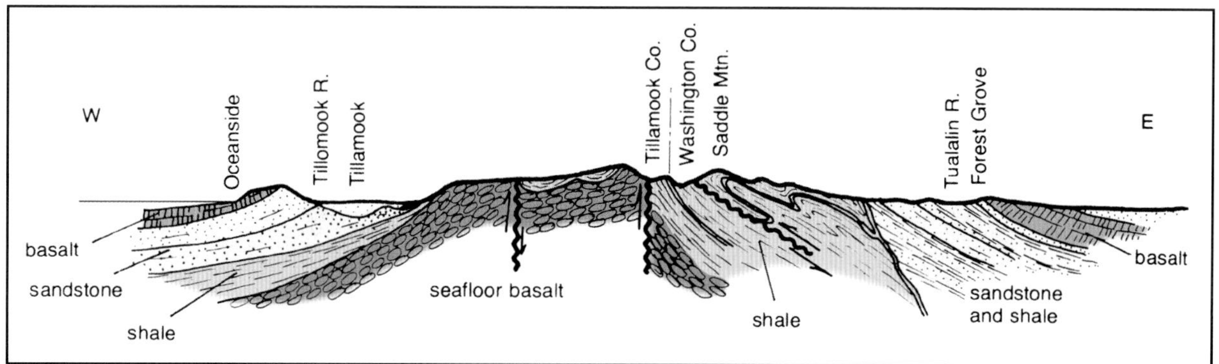

FIGURE 115. Geological cross-section across U.S. 101 at Tillamook Head. Modified from Alt and Hyndman (1978).

headlands, such as Hug Point, Arch Cape, and Cape Falcon, are sites where large amounts of basalt were intruded into the mudstone, much like Elephant Rock to the south.

General Description

Netarts Bay is a cresent-shaped embayment between two basalt headlands (Cape Lookout and Cape Meares) that has been almost closed off by a 5.7-mile-long sand spit (or bay barrier) that projects due north (Figure 116). Note that only small streams empty into the bay, so there was no river valley carved during the last low-sea-level period (when sea level was about 400 feet lower than present, about 12,000 years ago). Once sea level rose to its current level, the marine terrace on which the town of Netarts is built was eroded by waves before the spit built out and protected that shore. However, the structures built to the north of the spit are not protected, and this area has undergone periods of severe erosion during *El Niño* storms. The southern end of the spit, which is part of the Cape Lookout State Park, has also undergone long-term erosion, with a major breech in the dune during major winter storms in 1998-1999 that damaged park facilities. According to Allan and Komar (2002), instead of building a seawall, a cobble berm and dunes were placed behind the beach (a design with Nature approach that is well described in Allan and Komar, 2004). Obviously, very little sand is being added from the south, with only resistant basalt cliffs as the source. As sea level continues to rise and waves increase, breaches of the southern end of the spit are likely to increase over time. This area is another one where you can see the changes in the north and south ends of the spit from 1994 to present using the Google Earth historical imagery tool.

Basaltic rocks occur along the coast from Oceanside to Cape Meares, forming steep, high cliffs, wave-scalloped bays, and the sea stacks of Three Arch Rocks (Figure 117), Pillar Rock, and Pyramid Rock. To the north, the 5-mile-long Bayocean Peninsula projects to the north away from Cape Meares, nearly closing off Tillamook Bay. It is now undeveloped, but that wasn't always the case. Ruggiero et al. (2013) reported that construction of the jetties at the mouth of Tillamook Bay blocked off the seasonal north-south reversal in sand transport across the mouth, leading to severe erosion of the south side spit. There had been a large recreational resort on the spit, called Bayocean, built in the early 1900s (it was touted as the "Atlantic City of the West"). Erosion destroyed many structures over time, and in 1952, the spit was breached, and residents were evacuated in 1953. By 1960, there were no more structures on the spit, with the last house being claimed by the Ocean (The Oregon Encyclopedia, 2017a).

Tillamook Bay is typical of most estuaries in Oregon that have not filled with sediment since sea level stabilized about 5,000 years ago. Figure 118 shows how geologists think that the bay and spit developed over time: A) During the last sea level low stand, the rivers filled the valley with sediment all the way to the mouth of the bay and built wide coastal plains along the coast; B) At the start of the last sea level high stand, the valley was flooded and the coastal plain eroded; C) As sea level stabilized at its current level, spits start to build out across the bays; and D) Today, Bayocean Spit has migrated as far as it can because water flow in/out of Tillamook Bay carries the sand into the Bay or offshore, and the shoreline north of the north jetty has built out. Currently, Tillamook Bay has a beautiful bay-head delta of salt marsh and lots of

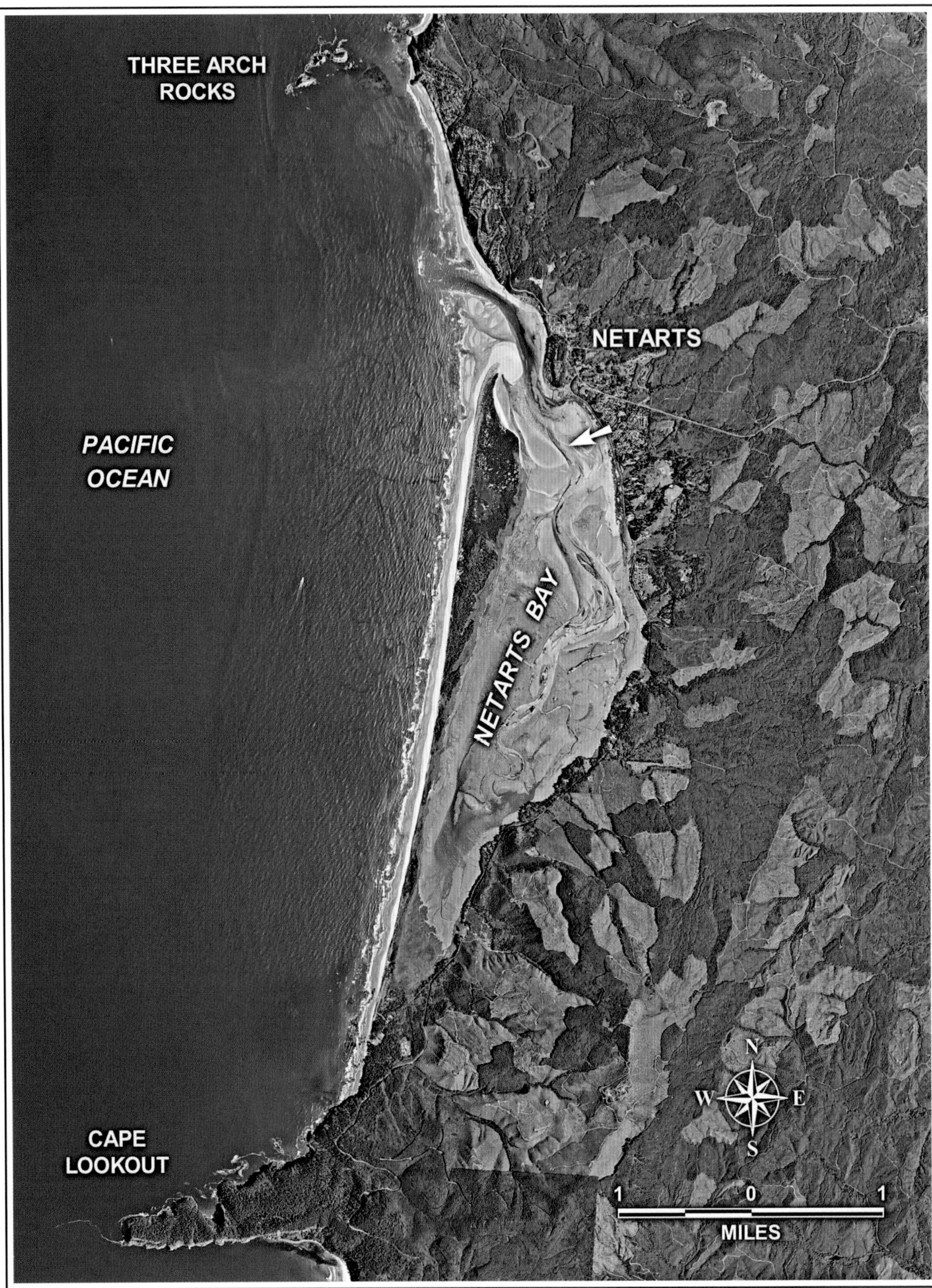

FIGURE 116. Vertical image of the entrance to Netarts Bay, OR. This bay is different than most others in Oregon and Washington because it formed by the spit building across a shallow embayment rather than across a valley carved by a river during the last sea-level lowstand. Note the well-developed flood-tidal delta (arrow). NAIP 2009 imagery, courtesy of Aerial Photography Field Office, USDA.

FIGURE 117. Oblique aerial view of Three Arch Rocks, OR. Public domain photo taken 18 April 2006, courtesy of Mary Beth Seibert.

open water. Nehalem Bay is similar in form and process, but the sand spit has migrated south across the bay, and the bay is much smaller.

Twin Rocks and Haystack Rock, shown in Figure 119, and other isolated sea stacks along this Zone are remnants of basalt, as discussed earlier. The sea arch at Twin Rocks is particularly impressive. In contrast, Cape Falcon and Tillamook Head are composed of large intrusions of basalt that form tall headlands with steep slopes; they border the shoreline of sand beaches backed by soft mudstones. These beaches are distinctive because they have quite a bit of gravel produced from landslides at the headlands. These gravels form impressive berms on the beaches just north of Tillamook Head (see example gravel beach in Figure 60) and gradually decrease in size and amount to the north. Gravel beaches are less likely to erode during storms; thus the bluffs behind the beach are vegetated and erosion rates are close to zero (Ruggiero et al., 2013).

General Ecology

There are over 60 seabird nesting colonies in this Zone. The most recent estimates (2003) are that there are over 368,000 nesting birds, including nearly 290,000 common murres, 25,000 Leach's storm petrels, and 3,900 tufted puffins (Naughton et al., 2007). The colonies on Three Arch Rocks are impressive. There are over 120,000 common murres nesting on Shag Rock, and 2,900 tufted puffins on Finley and Middle Rock. Tufted puffins are usually present on other colonies as a single pair, so it is amazing that these rocks have so many nesting together. The colonies also have large numbers of cormorants and gulls. We are glad that the basalt rocks in this area create such amazing sea stacks and islands where these birds can nest with reduced threats of predation.

All of the estuaries support thousands of migrating shorebirds and waterfowl. Snowy plovers, state and federally listed as threatened, have not been observed as nesting on beaches in this Zone from 2005 to 2015; however, they attempted to nest in 2015 and 2016, at least one chick was hatched on the Nehalem Spit in 2017, so there is hope that they will establish a nesting population there.

Hundreds of harbor seals haul out on the sand flat at the entrance to Netarts Bay and in the middle of Tillamook Bay, as well as on isolated beaches and rocky shores. The northernmost breeding site of Steller sea lions in the lower 48 states is on small Seal Rock, part of Three Arch Rocks, where a dozen pups are raised in mid-summer. Cape Meares, Neahkahnie Mountain, and Ecola State Park are among the best places to observe migrating whales.

Tillamook Bay is the second largest bay in Oregon; however, it has lost 85 percent of its historic tidal wetlands to development, which has degraded water quality, increased nutrients, bacteria, and sedimentation, and affected fish and shellfish. There are many efforts underway to restore habitats and improve water quality in this "Estuary of National Significance," one of only 28 in the country with this designation. Water quality is of particular importance to oyster farmers, who lease over 2,600 acres for shellfish cultivation in Tillamook Bay (the most in the state) and 400 acres in Netarts Bay, according to the Oregon Department of Agriculture, 2016 Shellfish Plat Production Annual Report.

Places to Visit

This Zone is loaded with great places to visit. The prominent Cape Meares area includes the Cape Meares State Scenic Viewpoint and the Cape Meares National Wildlife Refuge. Here you can see the largest Sitka spruce tree in Oregon, a 1890s lighthouse, and panoramic views north and south. Cape Meares is an Important Bird Area for marbled murrelets, with an estimated breeding population of 320 birds. Then there are all those nesting seabirds, migrating whales, wintering and migrating shorebirds and seaducks, seals and sea lions hauling out; something to see for every season.

Early and middle Pleistocene low sea level with river filling valleys with sediment and building coastal plains

Late Pleistocene high sea level with valleys flooded

Early stage in the building of Bayocean Spit and river delta in the bay

Present stage with river delta partially filling the bay and spit fully built

FIGURE 118. Geologist's conception of Bayocean Spit development. Modified from Terich and Komar (1973).

Nehalem Bay State Park, a place you definitely should visit, is located on the southward trending sand spit that has built across Nehalem Bay. Because of the recent attempts and success for nesting by the snowy plover, the southern half of the Park has been designated as "occupied" meaning that the area has restrictions from 15 March to 15 September, when dogs, vehicles, bicyles, camping, and kites are prohibited, and people in general are allowed only on the wet sand part of the beach.

Oswald West State Park, on the west flank of Neahkahnie Mountain, offers many trails to access secluded sand beaches, the best preserved coastal rain forest in Oregon, and some of the best views of the Ocean in northern Oregon. If you want to walk on a really secluded beach, head to Bayocean Peninsula County Park, on the seaward side of Tillamook Bay. As you walk along the beach, think of all those houses and people that once occupied this now natural setting.

Stop at Cove Beach to see the large gravel berms; they are pretty impressive. North of the village of Arch Cape you will encounter Hug Point State Recreation Site where you can walk out at low tide and explore the tide pools and the rocky point. Just be sure to check the tide tables and keep an eye on the water level. You could become stranded as the tide rises.

The Tolovana State Recreation Site is one of the best places to see tufted puffins (from April to August) because over 600 birds nest on Haystack Rock and it is so close to shore. It is a very popular site, as seen in the photograph in Figure 119B.

The last place we will recommend that you visit in this Zone is Ecola State Park. It offers amazing views, hiking trails, more than 50,000 nesting common murres, seals and sea lions offshore, whale migrations, and it is very photogenic. The lighthouse on an island 1.2 miles off the point was operated from 1881 to 1957, then,

FIGURE 119. (A) Twin Rocks, OR. Public domain photo taken 28 February 2015 by Linda Letters. (B) Haystack Rock at Cannon Beach, OR. Photo taken on 6 August 2012 by Walter J. Sexton.

strangely enough, it became a columbarium—a storage house for cremated remains in urns called "Eternity at Sea" until its license was revoked in 1999. The urns are still there in the building and the property remains in private hands (The Oregon Encyclopedia, 2017b).

16 COMPARTMENT 3 - CENTRAL ESTUARIES (Tillamook Head to Point Grenville)

INTRODUCTION

The coastal habitats in this Compartment are listed in Table 6. It has a variety of habitats along the shore, with six of them occurring along greater than 10% of its length. Of those six habitats, salt- and brackish-water marshes is the longest, occurring along 43.5% of the length of the shore (519 miles). Next comes sheltered tidal flats at 41.4%, swamps at 18.2%, exposed tidal flats at 11.7%, fine- to medium-grained sand beaches

TABLE 6. Coastal habitats in Compartment 3 (Tillamook Head to Point Grenville) (NOAA/RPI. 2015).

COMPARTMENT 3 - CENTRAL ESTUARIES (Tillamook Head to Point Grenville)			
ESI VALUE	ESI DEFINITION	LENGTH (MILES)	% ABSOLUTE LENGTH
1A	Exposed Rocky Shore	9.26	0.78%
1B	Exposed, Solid Man-Made Structures	7.90	0.66%
2A	Exposed Wave-Cut Platforms in Bedrock	6.19	0.52%
2B	Exposed Scarps and Steep Slopes in Clay	0.00	0.00%
3A	Fine- to Medium Grained Sand Beaches	133.94	11.22%
3B	Scarps and Steep Slopes in Sand	1.33	0.11%
4	Coarse-Grained Sand Beaches or Bars	0.36	0.03%
5	Mixed Sand and Gravel Beaches or Bars	14.87	1.25%
6A	Gravel Beaches or Bars	10.03	0.84%
6B	Riprap	49.95	4.19%
6D	Boulder Rubble	2.52	0.21%
7	Exposed Tidal Flats	140.06	11.74%
8A	Sheltered Rocky Shores and Sheltered Scarps	7.89	0.66%
8B	Sheltered, Solid Man-Made Structures	6.02	0.50%
8C	Sheltered Riprap	32.09	2.69%
8F	Vegetated, Steeply-Sloping Bluffs	0.00	0.00%
9A	Sheltered Tidal Flats	494.18	41.41%
9B	Vegetated Low Banks	12.44	1.04%
10A	Salt- and Brackish-Water Marshes	518.83	43.47%
10B	Freshwater Marshes	130.21	10.91%
10C	Swamps	217.12	18.19%
10D	Scrub and Shrub Wetlands	78.17	6.55%
	Total ESI Length (Miles)	1,873.35	156.97%
	Absolute Shoreline Length (Miles)	1,193.44	100.00%

at 11.2%, and freshwater marshes at 10.9%. The reason that these numbers are more than 100% is because, in several areas, the shore is composed of more than one habitat type. The abundance of sheltered habitats reflects the fact that the shore is dominated by three major bay/estuarine systems - Columbia River Mouth, Willapa Bay, and the Grays Harbor area - with their associated sheltered habitats.

For purposes of discussion, we have further subdivided this Compartment into two Zones (see the map in Figure 120 and vertical image in Figure 121):

1) Tillamook Head to Columbia River Mouth

2) Columbia River Mouth to Point Grenville

ZONE 1. TILLAMOOK HEAD TO COLUMBIA RIVER MOUTH

Introduction

The principal geomorphological features of this 18-mile-long Zone, illustrated in Figures 120 and 121, are:

1) A large, north-northwestward projecting sand spit, called Clatsop Sand Spit, that contains multiple parallel beach ridges;

2) Long and straight sand beach along the outer face of the main spit; and

3) Only two river outlets, the small Necanicum River in the south and the massive Columbia River at the north edge.

Geologic Framework

The entire Zone is composed of the Clatsop Sand Spit of recent (Holocene) age built on the southern side of the Columbia River estuary. As shown in Figure 26, this part of the Washington Coast is rising due to tectonic processes at a rate of +0.8 feet/century.

General Description

The Clatsop Sand Spit has built seaward ever since sea level rose to its present level, about 5-6,000 years ago, as evidenced by the multiple lines of beach ridges that are clearly seen in Figure 122. The sand dunes range in height from 25-50 feet. Construction of the south jetty started in 1885, which caused the north end of the spit to extend to the north for almost 3 miles, narrowing the river mouth (Ruggiero et al., 2013). They also point out that the northern tip of the spit is only several hundred yards wide and that there are concerns that it could be breached, which could *result in the formation of a second river mouth.*

As can be seen in Figures 121 and 122, the south end of the Clatsop Sand Spit abuts against the north end of Tillimook Head, where the beach contains a very well-developed cobble/boulder berm at the high-tide line. The rest of the spit consists of a beautiful sand beach that is mostly in an unaltered state, with tall, vegetated dunes.

The jetties at the mouth of the Columbia River (Figure 68) mark the entrance to a large salt-wedge type estuary (also called highly stratified; see illustration in Figure 77 - upper diagram). That is presumably due the combination of a large freshwater input by the Columbia River plus a relatively modest tidal range. By far the largest ebb-tidal delta on the Oregon/Washington Coasts occurs off the end of those jetties.

General Ecology

With mostly ocean-facing sand beach habitat, the wildlife in this Zone is dominated by shorebirds that stop on the ocean beaches in spring and fall by the thousands, and waterfowl that can be found in nearshore waters and inland lakes. Snowy plovers have not nested on the spit since 1984, thus there are efforts to improve nesting habitat at the northern end of the spit, in the hope that, "if we build it, they will come." Common murres and sooty shearwaters are highly abundant offshore.

The Columbia River Estuary is highly productive and supports an abundance of fish, marine mammals, and birds. The estuary is where the juveniles of the many listed species of salmon spend 1-2 years feeding and growing, in preparation for heading to the Ocean. Their survival during this period has been reduced due to many factors, including physical changes that have reduced habitats in the estuary from diking and filling of wetlands and changes in flow patterns due the operation of dams upstream. Predation, as discussed below, is also a major factor.

The tip of the Columbia River south jetty is a major haulout for hundreds to thousands of California and Steller sea lions. They arrive in peak numbers in fall and winter, after the breeding season on rookeries to the north and south, to feast on fish moving into the Columbia River. In fact, according to the Washington Department of Fish and Wildlife (2017), the feasting of California sea lions on endangered and threatened stocks

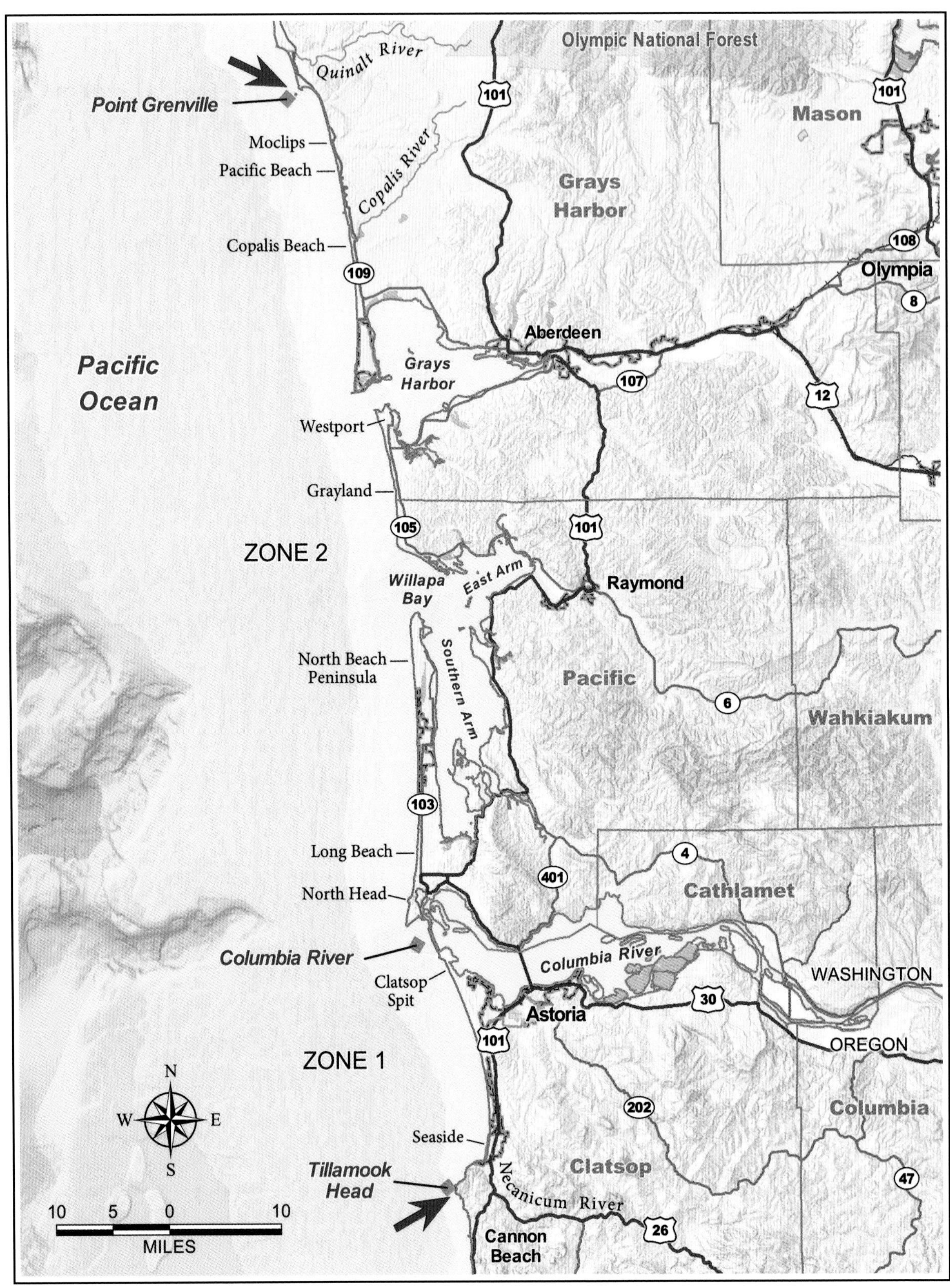

FIGURE 120. Map of Compartment 3.

FIGURE 121. Image of Compartment 3. Landsat imagery acquired in 2014-15, courtesy of the U.S. Geological Survey.

FIGURE 122. Clatsop Sand Spit, OR. Arrow points to the mouth of Necanicum River. NAIP 2011 imagery, courtesy of Aerial Photography Field Office, USDA.

of salmon and steelhead has been so severe (especially below the Bonneville Dam where the fish congregate before moving up the fish ladder), that federal agencies have authorized their removal (most are euthanized though some are placed in zoos and aquariums). Steller sea lions prefer sturgeon so they are not targeted for removal.

Several sites within the Columbia River estuary are recognized as Important Bird Areas, and most of the marsh islands and tidal flats have been designated as a regional Western Hemisphere Shorebird Reserve Network site. East Sand Island, a dredge spoil island located about 5 miles upstream from the mouth of the River, is the largest Caspian tern nesting site in the World (8,300 pairs in 2010). This colony has an interesting history; terns originally occupied the site in 1984, but by 1986, vegetation had overrun available nesting habitat and the birds had moved upstream to Rice Island. Concerns over increasing predation pressure on juvenile salmon prompted management to move to restore nesting habitat on East Sand Island in an attempt to move the colony further downstream (where their diet would rely less on juvenile salmon), and by 2001, all of the Caspian terns had shifted back to nesting on East Sand Island. Current management efforts are in place to reduce the colony size even further, and in 2016, only 5,915 pairs of terns were estimated to nest on East Sand Island (Bird Research Northwest, 2016). East Sand Island is also the largest breeding colony of double-crested cormorants in North America (~30,000) and the largest known roosting site for brown pelicans, with up to 17,000 birds. As the Columbia River enters the ocean, it forms the Columbia River plume, which is rich in nutrients and supports high densities of seabirds and marine mammals. It is not uncommon to observe 1,000s to 10,000s of sooty shearwaters in the plume between April and September.

The 18 miles of beach along the spit represent 95% of the razor clam harvest in Oregon because of such dense populations.

Places to Visit

Necanicum Estuary Natural History Park at Seaside is a good place to watch migrating shorebirds feeding on the tidal flats. Nearly all of the other places to visit are sand beaches.

Sunset Beach State Recreation Site provides great views north and south. But it also marks the western trailhead for the historic Fort-to-Sea Trail, connecting two units of the Lewis and Clark National Historical Park. The trail goes along part of the route that the "Corps of Discovery" used to explore after setting up Fort Clatsop on the Lewis and Clark River south of Astoria. Remember what a rough winter they had there in 1805-06, when it rained constantly.

At the north end of the spit, you will reach the Fort

Stevens State Park, which was the Oregon part of the harbor defense system for the Columbia River mouth. It was active from the Civil War to World War II, after which it became part of the state park system. There are many trails and an historic 1906 shipwreck, the *Peter Iredale*, that you can still see on the beach. And in fall and winter, you will be impressed by the numbers of sea lions hauled out on the south jetty, and year-round by the many thousands of harbor seals hauled out on sand flats at the entrance to the River.

ZONE 2. COLUMBIA RIVER MOUTH TO POINT GRENVILLE

Introduction

The principal geomorphological features of this 74-mile-long Zone, illustrated in Figures 120 and 121, are:

1) Two large estuarine complexes – Willapa Bay and Grays Harbor;

2) Long sand spits with multiple beach ridges on the landward side that partially close off both bays;

3) A conspicuous, large crenulate bay to the south of Point Grenville; and

4) Two primary river systems, the small Willapa River and the larger Chehalis River.

Geological Framework

A general map of much of the surficial geology in this Zone 3 is given in Figure 123. This Zone is in a geological unit called the Willapa Hills that was described as follows by Alt and Hyndman (1984): *The bedrock throughout the region is simply a large slab of oceanic crust still lying almost as flat as it formed, but about two miles above the normal elevation of the oceanic crust. The oldest rocks are greenish black pillow basalts, which erupted far offshore and once formed the bedrock crust of the Pacific Ocean. In large areas, a deep cover of oceanic sedimentary rocks still blankets the pillow basalts.* As shown in Figure 123, much of the land west of U.S. 101 is covered by a sand and gravel terrace consisting of glacial outwash deposits that are 12,000 to more than 1 million years old (Walsh et al., 1987), with modern sand spits along the bays. In fact, these sand and gravel terraces occur west of U.S. 101 all the way to the Hoh River, north of which the bedrock geology comes to the surface along the shoreline, creating the famous rock formations along the Olympic Peninsula. However, from about Copalis Beach to the Hoh River, the shoreline is backed mostly by steep cliffs of soft sediments, and the beaches have a lot of gravel eroded from the cliffs. As shown in Figure 26, this part of the outer Washington Coast is rising due to tectonic processes at a rate of +0.1 to 0.7 feet/century, with rates decreasing from south to north.

General Description

The southernmost end of this Zone is illustrated in Figure 124. There are two major segments of riprap jetties in that area: 1) To the east is a north-south oriented jetty 1,400 yards long that occupies the west side of the entrance to Baker Bay; and 2) The main north jetty at the mouth of the Columbia River that is oriented northeast/southwest and is 2,840 yards long. Cape Disappointment occupies the space between these two jetty segments. That bedrock cape rises to an elevation of 287 feet.

The sand beach north of the north jetty at the Columbia River mouth built out thousands of feet after the completion of the north jetty in 1917 (the pre-jetties shoreline was generally between North Head and Cape Disappointment; see Figure 124). Both headlands have lighthouses.

A massive sand spit called North Beach Peninsula extends for 25 miles from North Head to the entrance to Willapa Bay. The sand beach along the shore of the Peninsula measures slightly over two miles in width in places, and it contains as many as half a dozen preserved and vegetated beach ridges behind the beach front in places. This is striking evidence for the strong northerly alongshore sediment transport direction and the volume of sand that was added to the shoreline from the shoals and bars at the mouth of the River after jetty construction. The jetties were designed to narrow the River mouth and increase the water currents, to naturally scour the channel and keep it from shoaling in. The sand shoals of the massive ebb-tidal delta were pushed further offshore where they were no longer maintained by tidal currents, and the strong southwest waves pushed that very large volume of sand to the north. Also, some of the sand was transported inshore where it created Desdemona Sands (Sherwood et al., 1990). However, the beaches have undergone extensive erosional/depositional cycles in the last several decades, as they adjust to the changes caused by the construction of the Columbia River jetties.

One consequence of the northern longshore

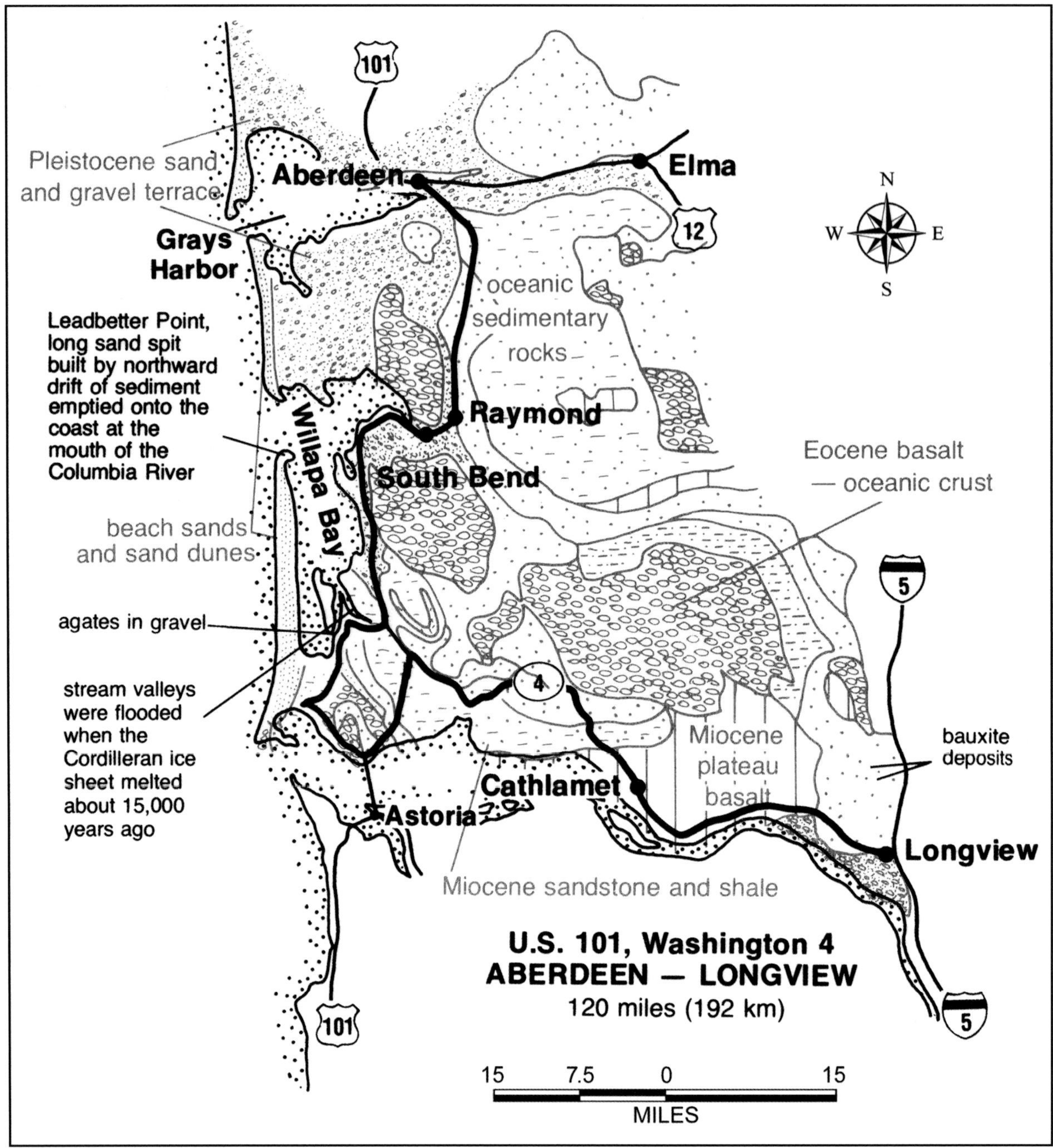

FIGURE 123. General map of the surface geology of much of Zone 2 of Compartment 3. Modified from Alt and Hyndman (1984).

transport of sand from the Columbia River is the formation of a heavy mineral lag deposit by the process shown in Fugure 57A. According to Li and Komar (1992), the heavy mineral content of the sand just north of the north jetty can reach 66%, but decreases to less than 10% 6 miles to the north. You can see how dark the beach sand looks in Figure 124.

The geologic histories of Willapa Bay and Grays Harbor since the last 10,000 years are very different (Twichell and Cross, 2001). Willapa Bay lacks a significant river, so no big valley was carved when sea level was at least 400 feet lower than present during the last ice age; it was a broad coastal plain built from sediments from the Columbia River. Grays Harbor does have a river, and

FIGURE 124. The Cape Disappointment area, north side of the Columbia River mouth, WA. Arrows point to the two segments of jetties (composed of riprap) present in the area. NAIP 2011 imagery, courtesy of Aerial Photography Field Office, USDA.

the ancient Chehalis River cut a valley that was 150-200 feet deep. As sea level rose quickly between 10,000 and 5,000 years ago, Willapa Bay was flooded but filled with sediment from the Columbia River, so it is shallow. The drowned river valley in Grays Harbor has partially filled in (Figure 125), also with sediment mostly from the Columbia River.

The shoreline between the two bays is called the Grayland Plains. On topographic maps and aerial imagery you can see a sharp break between the rolling hills to the east and the flat, 1-2 mile wide plain to the west. That break represents where a higher sea level during one of the inter-glacial periods once stood, with waves cutting a bench into the cliffs of sand and gravel. Once sea level dropped then rose to its current level, beach ridges built out, forming the landform now occupied by the towns from North Cove to Cohassett Beach. To the north, outside the influence of sand from the Columbia River, the width of these Holocene beach ridge deposits thins to just a few hundred yards wide at the mouth of the Copalis River. Further north, there are only some areas with low dunes behind the beach

FIGURE 125. Grays Harbor Estuary, WA. NAIP 2009 imagery, courtesy of Aerial Photography Field Office, USDA.

adjacent to inlets.

As shown in Figure 126, this part of the shoreline has developed a classic crenulate bay shape, anchored by the rocky headland at Point Grenville (see model in Figure 63). The geomorphology there is playing tricks on us. The crenulate bay indicates a southerly sediment transport direction whereas the spits to the south indicate a northerly sediment transport direction. Looks like some research is needed to solve that puzzle.

General Ecology

The distribution of seabird nesting colonies in this Zone reflect the scattered presence of isolated rocky cliffs and islands, their preferred nesting habitat. Thus, colonies are present only on the rock cliffs at Cape Disappointment, North Head, and Point Grenville, and on the jetties at the Columbia River and Grays Harbor, where pigeon guillemots and cormorants nest in the crevices in the riprap. The sand beaches and tidal flats are heavily used by migrating and wintering shorebirds, with tens of thousands of dunlins and sanderlings, and more than a million shorebirds in total. Because of these numbers, Grays Harbor is designated as a site of Hemispheric Importance by the Western Hemisphere Shorebird Reserve Network, one of six sites from Baja, Mexico to Bristol Bay, Alaska with this highest rank. At the Grays Harbor National Wildlife Refuge, you can take the Sandpiper Trail to visit the mudflats and watch for birds (it is best to plan a visit within two hours of high

FIGURE 126. Crenulate bay to the south of Point Grenville, WA. NAIP 2009 imagery, courtesy of Aerial Photography Field Office, USDA.

tide, when the birds will be closer to shore and easier to see).

Three sand beaches bordering Willapa Bay are the only ones in Washington that have consistently been used for nesting in the last decade by snowy plovers: the northern tip of North Beach Peninsula (a site that is actively managed by the Willapa Bay National Wildlife Refuge to improve nesting habitat), Midway Beach, and Graveyard Spit. Cormorants, terns, and gulls nest on the sandy islands in Grays Harbor. Nearshore waters are concentration areas for common murres, sooty shearwaters, and surf scoters, although many waterfowl are present. Northen spotted owls and marbled murrelets (both federaly listed as threatended) nest in the forests along southeast Willapa Bay.

The southern border of the Olympic Coast National Marine Sanctuary is just north of Copalis Beach, and the Copalis National Wildlife Refuge includes the offshore rocks starting at Point Grenville and extending north to the Raft River.

Some sea lions use the Columbia River north jetty as a haulout, but nowhere near the numbers on the south jetty. There are numerous harbor seal haulout sites throughout Willapa Bay and Grays Harbor that are used for pupping, nursing, and resting. Peak abundances occur during the pupping season (mid-April through June) and the annual molt (July through August). Gray, humpback, and killer whales are seasonally present, and harbor porpoises and seals are common.

The sand beaches from Long Beach to Moclips are popular for razor clam harvesting, because of their high density. Willapa Bay is an important nursery habitat for Dungeness crab.

Places to Visit

Cape Disappointment State Park (shown in Figure 124) occupies most of the southern tip of land from the north jetties to North Head and is very popular. It offers access to beaches, the north jetties to see birds and marine mammals, two lighthouses, and great trails. It is a good place to "ship watch" the vessel traffic in/out of the Columbia River. While there, you may also want to visit the Lewis and Clark Interpretive Center for a review of the history of the site as well as some great views.

Leadbetter Point State Park and Willapa National Wildlife Refuge share management of the north end of the North Beach Peninsula (Figure 120). There are great beaches on both sides of the spit, and clamming is very popular in season. There are numerous beach state parks to visit. Some wrap around spits, offering both ocean and bay beaches for fishing, bird watching, and, oh yeah, sunbathing and swimming.

If you want to see whales, your best chance is to book a whale watching cruise out of Westport in Grays Harbor. There are facilities to charter boats for fishing, either big boats for ocean fishing or smaller boats for the bays.

17 COMPARTMENT 4 - OLYMPIC PENINSULA (Point Grenville to East End of Dungeness Spit)

INTRODUCTION

The coastal habitats present in this Compartment are listed in Table 7. This Compartment has a variety of habitats along the shore, with six of them occurring along greater than 10% of its length. Of those six habitats, exposed wave-cut platforms in bedrock is the longest, occurring along 32.8% of the length of the shore (140 miles). Next comes mixed sand and gravel beaches or bars at 23.2%, gravel beaches or bars at 20.2%,

TABLE 7. Coastal habitats in Compartment 4 (Olympic Peninsula).

COMPARTMENT 4 – OLYMPIC PENINSULA (Point Grenville to East End of Dungeness Spit)			
ESI VALUE	**ESI DEFINITION**	**LENGTH (MILES)**	**% ABSOLUTE LENGTH**
1A	Exposed Rocky Shore	39.71	9.32%
1B	Exposed, Solid Man-Made Structures	1.13	0.26%
2A	Exposed Wave-Cut Platforms in Bedrock	139.84	32.81%
2B	Exposed Scarps and Steep Slopes in Clay	0.54	0.13%
3A	Fine- to Medium Grained Sand Beaches	5.07	1.19%
3B	Scarps and Steep Slopes in Sand	3.98	0.93%
4	Coarse-Grained Sand Beaches or Bars	46.24	10.85%
5	Mixed Sand and Gravel Beaches or Bars	98.67	23.15%
6A	Gravel Beaches or Bars	86.10	20.20%
6B	Riprap	17.22	4.04%
6D	Boulder Rubble	10.27	2.41%
7	Exposed Tidal Flats	66.05	15.50%
8A	Sheltered Rocky Shores and Sheltered Scarps	16.89	3.96%
8B	Sheltered, Solid Man-Made Structures	0.38	0.09%
8C	Sheltered Riprap	6.12	1.44%
8F	Vegetated, Steeply-Sloping Bluffs	0.00	0.00%
9A	Sheltered Tidal Flats	9.36	2.20%
9B	Vegetated Low Banks	3.66	0.86%
10A	Salt- and Brackish-Water Marshes	19.88	4.67%
10B	Freshwater Marshes	5.43	1.27%
10C	Swamps	64.61	15.16%
10D	Scrub and Shrub Wetlands	15.57	3.65%
	Total ESI Length (Miles)	656.73	154.09%
	Absolute Shoreline Length (Miles)	426.19	100.00%

exposed tidal flats at 15.5%, swamps at 15.2%, and coarse-grained sand beaches or bars at 10.9%. Clearly, the abundance of those types of habitats along the shore of this Compartment shows that, for the most part, they occur along the outer coast where they are exposed to considerable wave action. The reason that these numbers are more than 100% is the fact that in several areas the shore is composed of more than one habitat type.

For purposes of discussion, we have subdivided this Compartment into the four Zones located on the map in Figure 127:

1) Point Grenville to Mouth of Hoh River
2) Mouth of Hoh River to Cape Flattery
3) Cape Flattery to Agate Bay
4) Agate Bay to East End of Dungeness Spit

ZONE 1. POINT GRENVILLE TO MOUTH OF HOH RIVER

Introduction

The principal geomorphological features of this

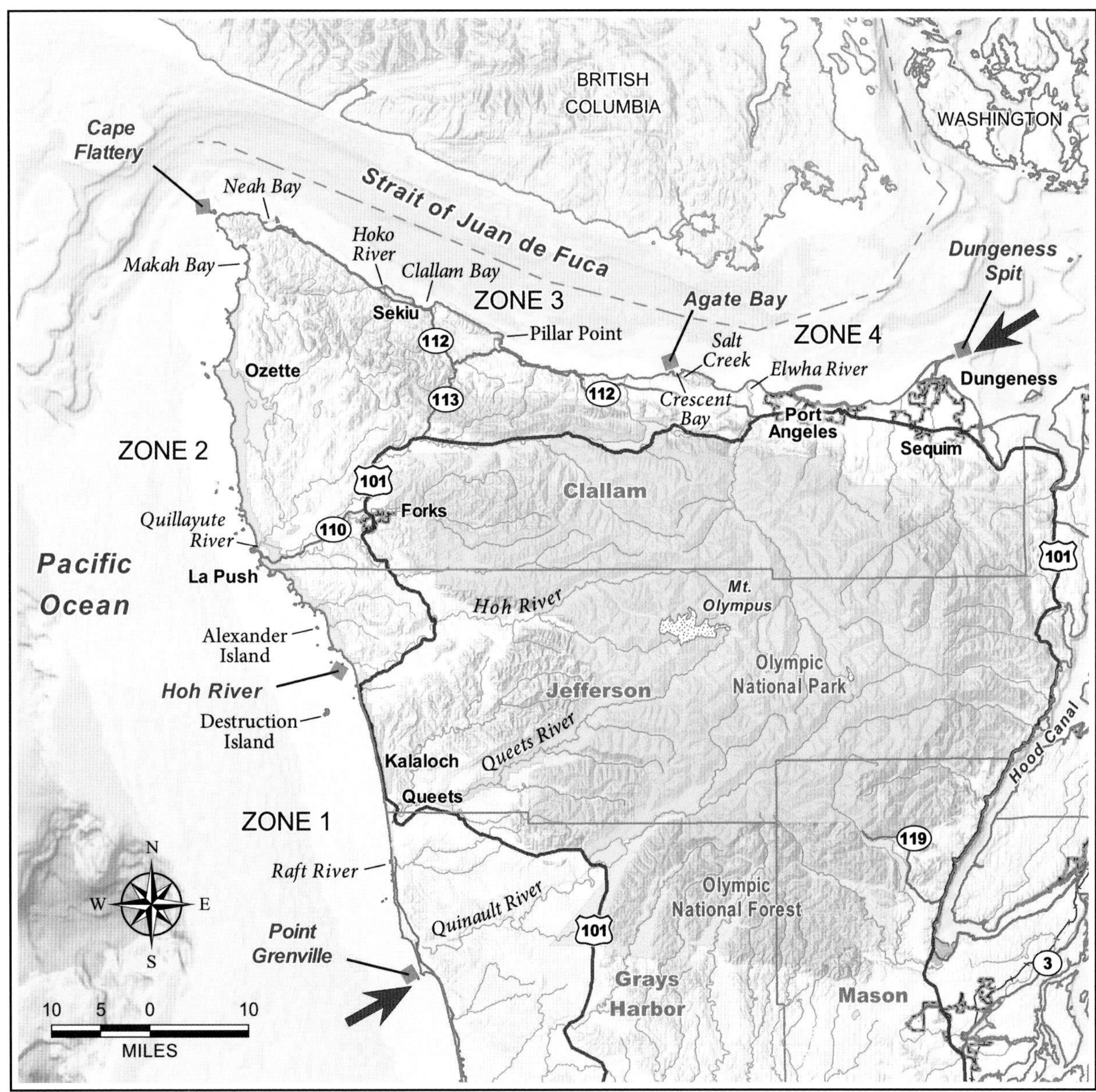

FIGURE 127. Map of Compartment 4.

32-mile-long Zone, illustrated in Figure 127, are:

1) Fairly extensive stretches of eroding shores backed by cliffs of semiconsolidated sediments, with scattered bedrock headlands and sea stacks, some of which shield tombolos;

2) Extensive zones of coarse-grained sand, sand and gravel, and gravel beaches, some of which are highly cuspate; and

3) Three moderate-sized rivers – the Quinault, Queets, and Hoh Rivers – as well as a number of smaller ones, a reflection of the abundance of rainfall in the area.

Geological Framework

A general map of the geology of this Zone is given in Figure 128. Alt and Hyndman (1984) pointed out that a rather flat terrain occurs between the mountains and the Ocean, much of which is *old shallow seafloor now above sea level.* They also noted that very little bedrock is exposed except in the sea cliffs, pointing out that these sea cliffs are the only easily accessibly exposures of the trench filling that forms the core of the Olympic Mountains. Many exposures reveal tightly folded layers of sandstone, and in places the rock consists of angular fragments stuck together – what geologists call a "breccia." That folding and breakage happened while the sinking oceanic crust scraped these rocks into the trench.

And at this point another geological factor enters the equation, deposits left by the glaciers during the Pleistocene epoch. The distribution of these deposits is outlined in the map in Figure 129. With regard to the glacial deposits, Alt and Hyndman (1984) reported:

A large valley glacier ended at the west end of Quinault Lake (shown on Figure 129), *and built a morainal ridge there, which now functions as a natural dam to impound the Lake. Meltwater pouring over that moraine spread over the countryside to the west, and dumped its load of sediment to build an enormous deposit of sand and gravel.*

General Description

The shore at Point Grenville is a half-mile-wide zone of eroded high scarps with numerous sea stacks located just offshore. It is one of the few places in this Zone where the resistant volcanic basalts are exposed along the shoreline (the other locations are the Hogsbacks south of the Raft River). From Point Grenville north to

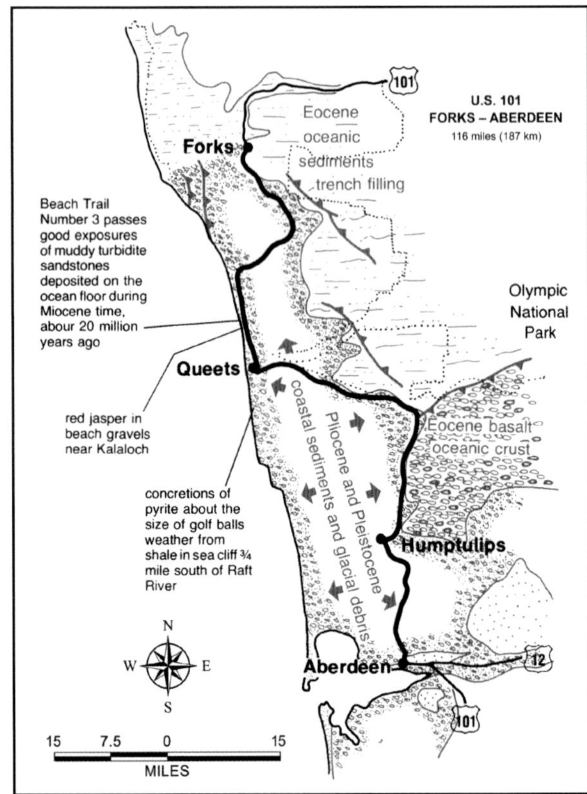

FIGURE 128. General geological map that includes Zone 1 in Compartment 4. Modified after Alt and Hyndmann (1984).

the village of Taholah, the shoreline is backed by scarps in a mix of layered sedimentary rocks overlain by sand and gravel deposits, both of which are seriously eroding. As those scarps eroded, large crescentic scallops (land slides) of the slumped materials have been deposited on the back beach; thus, the beaches contain a lot of gravel. The elevated scarp in light-colored material ends at the northern end of this two-mile section, near the southern side of the village of Taholah. The map of the vertical land movements in feet per century in Figure 26 shows that this region transitions from slightly rising to slightly subsiding due to tectonic processes. Thus, sea-level rise will likely increase the rates of cliff erosion in this area.

Just north of Taholah, Cape Elizabeth juts out into the Ocean, forming elaborate, nearly vertical, scalloped cliffs in a resistant conglomerate (sedimentary rock composed of sand and gravel deposits). These rocks are layered with a slight tilt to the east, forming extensive wave-cut bedrock platforms in the interidal zone and offshore sea stacks and shallow reefs. Softer sedimentary rocks extend north to the mouth of Camp Creek, thus

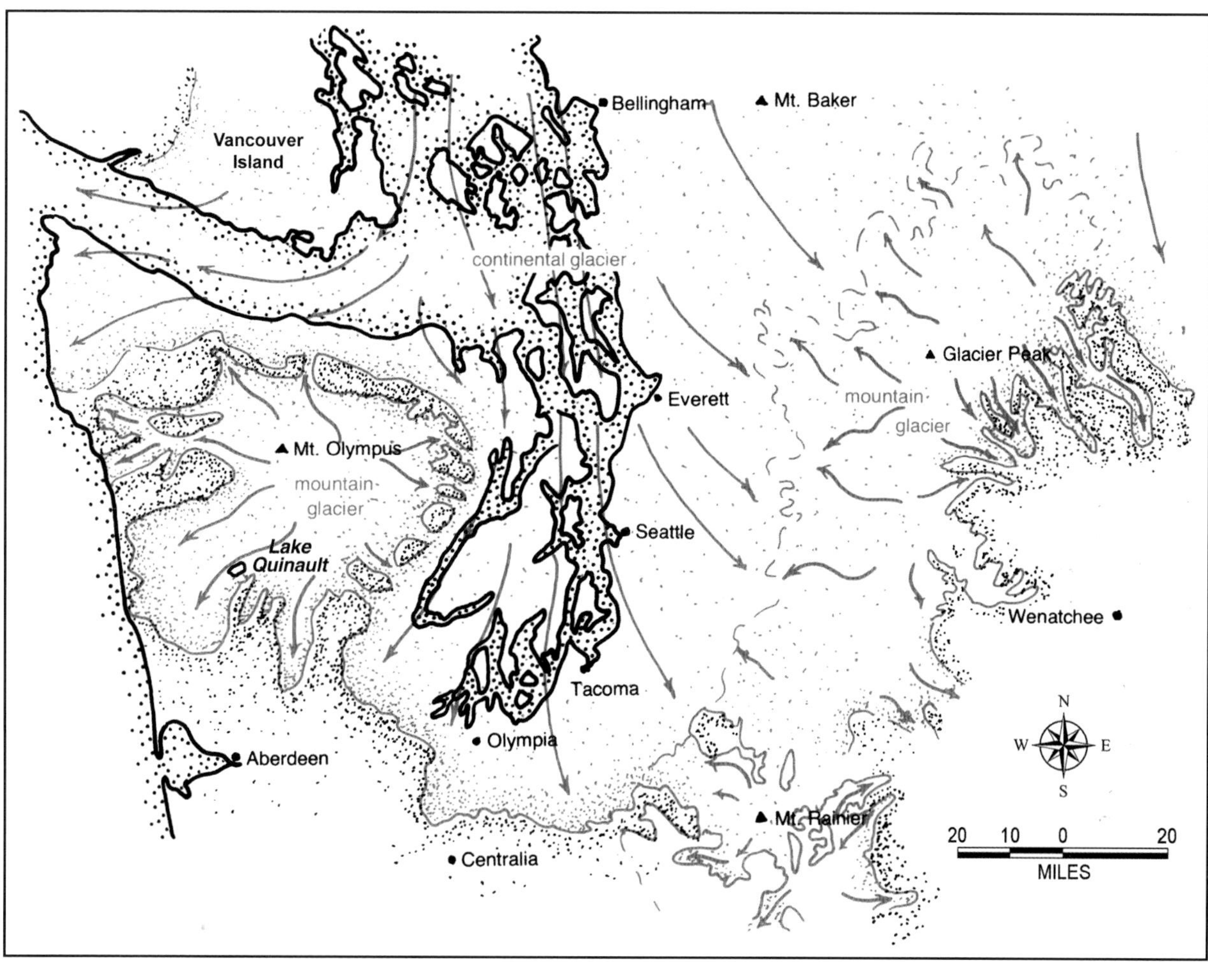

FIGURE 129. Approximate extent of Pleistocene glaciation in the Olympic Peninsula and Puget Sound lowland. Modified after Alt and Hyndmann (1984).

that entire section is backed by eroding scarps.

An 0.6-mile zone of high vertical scarps in sedimentary rocks called Pratt Cliff are located just to the north of the mouth of Camp Creek. These cliffs are inaccessible even at low tide. A little more than a mile offshore is a group of islands composed of resistant basalt called Split Rock and Willoughby Rock, that are important sites for seabird nesting and seal haulouts. The softer sedimentary rocks that were deposited on top of the basalt have been eroded over time, leaving the islands isolated offshore. That will be the fate over time for the two spectacular sea stacks in the intertidal zone to the north, called Little Hogsback and Hogsback. Both of those sea stacks provide shelters for the formation of text-book-style tombolos composed of sand and gravel. Tunnel Island, another amazing sea stack at the mouth of the Raft River, and the sea stacks north and west of it, are different in that they are composed of a relatively hard sandstone. Obviously some layers are harder than others, because waves have carved sea caves, arches, and tunnels. Some of the high cliff faces are nearly vertical, making ideal seabird nesting sites.

Eroding cliffs extend north to the Queets River, with spits building from both north and south to the mouth (Figure 130). The lower portion of the Queets River is a classic example of a meandering stream (see model in Figure 82), presumably as a result of the flat terrain in that area. Note that the northern spit is not vegetated, indicating that it is frequently washed over during strong winter storms. The River has changed its course over time. There can be very large accumulations of logs on the beach and extending north. The mouth of Kalaloch Creek and the wide beaches to the north of it also have large log accumulations.

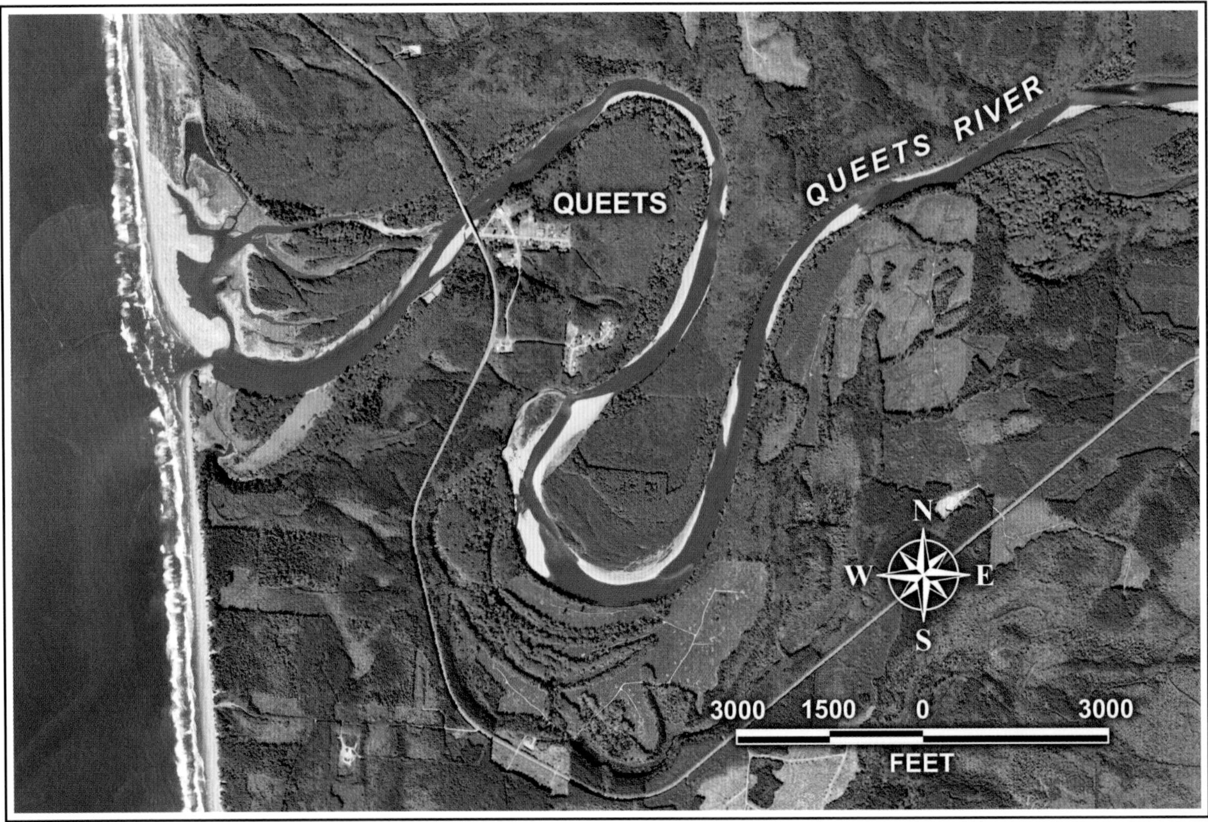

FIGURE 130. The lower portion of the Queets River, WA, a classic example of a meandering stream. NAIP 2013 imagery, courtesy of Aerial Photography Field Office, USDA.

The shoreline between Kalaloch Creek and the rocky outcrops known as Browns Point is backed by low-lying bluffs composed of silt and clay that were deposited between 70,000 and 17,000 years ago (Rau, 1973). Thus, they are soft and easily eroded, which is of further concern because this region of the coast is subsiding (see Figure 26). The sand and gravel beaches along that shore were derived from the glacial deposits that were laid down below the younger silts and clays and are being exposed as the cliffs recede.

Between Browns Point and Starfish Point, there are rocky outcrops of bedded sedimentary rocks similar to those further south that were discussed earlier, except these are steeply dipping to the east. According to Rau (1973), the youngest rocks are to the west, meaning that these rocks have been rotated beyond vertical, to an overturned position. The contact between the older sedimentary rocks and the overlying sediments is about 20 feet above high tide. It can be seen here and not further south at Kalaloch because that area has been downwarped and is now below high tide.

The area around Cedar Creek has three rocky islands, with Abbey Island in the intertidal zone, South Rock about ¾ miles offshore, and Destruction Island 3.5 miles offshore. The sandstone rocks on Destruction Island have been tilted to near vertical and folded, which can be seen in the horseshoe shape of the island, shown in Figure 131. Abbey Island is also interesting in that the rocks there and in the base of the adjacent cliffs are volcanic breccia (defined earlier).

North of Abbey Island is a zone of large landslides in the cliffs that are composed of sand and gravels overlying soft clay and silt. About halfway between Abbey Island and the mouth of the Hoh River, the cliffs change to colorful reddish-orange sands and gravels that exceed 100 feet in thickness. The geological history of this Zone is quite complex.

General Ecology

Much of the coastal areas in this Zone are part of the Olympic National Park and the Quinault Indian

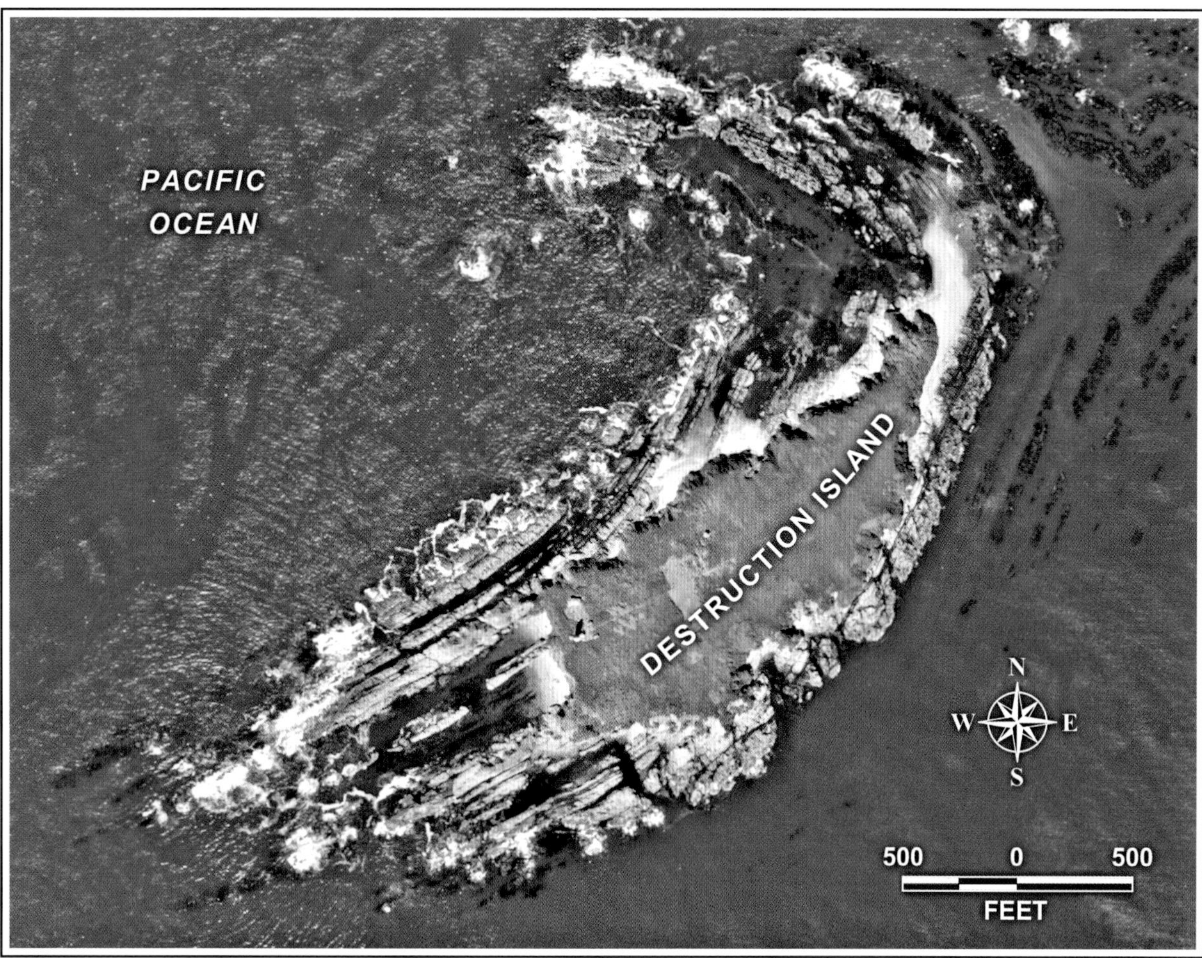

FIGURE 131. Vertical photo of Destruction Island showing the horseshoe shape of the island because of the rocks that have been tilted and folded. NAIP 2013 imagery, courtesy of Aerial Photography Field Office, USDA.

Reservation. The offshore waters are part of the Olympic National Marine Sanctuary. There are about 20 mapped seabird nesting colonies, mostly on offshore islands and a few on very isolated rocky headlands. Survey data on the number of seabirds at each colony are old (often dating prior to 2000), and new surveys being conducted by state biologists are not available. Thus, no counts are reported here, only the seabird species that have nested historically. The list of nesting seabirds includes double-crested, Brandt's, and pelagic cormorants, common murres, tufted puffins, pigeon guillemots, rhinoceros auklets, black oystercatchers, and gulls. The colony on Destruction Island includes the only location for nesting rhinoceros auklets in the Zone, though it is a big one, with 13,000 birds present March to October. Nesting cormorants there pale in contrast, with only hundreds of birds nesting. Some of the islands and cliffs have nesting peregrine falcons that prey on seabirds during nesting season, and waterfowl, shorebirds, and passerines year round.

Marbled murrelets (listed as threatened) nest in old-growth forests, with the highest densities south of Destruction Island to Kalaloch and the Raft River. Offshore waters support large numbers of wintering birds, including tens of thousands of red-throated loons, surf scoters, red-necked phalaropes, and sooty shearwaters.

Nearly every offshore rock reef and island in this Zone is used as a haulout site by harbor seals, with hundreds at many sites, particularly at Split Rock and Destruction Island. Large numbers of Steller sea lions and some California sea lions use the outcrops around Split Rock. A few individual northern elephant seals spend winter and spring around Destruction Island. You might see some of the introduced sea otters.

The few larger streams and rivers support runs of Chinook, chum, coho, and sockeye salmon, and steelhead. All of these salmon runs are federally listed as threatened. The Quinault River is one of two significant runs of sockeye salmon in this Compartment. Sockeye spawn in tributaries of large lakes, and the young fry rear in the lake. The Quinault Indian Nation has been restoring spawning habitat in the upper Quinault River above Lake Quinault to increase the population.

Steelhead are a type of rainbow trout that spawns and rears for a couple of years in freshwater, then heads to sea for 1-4 years before returning to their natal streams to spawn. There are summer and winter spawns. Not all fish die after spawning; some, particularly winter-run fish, can spawn as many as four times. Thus, these winter-run fish can be really big and fishing for them is really popular. However, habitat degradation, construction of dams, and overfishing has reduced the numbers of returning fish in the last few decades to about 10% of their historical numbers. As a result, the fishery is intensely managed, with many restrictions on the number of fish that can be taken by recreational fishers; all commercial fishing for steelhead is only allowed by Indian tribal fishers.

Places to Visit

The shore between Moclips and Queets is part of the Quinault Indian Reservation, and the mouth of the Hoh River is part of the Hoh Indian Reservation. So, if you are interested in visiting these areas, you should inquire about what activities are possible in these locations. Tribe members utilize the shoreline, marine waters, and upland areas within and surrounding Reservation lands for subsistence harvest of invertebrates, fish, birds, and other species. These areas are also of high cultural importance and sensitivity.

Public access to the coast is primarily along about 11 miles of U.S. 101 from just north of Queets to Hoh, within the boundaries of the Olympic National Park. There are two campgrounds, South Beach and Kalaloch, and six beaches with parking and/or overlooks. Kalaloch Beach is good for razor clamming, but check to make sure that it is open because it was closed in 2018. When visiting the beach, it is great to inspect the amazing intertidal sea stacks and rocky headlands. The Park Service warns visitors to never try to walk around headlands at high tide because you could be trapped. And you should to always carry a tide chart for that day so you know when high tide is. Good advice!

The beaches are mixtures of sand and gravel, sediments that have eroded out of the retreating cliffs over time, with the fine-grained sediments washed out to sea, leaving the coarser sediments to be reworked by waves. You can hike north or south, depending on how much privacy you want. North of Boulder Creek, very large boulders occur along the lower intertidal zone. They are only semi-rounded, because they were transported by ice rather than by streams (Rau, 1973). Figure 129 shows that one of the lobes from the glaciers that spread out from Mt. Olympus extends all the way to the current shoreline.

Ruby Beach is unique in that, in addition to spectacular sea stacks, you can hike the 3 miles north to the mouth of the Hoh River, which tends to be less populated. Ruby Beach got its name from the concentrations of garnet, a type of heavy mineral, in the sand (see Figure 57 and associated discussion on how these layers form). There is a viewpoint about 1 mile south of Ruby Beach where you see, weather permitting, 33-acre Destruction Island. Unfortunately, no visitors are allowed on the Island, where the lighthouse was operated from 1892 to 2008. The Island has quite a history. What an interesting stretch of shoreline that you surely won't want to miss!

ZONE 2. MOUTH OF HOH RIVER TO CAPE FLATTERY

Introduction

The principal geomorphological features of this 50-mile long Zone, illustrated in Figure 127, are:

1) Abundant exposed rocky headlands;
2) Numerous nearshore islands and sea stacks;
3) Randomly distributed tomboloid forms;
4) Numerous arcuate sand beaches; and
5) In addition to the Hoh River, there is one more good-sized river in this Zone, the Quillayute River (Figure 132) as well as a couple of smaller ones – the Sooes and Waatch Rivers.

Geological Framework

The general geology of the southern part of this Zone is illustrated in Figure 128. As shown in Figure

FIGURE 132. Mouth of the Quillayute River and the village of La Push, WA. NAIP 2013 imagery, courtesy of Aerial Photography Field Office, USDA.

129, this area was covered with ice during the last major glaciation of the Pleistocene epoch. Figure 133 is a cross-section of the geological units from Alexander Island to the mainland, north of the Hoh River. Rau (1980) described these units and the processes that formed them, from oldest to youngest, as:

1) Bedrock, composed of siltstones, sandstones, and conglomerates, that is 20-40 million years old. Many of these rocks have been so extremely deformed that they are not very resistant to erosion by wave attack (called melange rocks). Where these rocks are more resistant to erosion (often composed of less deformed massive sandstones and conglomerates), they form the headlands and offshore islands.

2) During the Ice Ages of the early Pleistocene epoch, glacial meltwater and even glacial ice deposited thick layers of sand and gravel over the bedrock, layers that were even deeper in the river valleys. These deposits originally extended many miles seaward; however, erosion by wave action removed much of these materials. Note that this wave erosion during higher sea levels also eroded the bedrock, creating the elevated wave-cut platforms illustrated in Figure 133. Also note the thick, older glacial deposits located further inland.

3) Near the end of the Pleistocene, a second and thinner series of sand and gravel sediments were

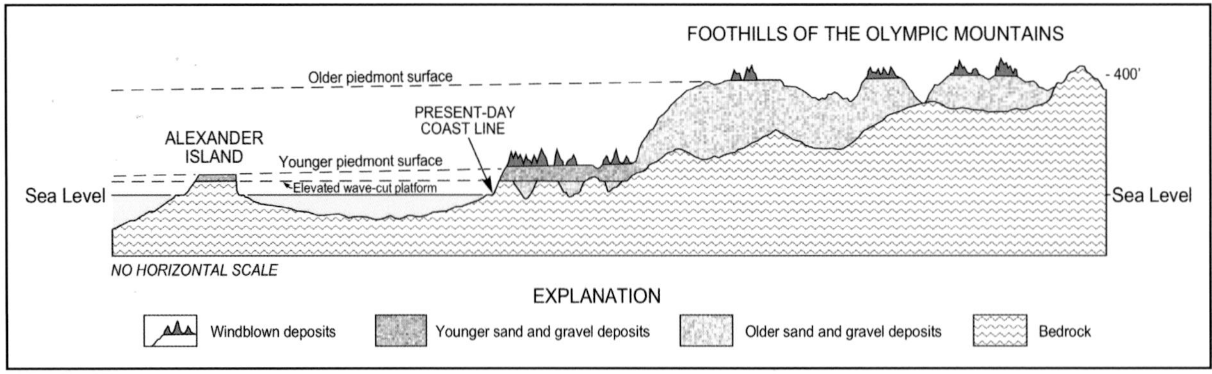

FIGURE 133. Cross-section of the geological units from Alexander Island to the mainland, North of the Hoh River. Modified after Rau (1980).

transported down major stream valleys and deposited on the now uplifted wave-cut platform. These younger deposits also extended seaward, but less so than the older deposits. They also were eroded together with bedrock by wave action back to the present-day coastline. Thus, Alexander Island (Figure 134) is a remnant of the mainland that existed farther west just a few thousand years ago. The many sea stacks are smaller erosional remants, and the flat rocky reefs are where even the sea stacks have been eroded away to the level of the waves.

4) Even later in the Pleistocene, windblown silt and sand covered much of the coastal area to a depth of as much as 15 feet. These deposits are very noticeable at the top of most cliffs, such as at Taylor Point.

General Description

Jefferson Cove, which can be easily accessed from the trail at the north side of the Hoh River, is typical of the many, less-accessible coves in this area: Rocky headlands in more resistant sandstones (and sometimes basaltic rocks such as at Portage Head and south of Point of Arches) that border arcuate sand or sand and gravel beaches backed by lower cliffs in the melange rocks with abundant landslides. This pattern is repeated over and over, all the way to Makah Bay. Sometimes the coves are small, less than 0.3 miles long, whereas others can be over 3 miles long. The headlands can be massive, such as at Hoh Head, which extends about 0.5 miles offshore (therefore, you hike over rather than around it). In contrast, often the coves are borderd by small sea stacks, erosional remanants of more impressive headlands.

On the western flank of Secret Cove, the contact of the darker deformed rocks with the lighter and resistant sandstone of Hoh Head can be seen just above the high-tide line (Rau, 1980). The melange rocks can include large blocks of sandstone that stand out in contrast to the surrounding area.

Offshore of Mosquito Creek, Alexander Island, with its flat top, displays the effects of waves cutting a platform in the rocks at higher sea levels, about 100 feet higher (relatively) than present. The top 20 feet of the island is covered by the younger Pleiestocene-aged unconsolidated sediments, as shown in Figures 133 and 134.

The shoreline from Mosquito Creek to Goodman Creek forms a deep, crenulate bay, and the cliffs have extensive landslides in the melange rocks there. Oddly, the mouth of Goodman Creek flows into a cluster of a dozen small nearshore bedrock islands and sea stacks. According to Rau (1980), the Goodman Creek valley is the location of a major fault that trends to the northeast, so that may be the reason it is located there. He also pointed out that you can see the rounded pebbles and boulders in the conglomerate that forms the cliffs north of Goodman Creek. At Toleak Point, there is a sandy, diamond-shaped tombolo about 300 yards across, and another one at Strawberry Point.

A very picturesque cluster of about 20 small islands and sea stacks, called Giants Graveyard, is located offshore Graveyard Point. The sea stacks have a wide range in shapes, sizes, and heights. This area is accessible from the trail to Third Beach. The Quillayute Needles at the north end of Second Beach, and the headlands at Teahwhit Head and Quateata have the same underlying geology as to the south, mostly massive sandstones and conglomerates.

The mouth of the Quillayute River at La Push

FIGURE 134. Oblique aerial photograph of Alexander Island, WA, a remnant of more resistant sedimentary rocks that used to be part of the mainland (see Figure 133). The flat top of the island is a wave-cut platform that is now covered by 20 feet of sand and gravel of Pleistocene age. Photo taken 18 February 2008 by Sam Beebe/Ecotrust CC BY-SA 3.0.

(Figure 132) had migrated north to the end of Rialto Beach as recently as 1909 (Rau, 1980). The inlet was stabilized in its current position in 1931 when the jetties were constructed. Figure 26 shows that the area around La Push is slightly subsiding.

The bedrock geology of the coastal zone from La Push to Makah Bay is similar to that discussed above, with rocky headlands and breathtaking sea stacks offshore bordering sand and gravel beaches, though the embayments between the headlands are much larger, several miles long. There are some spectacular sea arches, such as Hole in the Wall north of Rialto Beach and Point of the Arches (Figure 135), south of Shi Shi Beach, which has its own awesome sea stacks. However, the Pleistocene deposits on top of the bedrock include glacial till, a dense clay-rich matrix with gravel and scattered boulders mixed in, in addition to the outwash deposits. The till was laid down by the Juan de Fuca lobe of the continental ice sheet that covered the northern end of the Olympic Peninsula about 15,000 years ago (Gerstel and Lingley, 2000), rather than the mountain glaciers that spread out from Mt. Olympus (see Figure 129). As shown in Figure 26, there has been a change from a subsiding to a rising coast north of La Push.

Makah Bay has almost a "W" shape, bordered by headlands, but also with a sea stack right in the middle with a large tombolo connecting it to the beach (Figure 136). Two small rivers help bring in sand from the watershed as well as the eroding cliffs, so the sand beaches are wide and have small spits building to the north of each river mouth.

The massive headland of Cape Flattery, the "most northwestern point in the contiguous lower 48 states," has wide wave-cut platforms along the southwestern

FIGURE 135. Oblique aerial photograph of the Point of the Arches, WA. Photo taken 18 February 2008 by Sam Beebe/Ecotrust CC BY-SA 3.0.

corner, adjacent to the Waatch River (Figure 136), where the cliffs are composed of siltstones. However, north to and around the Cape, the cliffs are composed of basaltic rocks that is more resistant and thus forms vertical rocky shores with numerous "slot" bays along weaker zones in the rock (see Figure 42A). If you hike all the way to the Cape (which we highly recommend), you will see the crystal-blue water and waves crashing off the vertical cliffs and sweeping into those slot bays.

General Ecology

With so many sea stacks and offshore islands, there are over 100 locations mapped as having seabird nesting colonies and/or peregrine falcon nest sites. The species list is impressive: Brandt's, double-crested, and pelagic cormorants, pigeon guillemots, Cassin's and rhinoceros auklets, tufted puffins, Leach's storm petrels, glacous-winged gulls, and black oystercatchers. Nesting starts in April and ends by July to October.

Some of the locations have just one species. For example, some of the very steep sea stacks with no flat or vegetated top have mostly pigeon guillemots, because they prefer to nest in the small cavities in the rocks. Tufted puffins prefer burrows in grassy slopes. Some locations only have a little bit of land above high tide, and a pair of black oystercatchers can nest there. The larger islands, with a mix of habitats, have more species occupying their preferred nesting niche.

Marbled murrelets (listed as state endangered and federally threatened) nest in old-growth forests. Offshore waters support large numbers of wintering birds, sooty shearwaters, fork-tailed shearwaters, Leach's storm petrels, and common murres considered as abundant,

FIGURE 136. Vertical image showing the mouths of the Sooes and Waatch Rivers, WA. NAIP 2009 imagery, courtesy of Aerial Photography Field Office, USDA.

and up to ten thousand rhinocerous auklets, surf scoters, and red-throated loons.

Tatoosh Island, off the very tip of Cape Flattery, is famous for many reasons, but particularly for the long history of study by Dr. Robert Paine, University of Washington researcher, of intertidal communities and how predation affects their food webs, and also of the impacts of peregrine falcons preying at seabird nesting colonies. Paine (1966) did some interesting experiments that had never been conducted before in intertidal communities—he removed sea stars–voracious predators of mussels, barnacles, and limpets–then watched what happened. Pretty quickly, in the absence of being eaten by sea stars, the open spaces were occupied by barnacles, then later replaced by mussels, forming dense beds that crowded out other species such as algae and limpets. The bottomline was: predators increase diversity through constant disturbance. He later termed these "keystone" species (Paine, 1969), named after the central stone that prevents an arch from collapsing.

The return of American peregrine falcons from near extinction due to the use of DDT as a pesticide had caused concerns that these "duck hawks" were reducing seabird nesting success on breeding colonies. Paine et al. (1990) studied this issue for over ten years, from 1978 to 1988, with interesting results. Yes, peregrine falcons reduced the numbers of both Cassin's and rhinoceros auklets, their preferred prey. But, they also ate or scared off northwestern crows, who are voracious egg and nest predators themselves. So, the peregrine falcons reduced the crow-related egg predation, which resulted in increases in the number of common murres, pelagic commorants, and black oystercatchers. Good news if you are these species; bad news if you are the preferred prey for peregrine falcons.

There are over 50 sites in this Zone used as haulouts by seals and sea lions. Most of them are intertidal areas around islands, rocks, and reefs used by harbor seals for pupping and molting. The main haulout sites for Steller and California sea lions are at Carroll Island (2 miles offshore at the south end of Ozette Lake), Bodelteh Islands (off Cape Lava), Cape Alava, and Tatoosh Island. Each of these sites can have hundreds of animals present. Only male California sea lions head into Washington waters; the females remain near their rookeries off California and Mexico. Both sexes of Steller sea lions occur in Washington, but there are no breeding rookeries.

Gray, humpback, and killer whales are common depending on the season, and sea otters have been reintroduced to this area. Sea otters are more concentrated around Destruction Island, the mouth of the Hoh River, Perkins Reef, Cape Johnson, Sand Point, Cape Alava, and Duk Point.

Most every stream and river in this Zone support runs of Chinook, chum, and coho salmon, and steelhead. All of the salmon runs are federally listed as threatened. There is a small run of sockeye salmon in the Ozette River, one of only two rivers along the outer coast of Washington that have sockeye runs.

Places to Visit

There are three areas along the shore of this Zone that are easily accessible by driving at the mouth of the Quillayute River at La Push, Hobuck Beach at Makah Bay, and Cape Flattery. At La Push, you can see the sea stacks at Rialto Beach and hike trails along the Quillayute River. Along La Push Road, there are relatively short trails to Second and Third Beach where you can see sea stacks galore, including Giants Graveyard. We can't describe just how beautiful this area is.

The shoreline north of La Push to Ozette is a true wilderness, with no overland trails except from Ozette to Sand Point and Cape Alava. You get to see this amazing coast by walking the hiking trails that go along beaches where access is possible and overland over the rocky headlands. The trail from Rialto Beach to the Ozette trailhead is 20 miles, with designated campsites along the way. Once you reach the trailhead, you still have about 3 miles to go from the shore to the Ozette Ranger Station. It would be a 6- or 9-mile-long day hike from the Ranger Station, but well worth the effort. Reservations are required for backpacking along the Ozette Coast from May through September, so make your plans early if this is how you are visiting.

The Makah Indian Reservation extends from north end of Shi Shi Beach all around Cape Flattery to beyond Neah Bay. You need a recreation pass for any recreation on the Reservation, which is available in most businesses in Neah Bay. The road to the trailhead to Shi Shi Beach, Point of the Arches, etc. is located on the south end of Hobuck Road. It is only 2 miles to the beach and the wonderful sea stacks and wildlife to see.

You can drive a winding road through lush forests to the parking area at Cape Flattery, where there is a

1.5-mile roundtrip hike to stunning views of the Pacific Ocean and the Strait of Juan de Fuca. Just off the tip of Cape Flattery is Tatoosh Island and the historic Cape Flattery Light. As you might expect, the Cape is an excellent spot for viewing whales, birds, sea lions, sea otters, vessel traffic, etc.

ZONE 3. CAPE FLATTERY TO AGATE BAY

Introduction

The principal geomorphological features of this 50-mile-long Zone, illustrated in Figure 127, are:

1) Some exposed rocky headlands, with the ones in the Cape Flattery area being the largest;

2) Mostly an undeveloped shore backed by forested low bluffs. Only three sizeable villages occur along this shore of the Strait of Juan de Fuca – Neah Bay, Sekiu, and Clallam Bay;

3) Only one modest sized river, Hoko River, and a number of smaller ones, including the Clallam and Pysht Rivers; and

4) Eight small rivers, none of which have jetties.

Geological Framework

The general geology of this Zone is illustrated in Figure 137. This map shows that, from Cape Flattery east to about the town of Twin, the bedrock along the shore is Oligocene-Miocene sandstone, shale, and conglomerate, and from Twin to the east the coast is coverd by glacial deposits of sand, gravel, and mud, except for the basalt rocks just west of Crescent Bay. As shown in Figure 129, the glacial deposits represent the edge of the continental ice sheet from the north during the latest Pleistocene glacial event (25,000 to 10,000 years ago). The glacial deposits get thicker to the east, and there is a transition zone where waves have eroded the glacial deposits, exposing the underlying bedrock along the coast. The presence of glacial deposits dramatically changes the nature of the coast, with abundant gravel, including large boulders, on the beaches as the steep cliffs in the loose sediments erode. Many of the boulders are composed of granite, further evidence of the effects of glaciers because there are no granitic rocks in the Olympic Peninsula. The granite was transported by the continental ice sheet from granitic rocks in British Columbia.

General Description

East of Neah Bay, four parallel vertical bedrock ledges of a very hard sandstone (with softer sedimentary rocks between them) oriented to the northwest create straight rocky shores as shown in Figure 138, which are very unusual for rocky shorelines. The westernmost sandstone ledge extends to form the east shore of Waadah Island. The rocky ledges extend into the intertidal zone as low ridges. These ledges act as natural groins, trapping

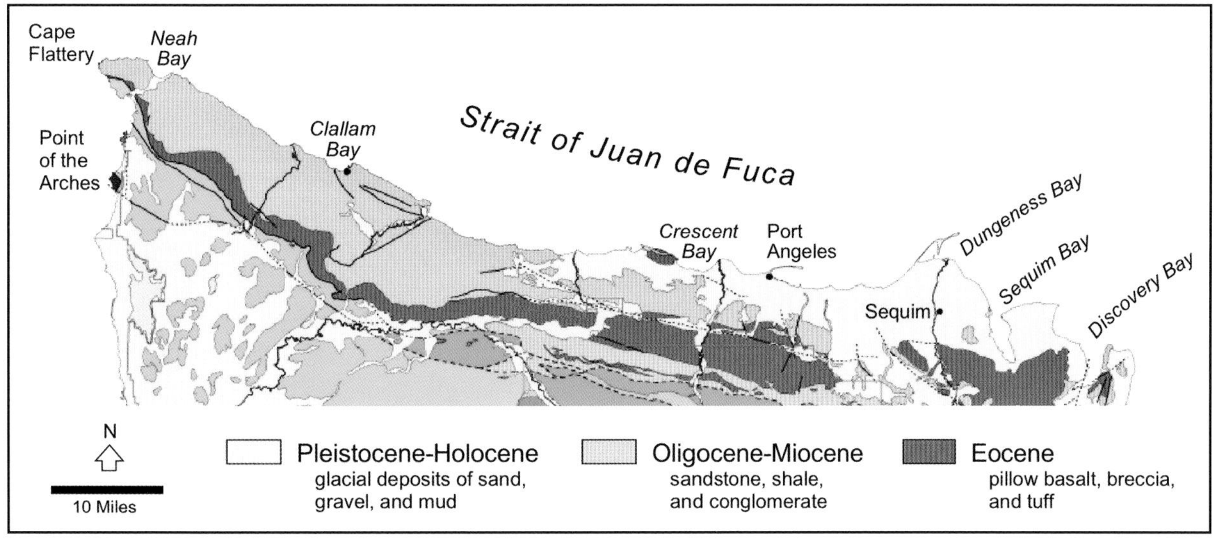

FIGURE 137. General geology of Zones 3 and 4 of Compartment 4. Modified from Geologic Map of the Olympic Peninsula, Washington, courtesy of USGS (1978).

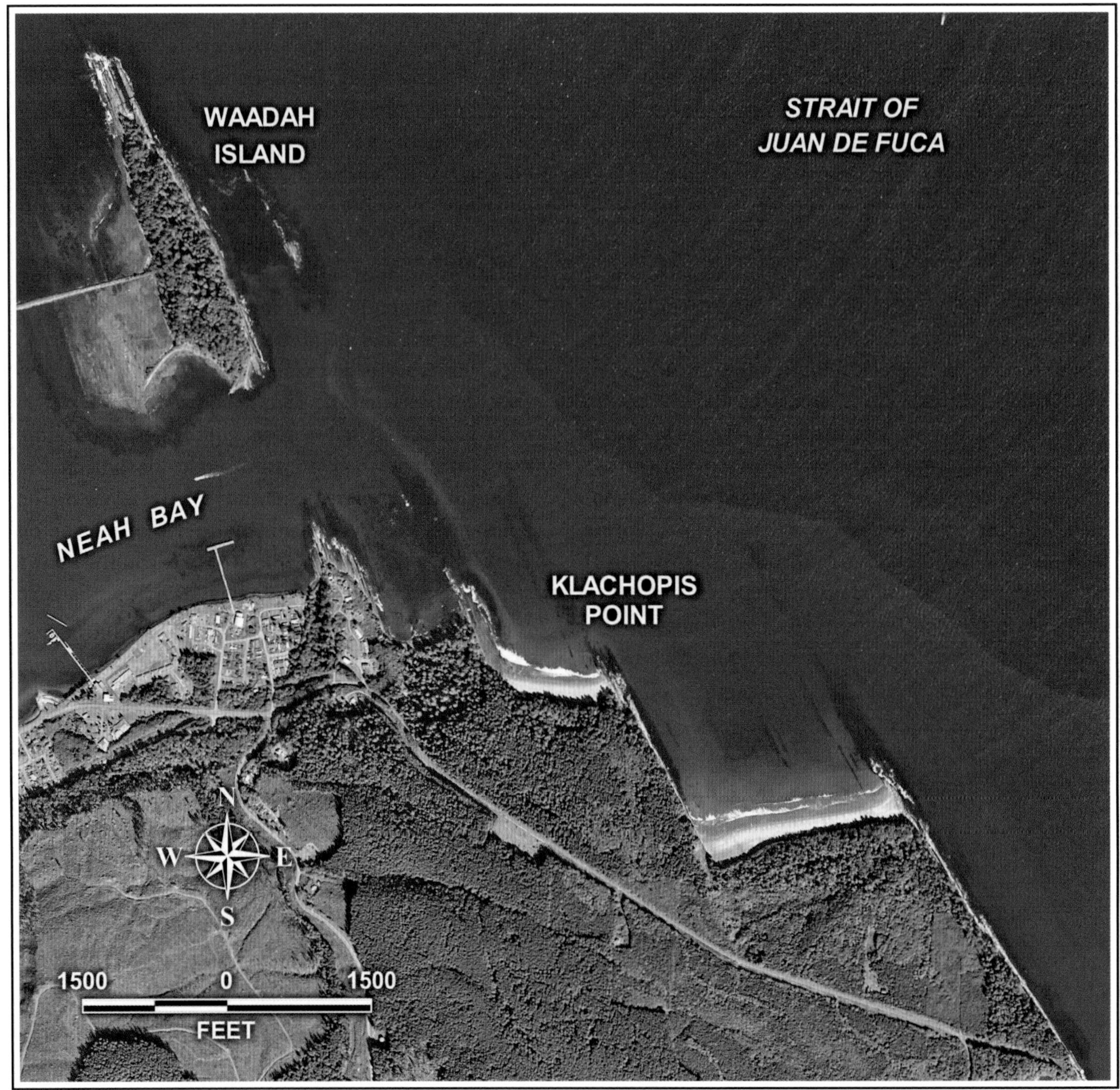

FIGURE 138. Two natural groin systems located just east of the village of Neah Bay, WA. NAIP 2009 imagery, courtesy of Aerial Photography Field Office, USDA.

sand in the pockets between them. Those are the most clear-cut examples of natural groins that we have seen anywhere.

[NOTE: A **groin** is a linear, relatively short, shore-protection structure built perpendicular to the shore for the purpose of widening the beach by trapping a portion of the sand in the littoral drift. See Figure 85B. In a few cases, a projection of bedrock along the shore can also trap sediment on its updrift side in a fashion similar to the effect of a man-made groin. These are called **natural groins**.]

The shoreline from Neah Bay to Clallam Bay is exposed to some pretty large waves, as evidenced by wide wave-cut platforms, abundant gravel beaches, and lots of very large boulders both against the cliffs and lower on the rock platform. Kelp beds are abundant. There are low sea stacks and rock reefs offshore. Sections of State Highway 112 built too close to the shore now have riprap for protection from erosion.

The creeks drop their load of sediment when

they empty into the Strait, forming wide flats and bars composed of sand and gravel at the mouth. The Hoko and Clallam Rivers carry larger volumes of sediment that have been reworked by waves to form bays with a rocky headland on the east and spits built across the river mouths. Interestingly, the spit across the Hoko River is built to the east, whereas the spit across the Clallam River is built to the west, reflecting the bimodal wind patterns in the Strait.

From Slip Point east of Clallam Bay to the mouth of the Pysht River, the shoreline is composed of steep rocky cliffs. There can be large boulders on the shore, and there is very little sand. Pillar Point headland provides shelter from waves and thus the Pysht River mouth area includes salt marshes, extensive sand flats, and a nice sand spit (shown in Figure 139).

The shoreline east of the Pysht River to just west of the Lyre River has wide intertidal platforms composed of sand, gravel, and low rock outcrops. The bedrock there is obviously less resistant to erosion by waves, thus the bedrock is eroded into flat platforms at sea level. And remember that sea level has been at its present height for only about 5,000 years, and the land is rising (Figure 26). You can start to see the glacial deposits exposed in the cliffs, overlying the folded and faulted bedrock melange just west of the Lyre River.

From Murdock Creek east, sand and gravel beaches are fronted by wide platforms until Agate Bay, where a sand and gravel pocket beach has formed in front of the basaltic headland.

General Ecology

East of Neah Bay, the bird species reflect the paucity of isolated and steep rocky shores and offshore sea stacks, meaning that there are very few seabird nesting colonies, and mostly pigeon guillemots. However, there are many ducks, shorebirds, and post-breeding murres. And hundreds of bald eagles move down from British Columbia to join the resident population between January and April along the shoreline between Neah Bay and Clallam Bay.

The Southern Resident killer whale is a distinct population that spends several months in summer and fall in the inland waters of Washington and British Columbia, Canada (called the Salish Sea). There are three family groups, called J, K, and L pods, with a population in December 2017 of 76. They were listed as federally endangered in 2005. They come to feed on runs of Chinook salmon, mostly those heading to the Fraser River in British Columbia.

Humpback whales come in summer to feed and are often seen on whale watching tours. These whales are part of the Mexico Distinct Population Segment (because they spend winter and calving there) that is federally listed as threatened. Harbor seals haulout on the wide intertidal rock platforms; abundances peak during pupping season (mid-June to August) and the annual molt (late August to October).

The salmon fishery in the Strait of Juan de Fuca and Puget Sound is tightly regulated and managed, particularly Chinook salmon because of their importance to the survival of the endangered Southern Resident killer whales. All salmon returning to spawn in their natal streams must pass through the Strait. The timing of these "runs" varies by species, thus it is possible to set harvest limits by species and runs, to ensure that sufficient fish survive to spawn. There is a recreational fishery for Chinook, coho, and chum salmon, and steelhead. Dungeness crab fishing is also carefully managed, and all crabbers are required to complete and submit catch record cards. The sport fishery is opened/closed as needed to maintain sustainable harvests.

Places to Visit

The town of Neah Bay is home to the Makah Museum where you can learn about Makah life prior to European contact. State Highway 112 runs along the coast a part of the way, and there are pulloffs at scenic or beach access areas. Pillar Point County Park provides access to the Pysht River estuary and tidal flats with good birdwatching and crabbing. Murdoch Beach near the Lyre River is supposed to be a good place to search for agates, round concretions, and other interesting types of rocks that have eroded out of the basalt headlands (its nickname is Round Rock Beach).

ZONE 4. AGATE BAY TO EAST END OF DUNGENESS SPIT

Introduction

The principal geomorphological features of this 33-mile-long Zone, illustrated in Figure 127, are:

1) A couple of perfectly formed crenulate bays

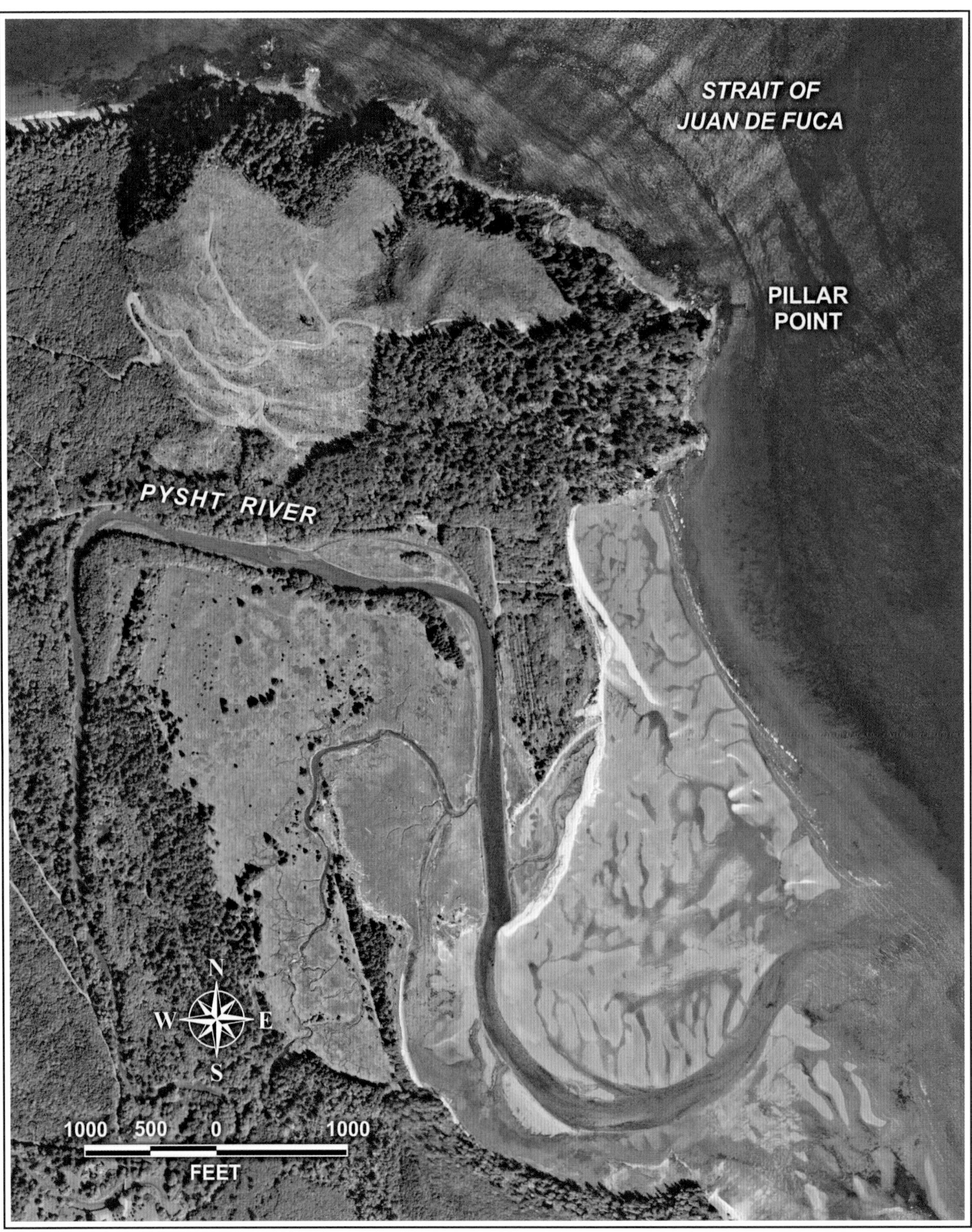

FIGURE 139. Coastal area in the vicinity of Pillar Point, WA on the Strait of Juan de Fuca. NAIP 2015 imagery, courtesy of Aerial Photography Field Office, USDA.

created by dominant waves from the northwest;

2) Basalt rocky headland at Striped Peak and Elwha River delta;

3) Two large spit complexes, also showing the effects of waves from the northwest, including the fam**ous Dungeness S**pit;

4) Three tidal inlets on this outer coast – Mouth of Salt Creek at Salt Creek Recreation Area; Mouth of Elwha River at Angeles Point; and Entrance to Port Angeles Harbor at Port Angeles. None of those inlets have jetties.

Geological Framework

The general surficial geology of this Zone is illustrated in Figure 137. Most interesting are the basaltic rocks that form the headlands on either side of Crescent Bay and the rocky peninsula along Striped Peak. These are the only areas where this rock type is exposed at the shoreline. This resistant rock forms steep cliffs, rocky ledges, lots of small bays with pocket gravel beaches, and even a sea stack at the southern edge of the peninsula, Observatory Point.

From there to Dungeness Spit, the shoreline is backed by thick glacial deposts that form steep cliffs fronted by mixed sand and gravel beaches, with some very large boulders that were transported by the ice, rather than by streams. Two deltas build out into the Strait: Elwha and the small Twin River; and there are two large spits: Ediz Hook and Dungeness. These large, elongated spits form when: 1) There is an abundant sand and gravel supply (such as the eroding cliffs of glacial deposits); 2) Waves approach the shoreline at an oblique angle (large waves in the Strait come for the WNW); 3) There are two sets of wave that approach from different directions (there is a secondary wind pattern, and the waves they generate are out of the southeast); and 4) There is a shallow area offshore on which the sediments can build on. All of these conditions come together to form these large spits along the eastern shore of the Strait de Juan de Fuca.

General Description

The beach at Crescent Bay, with its classic crenulate bay pictured in Figure 64, is composed of sand, something of an exception along the Strait where there is abundant gravel. The sand comes from Salt Creek, which drains a relatively flat area underlain by glacial sediments, so it carrys mostly sand. The creek has obviously migrated east and west, between the rocky headlands, and the small estuary has filled with sediments and salt marsh.

The basalt bedrock forms the shallow peninsula between Tongue Point and Observatory Point. Pillow basalt is exposed in some of the cliffs.

Freshwater Bay is borderd by the basalt headlands to the west and the Elwha River delta to the east. The eastern half is a well-defined crenulate bay, also with a gravel beach. The Elwha River delta is illustrated by the vertical image in Figure 75. Note that it has the shape of a wave-dominated delta (outer shoreline shaped and molded by waves), although the number of beach ridges is limited.

There is a great story about the Elwha River, starting with the construction of two dams in the early 1900s that blocked migration of salmon upstream, trapped sediment and wood behind the dam that caused the river mouth to erode, and flooded the historic homelands and cultural sites of the Lower Elwha Klallam Tribe. After much study and collaboration among the stakeholders, funds were approved by Congress to remove the two dams to restore both ecological and cultural resources and use. This was the largest dam removal project in the U.S. to date. The Elwha Dam was removed September 2011-March 2012; the Glines Canyon Dam was removed in 2014. There are some great videos of the actual dam removal operations and of the river reworking the sediments that were trapped behind the dams into suitable salmon spawning habitat. And you can use the Google Earth historical imagery tool to see the pulse of sediment flowing into the delta and being reworked by waves. The Elwha River is now running free again. Very exciting stuff and quite the success story.

The heavily engineered Ediz Hook is 3.5 miles long and protects the port and town of Port Angeles. The beach on the northern side of the Hook has been stabilized with extensive riprap and, as a result, it has not changed much over recent time. Not too interesting in terms of coastal geomorphology, expecially compared to the wild and free Dungeness Spit.

But, before we get there, the shoreline between the two spits has high, steep bluffs in glacial deposits that are having some significant erosion issues for the houses built too close to the edge. There is a lot of active slumping in some locations.

The northwest limb of Dungeness Spit is attached

to the end of the bluffs. Figure 140 shows the eastern part of the spit, with its double recurved spits of sand and gravel. You can see that the inner spit was built earlier and that the waves from the east have reworked both spits, competing with the larger waves from the WNW. The far northeast end of Dungeness Spit, a geomorphological wonder if there ever was one, marks the eastern boundary of this Zone.

General Ecology

In addition to the species described for Zone 3 in the previous section, pigeon guillemots nest in burrows in the high bluffs of glacial sediment, and waterfowl and shorebirds are more abundant, especially around Dungeness Spit.

Harbor seals haul out on the rock ledges at Striped Peak and east of the Pysht River, on the sand bars at the mouths of the Lyre River and Twin Rivers, and on the log booms in Ediz Hook.

Nearly every stream has runs of salmon and steelhead. The Elwha River was once one of the most productive rivers in the region; construction of the dams restricted spawning to the lower 5 miles. Since removal of the dams, fish have access to over 70 miles of river and tributaries. Chinook salmon are returning, but in small numbers as of 2017, based on counts of spawning depressions called "redds" in the river and tributaries above the dams. coho and chum salmon, and steelhead are recovering faster; however, full recovery will take decades.

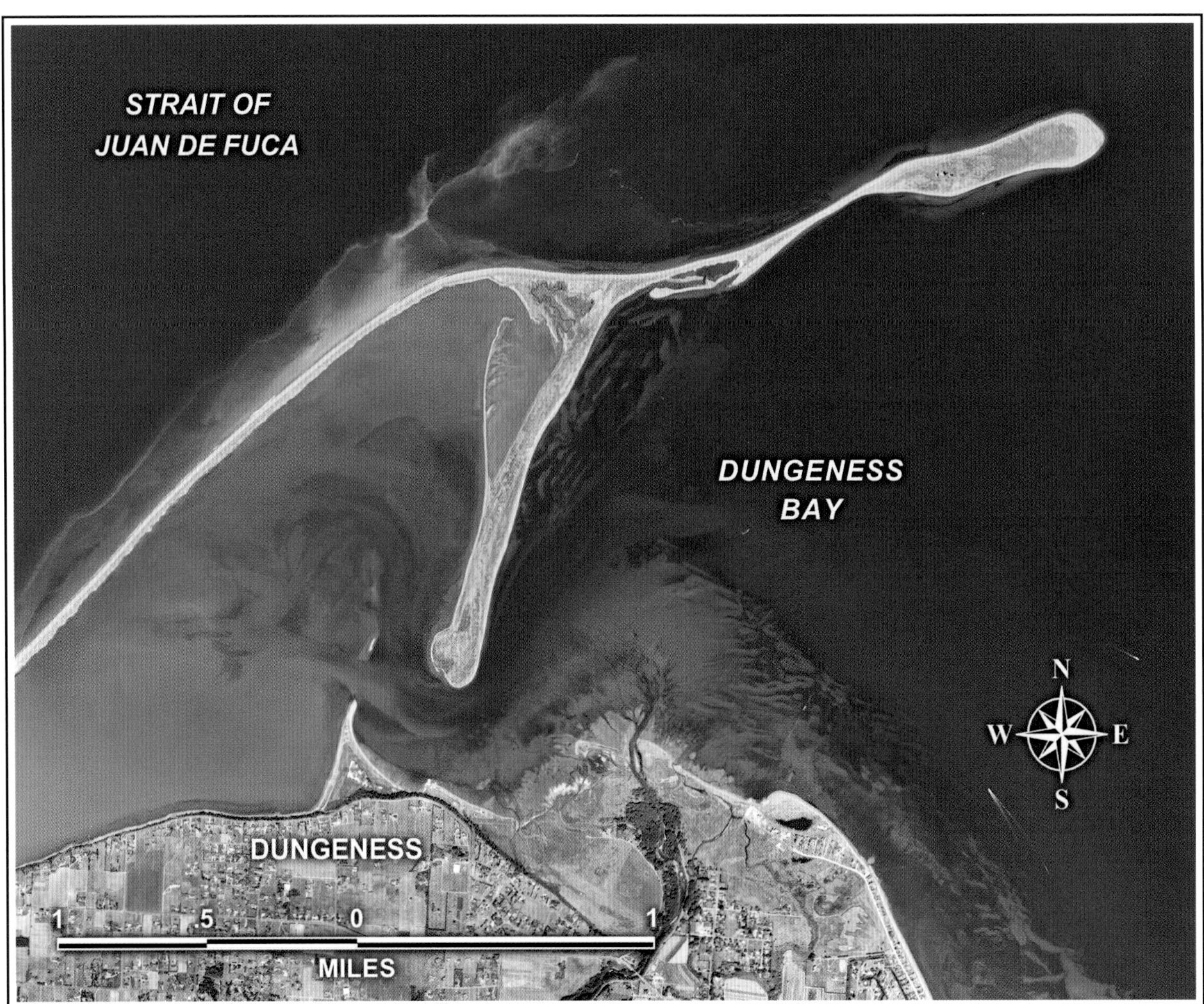

FIGURE 140. Dungeness Spit, WA. NAIP 2013 imagery, courtesy of Aerial Photography Field Office, USDA.

Places to Visit

There are two primary locations that we suggest you visit in this Zone – the Salt Creek Recreation Area and Dungeness National Wildlife Refuge. You can also consider going on a whale watching tour; there are charters out of Port Angeles that guarantee that you will see whales, and a lot more.

Salt Creek has sand beaches, great birding, and some exceptional tide pools on the rocky ledges in the basalt rocks at Tongue Point Marine Life Sanctuary. If you plan to explore the tide pools, remember to be there during low tide.

At the Dungeness National Wildlife Refuge you can see lots of different types of birds, depending on which side of the spit you look and when you visit. The tidal flats behind the spit can have large numbers of migrating shorebirds in spring and fall, and different species of waterfowl all year. Eelgrass beds attract large numbers of brant. Harbor seals haul out on the tidal flats. Along the front of the spit, you will see pigeon guillemots that nest in burrows in the bluff to the west of the spit, as well as lots of terns, gulls, and waterfowl. Rhinoceros auklets are common in summer and fall, and black oystercatchers are common most of the year. The species list for the refuge includes 244 birds.

RECOMMENDED ADDITIONAL READING

Collier, M. 1999. A Land in Motion: California's San Andreas Fault. Univ. Cal. Press, Berkeley. 118 pp.
Thorough discussion of the San Andreas Fault, a topic of special concern on Pacific Coast of North America. Excellent maps and photographs to go with a very interesting text.

Davis, R.A., Jr. and D.M. FitzGerald. 2004. Beaches and Coasts. Blackwell Publishing, Malden, MA. 419 pp.
The best modern textbook for giving the novice an introduction to the physical processes of the coastal zone.

Hyndman, D. and D. Hyndman. 2006. Natural Hazards and Disasters. Thomson Brooks/Cole, United States. 490 pp.
Superbly illustrated treatment of plate tectonics, earthquakes, volcanoes, and other natural hazards. Possibly the best place for a non-geologist to begin to learn about those processes.

Komar, P.D. 1976. Beach processes and sedimentation. Prentice-Hall, Inc., Englewood Cliffs, NJ. 429 pp.
Comprehensive coverage of all aspects of beach dynamic processes. Details are given on waves and sediment transport, as well as the beach cycle. This book was the first major work published on this topic.

Komar, P.D. 1998. The Pacific Northwest Coast: Living with the Shores of Oregon and Washington. Duke Univ. Press. 216 pp.
Anyone who owns property on this coast, or frequently visits it, should have a copy of this book. Details are given on the morphology and character of the entire outer coast, as well as a discussion on what hazards you might experience.

McConnaughey, B.H. and E.M. McConnaughey. 1985. Pacific Coast: The Audubon Society Nature Guide. Alfred Knopf, New York. 633 pp.
A useful field guide for identifying birds, marine mammals, fish, invertebrates and seaweeds of the Pacific Coast. Many excellent colored illustrations.

Ricketts, E.F and J.C. Calvin. 1939. Between Pacific Tides. Rickets, An earlier revision by Ricketts, E.F, J. Calvin, and J. W. Hedgepeth. Revised later by D.W. Phillips in1985. 5th/Revised edition, Stanford Univ. Press, Stanford, 652 pp.
Still the "Bible" for marine invertebrates on the U.S. West Coast. It has gone through numerous editions since its first publication with several revisions.

Schwartz, M.L., ed. 2005. Encyclopedia of Coastal Science. Springer, The Netherlands. 1211 pp.
The answer to any question you might have on this subject will probably be found in this phenomenal tome.

GLOSSARY

Amphibole – An important group of generally dark-colored, rock-forming inosilicate minerals, composed of double chain SiO4 tetrahedra, linked at the vertices and generally containing ions of iron and/or magnesium in their structures. Hornblende is a common example of this mineral group.

Amphidromic system – Separate, more-or-less circular, gyrating tidal systems (standing waves) with a nodal point in the middle. They are a large rotary tidal system created by the behavior of tides as giant low waves that are effected by the shape of the World Oceans and the rotation of the Earth. The height of the tide depends on its location within the system. Tides at the axis of the system (nodal point) are nil, but they increase outward with distance from the axis. In the Northern Hemisphere, the tides rotate counterclockwise about the axis; in the Southern Hemisphere, the rotation is clockwise. Both diurnal (one complete rotation of the tides daily) and semi-diurnal (two complete rotations of the tides daily) can coexist in the same ocean basin. The Pacific Ocean contains two diurnal systems and eight partial or complete semidiurnal systems. (Ed Clifton, pers. comm.). See examples in Figure 29.

Anadromous fish – Fish that live in salt water but migrate to freshwater habitats to spawn, such as salmon, steelhead (a unique form of rainbow trout), and Pacific lamprey.

Angular unconformity – A nearly horizontal, planar surface carved by the erosion of a series of tilted strata. This surface is later covered by layers of younger horizontal strata. Such an erosional surface indicates that a significant amount of geological time is missing from the rock record.

Anticline – The condition where layers of rocks are folded into a form similar to what you would get if you pushed one edge of a rug along the floor, folding it into parallel bands. Where the folded layers bend up, the fold is called an anticline.

Antidunes – Linear and parallel, trochoid-shaped features generated by rapid flow in the upper flow regime. In the process, a relatively thin layer of rapidly moving water flows as a standing wave across the top of the antidune surfaces. The standing waves in the water are in phase with the sediment surface. The form itself moves up current. But the individual sand grains move down current in a series of stops and starts. They have a typical spacing of about 15-20 inches on beaches, where they are commonly associated with flat beds. See Figure 56.

Asthenosphere – Part of the Earth's mantle below the lithosphere that behaves in a plastic manner. The rigid and more brittle lithosphere rides over it (Hyndman and Hyndman, 2006).

Astronomical tides – Periodic changes in water levels caused primarily by the gravitational attraction of the moon and sun on the World Ocean. See Figure 27.

Backwash of a wave – That portion of the water that rushes up the beachface under the impetus of a breaking wave, doesn't percolate into the beachface, and returns back down the slope of the beachface under the influence of gravity.

Barchan sand dunes – A wind-formed, crescent-shaped sand ridge, which has two "horns" that face downwind. A steep slip face is present on the convex side of the dune, which dips at the angle of repose of sand in air (30+ degrees). See Figure 72.

Barrier islands – Elongate, shore-parallel accumulations of unconsolidated sediment (usually sand), some parts of which are situated above the high-tide line (supratidal) most of the time, except during major storms. They are separated from the mainland by bays, lagoons, estuaries, or wetland complexes and are typically intersected by deep tidal channels called tidal inlets. They do not occur on coasts with a tidal range greater than 12 feet (macrotidal shores).

Barycenter – The center of mass of the combined Earth-Moon system. The Earth and the Moon revolve together around the barycenter, which is displaced a distance from the center of the Earth toward the moon, and is always located on the side of the Earth turned momentarily toward the Moon, along a line connecting the individual centers of mass of the Earth and Moon (Wood, 1982). See Figure 28.

Basalt – A dark-colored, very fine-grained igneous rock, the most common type of solidified lava. It is composed mostly of the minerals plagioclase feldspar, pyroxene, and olivine. It is a very prevalent component of the seafloor near mid-ocean ridges.

Batholith – A word derived from the Greek language (*bathos* = deep; *lithos* = rock) that is defined as a large emplacement of igneous intrusive (also called plutonic) rocks that form from cooled magma deep within the Earth's crust. They are almost always composed of feldspar-rich rocks such as granite, quartz monzonite, or diorite (Wikipedia Encyclopedia).

Beach cusps –The scalloped seaward margin of a beach berm consisting of a regularly spaced sequence of indented bays and protruding horns of a relatively small scale (rarely more than a few tens of feet between horns). They are thought to be primarily depositional features (i.e., on the seaward flank of a prograding depositional berm) that form as a result of the interaction of shore-normal standing waves (edge waves) with shore-parallel, reflective waves, the same general process that produces rip currents (but at a smaller scale) (Guza and Inman, 1975). See Figure 37.

Beach drift – The transport of sediment along the beachface by the combined processes of uprush and backwash. The uprush from a wave that approaches the beachface obliquely will move sand grains (or pebbles) at an angle across the beachface. Once the approaching uprush loses its momentum, the sand grains (or pebbles) are transported in a perpendicular direction back down the beachface by the backwash under the influence of gravity. The ultimate result is that the sediment particles move along the beachface in a sawtooth pattern away from the direction of wave approach. See Figure 36 (lower left).

Beach nourishment – Artificially adding sand to a beach to counter beach erosion. The sand can come from a variety of sources, with offshore sand bars or ridges being a common one.

Beachface – The zone of wave uprush and backwash during mid- to high-tide. It usually slopes seaward at an angle of 5-10 degrees and is commonly the seaward face of the berm. See Figure 51.

Beachrock – Formed by rapid solidification of beach sand through the precipitation of calcium carbonate from sea water into the pore spaces between the sand grains, a process more likely to occur where the water is warm.

Berm on a sand beach – A wedge-shaped sediment mass built up along the shore by depositional wave action in between storms. It typically has a relatively steep seaward face (5-10 degrees; the beachface) and a gently-sloping landward surface (2-3 degrees; the berm top). See Figure 51.

Boulder – A sediment particle that is greater than 256 mm (~ 10 inches) in size. See Figure 43C.

Brackish marsh – A wetland containing grassy vegetation that is regularly flooded by brackish waters during mid to high tides. Salinities average less than 15 parts of "salts" per thousand parts of water.

Breakwaters – See offshore breakwaters in this glossary.

Chert – A fine-grained, smooth-textured rock composed of SiO_2 and formed as a sedimentary deposit.

Chiton – A type of marine mollusc with only one shell that lives attached to rocks. Its shell is made up of several plates like a suit of armor.

Closure point – A term defined by coastal engineers as the most seaward distance for sand to move into the nearshore zone (off the intertidal beach in an offshore direction) during storms. In effect, it is the seaward limit for sand to move during the normal beach cycle that takes place in response to the passage of storms and intervening periods of relative quiescence. Many coastal geologists and some engineers dispute the validity of this concept. See Figure 50.

Cobble – A sediment particle that is between 64-256 mm (2.5-10 inches) in size.

Collision coasts – Coasts defined by Inman and Nordstrom (1971) as those that occur along the leading edges of continental plates, such as the ones along much of the west coasts of North and South America. They usually are mountainous and commonly contain rocky shores with short, steep rivers emptying into the sea.

Competency of flow – The ability of the moving water

to transport sediment.

Concretions – Local concentrations of chemical compounds, such as calcium carbonate or iron oxide, within a sedimentary rock layer that usually take on the form of a spherical nodule.

Conglomerates – Sedimentary rocks made up mostly of gravel-sized particles or clasts.

Continental drift – A theory proposed by Alfred Wegener (1912) that states that the present continents on the Earth's surface were joined together and later moved apart, based to some extent on the almost perfect fit of the east coast of South America with the west coast of Africa, plus other comprehensive evidence. Lacking in his theory was a provable mechanism to move the continents apart. See Figure 14B.

Continental shelf – The gently sloping seabed between the high-tide line and the shelf break, where there is an abrupt increase in slope, at depths measured in hundreds of feet. See Figure 83.

Continental slope – A steeply descending planar surface that adjoins the outer margin of the continental shelf. See Figure 83.

Core of the Earth – The central, interior portion of the Earth that is composed mostly of iron and makes up approximately 30% of the Earth's volume. It consists of two layers, an outer extremely hot zone, which is molten, and in inner layer, also very hot, that is solid because of the intense pressure at those depths within the Earth. See Figure 11.

Coriolis effect – An effect whereby a body of air, water, etc. moving in a rotating frame of reference experiences a force that acts perpendicular to the direction of motion and to the axis of rotation. On the rotating Earth, this effect deflects moving bodies to the right in the northern hemisphere and to the left in the southern hemisphere. It was named after a French engineer G. G. Coriolis (1792-1843).

Crenulate bay – An embayment along the shore that is usually separated by two rocky headlands. They typically have a log-spiral, or fishhook, shape, with the shank of the hook pointing in the direction of sediment transport. See Figures 63 and 64.

Crust of the Earth – The solid upper layer of the Earth composed of a variety of igneous, metamorphic, and sedimentary rocks. The crust under the oceans, which averages about 6 miles thick, is composed mostly of rocks that contain an abundance of iron and magnesium-bearing minerals (e.g., basalt), whereas the crust on the continents, which averages 20-30 miles thick, is composed of rocks made up of lighter minerals (e.g., granite) (Wikipedia encyclopedia).

Debris flows – Rapid movement of a partially consolidated mass of sediment and broken up rock fragments, some up to boulders in size, down a steep slope. Debris-flow deposits are commonly associated with turbidites, which are deposited by more fluid plumes, or avalanches, of sediment-laden water.

Depositional coasts – Type of coast defined by Hayes (1965) as one characterized by: 1) Coastal zone made up chiefly of broad coastal and deltaic plains; 2) Bedrock composed generally of Tertiary and Quaternary sediments (young geologically speaking); 3) Tectonically subsiding areas; and 4) Many large and long rivers emptying into the sea. They most commonly occur on the trailing edges of continental plates or along the shores of epicontinental seas, such as the Black and Caspian Seas.

Diatoms – One-celled marine algae that are roughly circular in shape and secrete intricate skeletons of silica.

Dike – A relatively narrow rock band (or zone) of distinctly different composition (usually) that has been intruded into and cuts across an older rock unit.

Diorite – A rock similar in texture to granite that is made up largely of white to light gray plagioclase feldspar (ranging from $NaAlSi_3O_8$ to $CaAl_2Si_2O_8$) and a certain amount of hornblende (a range of compositions including $CaNa(Mg,Fe)_4(Al,Fe,Ti)_3Si_6O_{22}(OH,F)_2$) with <5% quartz.

Diurnal inequality of the tides – A condition where, although the tides are roughly semi-diurnal, four significantly different levels (rather than two) occur during one tidal day (referred to as higher-high, lower-high, higher-low and lower-low tides). One possible cause of this is the tilt of the Earth's axis, because the depth of the water under the tidal bulge will be different on opposite sides of the Earth when the moon is not directly over the equator (Figure 28C). Another probable cause is a combination of tides from separate diurnal and semidiurnal amphidromic systems, which appears to the case on the California Coast (Ed Clifton, pers. comm.). See Figure 30.

Downdrift – The direction in which the individual

sediment grains are moving along the shore within the sediment transport system.

Ebb-tidal currents – Currents generated during the falling tide.

Ebb-tidal delta – A lobe of sediment, usually composed of sand, much of which has been transported from the scoured-out throat of a tidal inlet by ebb-tidal currents. The sand is deposited on the seaward side of the inlet throat as the ebb currents slow down upon entering the open ocean. The morphological components of ebb-tidal deltas include a main ebb channel (with slightly stronger ebb currents than flood currents) flanked by linear sand bars on both sides and a terminal lobe at the seaward end. See Figure 66.

Echinoderms – Radially symmetrical marine invertebrates of the phylum Echinodermata that have an internal calcareous skeleton and are often covered with spines. Examples include sea stars, sea urchins, and sea cucumbers.

Eelgrass – A common name for a group of plants that grow under water in estuaries and in shallow coastal areas. It is neither a grass nor a seaweed. It is an angiosperm, or flowering plant, that can live for many years.

***El Niño* event** – A globally coupled, ocean-atmosphere phenomena that results in: 1) Relatively low pressure over the Pacific Ocean; 2) A weakening of the trade winds (during normal conditions, these strong winds generate a very strong westerly flow of the surface waters to the west in the equatorial region) that contributes to an increase in water level on the California, Oregon, and Washington Coasts (as much as 8-12 inches); and 3) The thermal expansion of the warmer water because of a decrease in upwelling along the coast. During these events, unusually heavy rains typically accompany an increased number of winter storms. These facts, along with increases in water levels, leads to significant coastal inundation and sea cliff erosion during coastal storms (Inman and Jenkins, 2005b).

Ensoulment – A concept that suggests that humans may participate with the Earth as if it were a living being (Cajete, 1995).

Entrainment – The selective transport of sediment particles by water or air.

Environmental Sensitivity Index (ESI) – A system for mapping coastal environments that ranks coastal habitats on a scale of 1-10 based upon their sensitivity to oil-spill impacts. Exposed rocky shores are ranked 1 (least sensitive) and sheltered salt marshes are ranked 10 (most sensitive). ESI maps, which are used for oil-spill response and other types of environmental assessments on a world-wide basis, also include detailed biological and human-use information.

Epicenter of an earthquake – The point on the surface of the Earth that lies directly above the hypocenter, the place at depth where the fracture that causes the earthquake is located.

Estuary – A coastal water body that, according to Pritchard (1967), has three defining qualities: 1) A flooded river valley that was formed during the lowstand of sea level that terminated about 12,000 years ago; 2) A water body with a substantial freshwater influx; and 3) A water body subject to tidal fluctuations. See Figures 77 and 78.

Evaporates – A class of sedimentary rocks usually, but not always, formed as the result of evaporation of sea water, typically under high surface temperatures. Evaporite deposits are most commonly composed of halite (NaCl) and gypsum [$CaSO_4.(2H_2O)$].

Fault – A fracture along which there has been slipping of the contiguous rock or sediment masses against one another. Points formerly together have been dislocated or displaced along the fracture (Lahee, 1952). See Figure 13.

Fetch – The length of water surface over which the wind blows to form waves. See Figure 32B.

Flocculation – A process that takes place in estuaries when colliding clay particles tend to adhere together under the influence of the increased concentration of cations, such as Na+, Ca++ and Mg++, derived from the sea water. As a result, the clay particles, which may have diameters as little as one micron, cluster into groups (flocs) that may have diameters measured in hundreds of microns. As a result, many of the flocs sink to the bottom at slack tide. See Figure 78B.

Floodplain of a river – Many river channels run through valleys of different widths that were carved during the last lowstand of sea level during the last major glaciation, which terminated about 12,000 years ago. During floods, sediment rich (muddy) water spills out of the channel and builds up a flat plain between the channel and the sides of the valley. These flat-lying deposits occupy a zone known as the floodplain.

Flood-tidal currents – Currents generated by the rising tide.

Flood-tidal delta – A lobe of sediment, usually composed of sand, most of which has been transported from the scoured-out throat of a tidal inlet by flood-tidal currents. The sand is deposited on the landward side of the inlet throat as the flood currents slow down upon entering the more open bay on the landward side of the inlet. See Figure 66.

Flow regime – A concept worked out by hydraulic engineers to distinguish among features that develop in sediment (most appropriate for sand) along the bed of a stream or beach under the influence of flowing water. When the water is flowing slowly (or is relatively deep), the bed changes from small ripples to larger megaripples and sand waves under increasing flow strength. This first part of the sequence, during which the value of the Froude number (the ratio of water velocity to the square root of the gravitational constant times water depth) is <1.0, is called the lower flow regime. As the Froude number increases to values >1.0, by either increasing the velocity of flow or decreasing the water depth, the sand bed goes into the upper flow regime, where either flat beds or antidunes are formed.

Forearc basin – A depression in the seafloor located between a subduction zone and an associated volcanic arc.

Foredunes – Wind-blown sand dunes that form immediately adjacent to the backbeach. They are most commonly vegetated.

Garnet – A member of a complex suite of silicate minerals that are commonly reddish in color, one example being glossular garnet [$Ca_3Al_2(SiO_4)_3$]. Common constituent of metamorphic rocks, being present in both schists and gneisses.

Gastropod – A class of mollusks typically having a one-piece coiled shell and flattened muscular foot with a head bearing stalked eyes.

Geomorphology – A scientific discipline devoted to understanding the origin and three-dimensional shape of the landforms of the Earth.

Glacial rebound coasts – Coastal areas that continue to rise even to this day as a result of the removal of the vast ice sheets associated with the last Pleistocene glaciation.

Gondwana – A super continent that existed between 300-400 million years ago. It consisted of the present areas of South America, Africa, India, Australia, and Antarctica. It was separated from another supercontinent, Laurasia, by the Rheic Ocean. Figure 14A.

Graded bed – A sediment layer that is coarse-grained at the bottom and fine-grained at the top. A common mechanism for them to form is to be dropped out of suspension from the sediment-laden plume of a turbidity current. See Figure 84.

Granite – A light colored, coarsely crystalline igneous rock that is composed of 20-60% quartz, with most of the remaining component being between 10 and 65% of a mixture of alkali feldspar and plagioclase. It may also contain a few dark-colored minerals, such as hornblende and biotite.

Granodiorite – A light colored, coarsely crystalline igneous rock that is composed of 20-60% quartz, with the remaining mostly feldspar component being between 65 and 90% plagioclase. It may also contain a few dark-colored minerals, such as hornblende and biotite.

Granule – A sediment particle that is 2-4 mm in size. See Figure 6.

Gravel – Sediment with a median diameter greater than 2 mm. Gravel is defined according to the Wentworth (1922) scale, which includes four classes – *granules* (median diameter = 2-4 mm), *pebbles* (median diameter = 4-64 mm), *cobbles* (median diameter = 64-256 mm), and *boulders* (median diameter >256 mm). See Figure 6.

Graywacke – A type of sandstone that is usually dark-colored, made that way by either containing a lot of clay matrix between the sand grains or having abundant sand particles composed of dark-colored rock fragments from a variety of rock sources, such as metamorphic slates or schists. A common component of turbidite deposits.

Groin – A linear, relatively short, shore-protection structure built perpendicular to the shore for the purpose of widening the beach by trapping a portion of the sand in the littoral drift. Commonly composed of riprap, but other materials, such as sand bags and wooden timbers, are sometimes used. See Figure 85B.

Half-life – The amount of time it takes half of a radioactive parent element to decay to a stable daughter element. For example, it takes 1.25 billion years for half of Potassium 40 (the radioactive parent) to decay to Argon 40 (its stable daughter).

Halophytes – Plants that are adapted to grow in salty

soil.

Heavy minerals – Minerals with specific gravities significantly greater than quartz, which has a specific gravity of 2.65 (the specific gravity of water is 1.00). Examples include magnetite, garnet, and hornblende. See Figure 57.

Highstand – The period when sea level was high for a significant amount of time, measured in thousands of years, during intervals between the four major glaciations of the Pleistocene epoch, called interglacials. The last highstand before the present one occurred around 125,000 years ago, and it is considered to have been considerably higher that the one at present (several feet in areas with no tectonic uplift).

Holocene time – The segment of geological time that has occurred since the end of the last major glaciation, starting about 12,000 years ago.

Hornblende – A dark-colored silicate mineral of the ferromagnesium class that is a common constituent of both igneous and metamorphic rocks with a range in compositions including $Ca_2(Mg,Fe,Al)_5(Al,Si)_8O_{22}(OH)_2$ and $CaNa(Mg,Fe)_4(Al,Fe,Ti)_3Si_6O_{22}(OH,F)_2$.

Hydraulic equivalence – A size-density relationship that governs the deposition of mineral particles from flowing (or standing) water. Two particles of different sizes and densities are said to be hydraulically equivalent if they are deposited at the same time under a given set of conditions (e.g., if dropped into a cylinder of water, the different particles would settle to the bottom at the same speed despite their differences in size).

Hydrodynamic regime (*hydro* = water; *dynamics* = kinetic energy of the water) – It is commonly expressed as the ratio of wave energy to tidal energy (i.e., how big the waves are versus how large the tidal range is). It exerts a major control on the geomorphological makeup of depositional coasts.

Hypocenter – The focus, or location of the fracture, where the movement starts during an earthquake. Usually at some depth below the Earth's surface.

Igneous rocks – Rocks that crystallized from an extremely hot molten mass of material (magma). One class of igneous rocks, called granitic rocks, develops by slow cooling of the molten mass at great depths within the Earth's crust. As a result of the slow cooling, large crystals of individual minerals form that can be as much as an inch or so in diameter. A second class of igneous rocks, called volcanic rocks, crystallize rapidly as a result of molten lava being extruded suddenly onto the Earth's surface. The crystals of these minerals that form so rapidly are, for the most part, too small to be seen by the naked eye.

Imbrication of gravel clasts – A process whereby the maximum projection areas of the clasts face (dip toward) the strongest force acting upon them, usually at angles of 20-25 degrees. It is best developed by platy or disc-shaped clasts. See Figure 58.

Invertebrates – Animals that do not have an internal skeleton made of bone, such as anemones, sea stars, jellyfish, clams, snails, crabs, shrimp, and worms.

Island arc – An arc-shaped zone of volcanic islands built above a subduction zone.

Jetties – Long parallel structures built at a river mouth, a tidal inlet, or even an artificially dug harbor, to stabilize the channel, prevent its shoaling by sediments, and protect its entrance from waves. They are commonly composed of large pieces of rock called riprap. See Figure 70.

Joints in the rocks – Planar surfaces along which the rocks have split apart, usually during one or more of the ever present tectonic movements along this coast.

Kelp – A marine algae that consists of three components: a holdfast that attaches to the rocky seafloor; a stipe that is similar to a plant stalk; and fronds that attach to the stipe. Kelp forests require wave energy and currents to bring in nutrients, clear water to allow photosynthesis, and cool water temperatures. Kelp forests also require an appropriate water depth (typically shallower than 60-80 feet unless the water is unusually clear) and a hard substrate for attachment.

Krill – Shrimp-like planktonic crustaceans, a major food source for baleen whale species.

***La Niña* event** – A globally coupled, poorly understood ocean-atmosphere phenomena that generally shows effects opposite to those of the *El Niño*. For example, a decrease in water levels along the Oregon and Washington Coast is possible. Rainfall is greatly diminished, and the trade winds are usually strengthened, resulting in the intensification of upwelling along the coast. A high-pressure anomaly over the eastern North Pacific that hugs fairly close to the North American coast typically accompanies these events. This forces storm tracks to come ashore further north than those that occur during *El Niño* events (Inman and Jenkins, 2005b).

Landslide – A mechanism whereby loosened soil and rock materials on a hill/slope are moved from their original site of formation to lower lying areas under the influence of gravity. Earthquakes and unusually high concentrations of water in the soil are two of the primary causes of landslides. They can occur both above or below the sea. See Figure 87.

Laurasia – A supercontinent that existed between 300-400 million years ago. It consisted of the present areas of North America, Greenland, Europe, and part of Asia. It was separated from another supercontinent, Gondwana, by the Rheic Ocean. See Figure 14A.

Limestones – Sedimentary rocks composed of calcium carbonate (CaCO3), which usually develop in marine waters by a combination of chemical precipitation and accumulation of the hard parts of marine organisms, such as sea shells.

Limpet – A variety of mollusk, which have a characteristic conical shell and a suckerlike foot. Limpets (like chitons) are well adapted for life on rocky surfaces exposed to wave action and in the higher levels of the intertidal zone.

Lithosphere – The rigid outer rind of the Earth approximately 40-60 miles thick. It composes the tectonic plates (Hyndman and Hyndman, 2006).

Littoral cell – A segment of the shore with updrift and downdrift boundaries that do not allow sediment in the longshore sediment transport system to pass into or out of the segment in a longshore direction. On the Pacific Northwest Coast, these boundaries are rock headlands. Sediment is added to the littoral cells from rivers and bluff erosion, and sediment is lost from the cells primarily down submarine canyons and into onshore dune fields. See Figure 61.

Littoral drift – Volume of sand moving along the shore in the sediment transport system.

Littoral current – Current formed when an obliquely approaching wave piles up water in the surf zone that flows away from the direction of the approaching wave. The resulting current runs parallel to the shore.

Longshore sediment transport – The movement of sediment along the beachface and within the nearshore zone when waves approach the beach at an oblique angle. This movement of sediment is usually produced by the combined processes of beach drift and longshore currents. See Figure 36 (Lower Left).

Lower flow regime – See definition of flow regime.

Lowstand – Term that refers to the period when sea level was low for a significant amount of time, measured in thousands of years, during the four major glaciations of the Pleistocene epoch. For example, the lowstand during the last glaciation, the Wisconsin, which occurred around 20,000 years ago, was about 350-400 feet lower than it is today.

Low-tide terrace – A relatively flat surface that is located seaward of the toe of the beachface and slopes offshore at a small angle (2-3 degrees). The landward margin is usually near mean sea level and the seaward margin near mean low water.

Macrotidal – Term originally used by Hayes (1976) to define types of coasts from the perspective of their coastal geomorphology. Davies (1973) defined macrotidal coasts as those with a tidal range (vertical distance between high and low tide) of approximately 12 feet, whereas Hayes (1979) proposed the same >12 feet boundary. Davies did not relate tidal range to coastal geomorphology in the fashion Hayes did.

Magma – A molten mass of material that usually forms deep within the Earth's crust.

Magnitude-Moment (M) scale – A logarithmic scale (base 10) used to determine the amount of energy released by an earthquake. It is based on several factors, including: a) How far the fault slipped (i.e., how far a fence line was offset); b) Length of the fault line that was affected; c) Depth at which movement was initiated; and d) The "stickiness" of the surfaces being faulted (rock rigidity). This is a more complex way to measure earthquake magnitude than the Richter scale, which is based on the amplitude of shear waves as measured by a seismograph. The Loma Prieta earthquake on 17 October 1989, which resulted in severe damage in San Francisco, measured 6.9 on the magnitude-moment scale and 7.1 on the Richter scale.

Mantle – The thick layer of material below the relatively thin Earth's crust. It contains more of the heavier elements, such as iron and magnesium, than the crust, is about 1,800 miles thick, and makes up approximately 70% of the Earth's total volume. It is composed mostly of the igneous rock peridotite. See Figure 11.

Mass wasting – A mechanism whereby weathered

materials are moved from their original site of formation to lower lying areas under the influence of gravity.

Meandering channel – Channel with a highly winding path, like the loops one throws into a rope. Such channels tend to have: 1) relatively flat slopes; 2) a high ratio of suspended to bedload sediments; and 3) a steady discharge. Meandering tidal channels are by far the most common channel type in the coastal water bodies of Oregon and Washington. See Figure 82.

Megaripples in sand – Asymmetric, linear, or cuspate features moved along a sandy sediment surface by flowing water with a spacing between their crests of 2-18 feet. They are seen more commonly in tidal channels, where the water is deeper and the currents are stronger. They can also be produced by wave surge beneath or just seaward from the breaker zone. See Figure 52.

Melange – A prevalent rock assemblage that formed under intense tectonic forces at great depths within a trench. It consists of a soft crushed shale or sepentinite matrix with blocks of other rocks "floating" in it (Sloan, 2006). These foreign blocks vary in composition and size.

Mesotidal – Term originally used by Hayes (1976) to define types of coasts from the perspective of their coastal geomorphology. Davies (1973) defined mesotidal coasts as those with a tidal range (vertical distance between high and low tide) of 2 6-12 feet, whereas Hayes (1979) proposed 3-12 feet, with low-mesotidal being 3-6 feet. Davies did not relate tidal range to coastal geomorphology in the fashion Hayes did.

Metamorphic rocks – Rocks that result from dramatic changes in igneous and sedimentary rocks affected by heat, pressure, and water that usually creates a more compact and more crystalline condition. These changes usually take place at significant depths below the Earth's surface.

Microtidal – Term originally used by Hayes (1976) to define types of coasts with respect to their coastal geomorphology. Davies (1973) defined microtidal coasts as those with a tidal range (vertical distance between high and low tide) of 6 feet, whereas Hayes (1979) proposed the boundary of <3 feet. Davies did not relate tidal range to coastal geomorphology in the fashion Hayes did.

Mid-ocean ridge – A long ridge with a deep rift along its crest that usually runs the length of, as well as along the middle of, one of the major oceans (with the Pacific Ocean being an exception; Alt and Hyndman, 2000). These ridges mark the lines where tectonic plates pull away from each other. As new lava flows spill out onto the ocean floor along fissures in these ridges, new oceanic crust is created. See Figure 4.

Mixed-energy coasts – Term used by Hayes (1976) to define a specific type of geomorphic assemblage on depositional coasts. Most mixed-energy depositional coasts are characterized by short, drumstick-shaped barrier islands with complex tidal flats and coastal wetlands on their landward flanks. They usually have tides in the mesotidal range (3-12 feet). See Figure 5B.

Mollusc – An invertebrate having a soft unsegmented body usually enclosed in a shell.

Mud – Sediment with a median diameter finer than 0.0625 millimeters (mm). It includes two classes – silt (median diameter 0.00391-0.0625 mm) and clay (median diameter <0.00391 mm. Silt rubbed between your teeth feels gritty, whereas clay feels smooth, a practice recommended only for pristine areas.

Mudflats – Tidal flats composed primarily of muddy sediments. They occur mostly in the more sheltered (from wave action) parts of coastal water bodies.

Mudstone – Sedimentary rock made up mostly of silt and/or clay that has a massive character.

Natural groin – A natural, usually rocky protrusion that sticks out into the littoral sediment transport system that can serve the same function as a man-built groin, that is, trap sand and form a relatively permanent sand beach. See Figure 138.

Neap tides – The minimum tides of the month that exist when the gravitational attraction of the Moon and Sun on the water bodies of the World Ocean are working in opposition, during the first and third quarter of the Moon (condition called quadrative). The maximum tides of the month (spring tides) occur during full and new moons. See Figure 27.

Nonconformity – An unconformable erosional surface along which sedimentary rocks, or semiconsolidated sediments, overlie igneous or metamorphic rocks that are massive in character (i.e., not bedded). This surface represents a loss of a significant amount of time from the rock record.

Normal, or gravity, fault – The type of fault that occurs when a rock or sediment mass located on the upper side of a sloping planar surface (a fracture in the rocks) slides down the fault relative to the lower side under the influence of gravity (and/or tension). See Figure 13 (Top).

Offshore breakwaters – Offshore segments of riprap or heavy concrete blocks of miscellaneous shapes several tens of feet long that are detached from and usually oriented parallel to the beach (but not always). They reduce wave energy on their landward sides, which causes sand moving along shore in the longshore sediment transport system to accumulate in their lee.

Pangaea – A single, large mass consisting of all the land area on the Earth's surface that formed about 300 million years ago, during the Carboniferous Period of the Paleozoic Era. It was created when two supercontinents, Gondwana and Laurasia, collided and welded together. See Figure 14A.

Parabolic sand dunes – A wind-formed, scoop-shaped sand ridge with the plan-view shape of a parabola. Two linear arms that trail along side of the "scoop" point upwind. See Figures 72.

Pebble – A sediment particle that is 4-64 mm in size. See Figure 6.

Pegmatite dike – A tabular body of extremely coarse, crystalline granitic rocks that cuts across adjacent rocks.

Peridotite – A coarsely crystalline, dark-colored igneous rock that consists mainly of olivine and other ferromagnesian minerals. It is the principal rock type in the mantle.

Period of a wave – The amount of time it takes two succeeding wave crests to pass a single point.

Permafrost – Ground that remains frozen for at least two consecutive years.

Permeability of sediments – A measure of the ease with which a fluid flows through open pore spaces (e.g., between sand grains or gravel clasts).

Pillow basalt – A rock type that typically forms when basaltic lava flows onto the seafloor in deep water, as is common along the mid-ocean ridges. The cold water in those depths quenches the surface of the flow and the lava commonly takes on a pillow shape.

Pinnipeds – A group of carnivorous marine mammals with fin-like limbs. Some common species occurring on the Oregon and Washington Coasts include California sea lion, harbor seal, and northern elephant seal.

Placer deposits – Layers of the heavier grains of sediment (mostly sand; typically called heavy minerals) that were deposited primarily as the result of the selective sorting out and transport to elsewhere of most of the coarser-grained, light-weight minerals in the original sediment source. This may take place by a variety of processes, with increased pivotability and transportability of the larger, lighter sand grains (and granules in some cases) in the shallow, sheet-flow of the wave swash being the primary process creating placers on beachface or berm top surfaces. Accumulation of individual grains of gold within sand bars in a stream bed is a another type of placer deposits.

Plate tectonics – The theory named by J. Tuzo Wilson, which states that individual blocks of the lithosphere, called plates, compose the outer portion of the Earth's surface. The individual plates either move apart (as along mid-ocean ridges), into each other (as in subduction zones), or parallel to each other (as along the San Andreas Fault). The plates move slowly (2-4 inches per year) above (riding on) the portion of the Earth's mantle called the asthenosphere, which behaves in a plastic manner.

Plateau-shield coast – Coastal type defined by Hayes (1965) as one characterized by plateaus and relatively narrow continental shelves. The bedrock along the shore is typically composed of ancient basement complexes of granite and gneiss (in shield areas) and Paleozoic and Mesozoic sedimentary rocks (in plateau areas). They are tectonically stable and some moderately long rivers may be present, depending upon the climate.

Plunging waves – Waves that take on a cylindrical shape (Hawaii Five-O type) when they break, falling abruptly down (collapsing) with considerable force. See Figure 34A.

Pocket beach – Usually a small beach, between two headlands. In an idealized setting, there is very little or no exchange of sediment between the pocket beach and the adjacent shore.

Porosity of sediments – The ratio of the open pore spaces between the individual sediment grains or gravel clasts to the space taken up by the sand and gravel clasts themselves (when they are stacked together).

Porphyritic igneous rock – An igneous rock that has exceptionally large, separate, and distinct individual crystals imbedded within a finer-grained matrix. The larger crystals were formed first, during which time they "floated around" within the molten magma as they slowly grew larger. The matrix, which is finer-grained, crystallized more abruptly when a cooler temperature of the molten magma was reached that promoted the more rapid crystallization of the matrix minerals.

Precocial chicks – The chicks of shorebirds, such as the snowy plover, that are hatched with their eyes open, are covered with down, and feed themselves immediately after hatching.

Pyroxene – A silicate mineral that occurs in a variety of igneous and metamorphic rock types and is an important constituent of the upper mantle. Pyroxenes have the general formula $XY(Si,Al)_2O_6$ (where X represents calcium, sodium, iron^{+2} and magnesium and more rarely zinc, manganese and lithium, and Y represents ions of smaller size, such as chromium and aluminum).

Quadrative – The times when the gravitational forces of the Moon and Sun work in opposition to each other with regard to generating tides in the World Ocean (first and third quarter of the Moon). This condition produces the smallest tides of the month (neap tides). See Figure 27.

Radioactive decay – A process whereby naturally occurring radioactive materials break down into other radioactive materials at fixed rates, or half lives.

Radiolaria – Microscopic-sized, unicellular marine protozoans. Protozoans are a large group of single-celled eukaryotic organisms (i.e., their cells have a nucleus that contains its chromosomes). Radiolaria have silica skeletons that accumulate on the seafloor in phenomenal numbers to form a "radiolarian ooze."

Raised marine terrace – A flat, or slightly seaward dipping, wave-cut rock terrace usually at least a few hundred yards wide that is flanked on its landward side by a somewhat degraded former wave-cut rock cliff and on its seaward side by a high, more recently active wave-cut rock cliff. The youngest ones originally formed during the last highstand of sea level between 80-125,000 years ago; then it has been raised well above the present high-tide level by continuing tectonic uplift. There may be a series of them up the sides of the mountains that face the shore, like a giant's staircase. Where such multiple terraces exist, the highest ones are the oldest and they get progressively younger as you move down toward the sea. See Figures 48 and 49.

Revetment – A structure built to protect a shoreline from erosion. It can be composed of a variety of materials including concrete, riprap, wood, and sand bags.

Rhythmic topography – Bulges of sand on the beach that move down the shore like the sinusoidal loops one might throw along a rope. They are created by waves that approach the beach at an oblique angle and are typically spaced 10s to 100s of feet apart. See Figure 36 (Upper Right).

Rhyolite – A fine-grained, extrusive volcanic rock, similar to granite in composition.

Richter scale – A scale that reflects the magnitude of earthquakes based on the calculation of the maximum amplitudes of the shear waves as recorded on a seismograph. The scale is logarithmic (base 10). For example, a magnitude (M) 5 earthquake would result in ten times the level of ground shaking as an M 4 earthquake (and 32 times as much energy would be released). Quakes with Richter scale values of 3.5 to 5.4 are often felt, but rarely cause damage. Those above 7.0 on the scale are classified as major earthquakes and can cause widespread damage.

Right-lateral fault – A type of strike-slip fault, of which the San Andreas Fault is a prime example. If you are standing on one side of the fault line looking across it, the *terra firma* on the other side of the fault line has moved to the right during an earthquake. The same thing is true if you step across the fault line and look back at the side you were standing on (i.e., that side also has moved to the right in a relative sense).

Rip currents – Wave-generated currents that flow straight off the beach and tend to be very regularly spaced. They form most commonly when the waves come straight on shore (not at an angle usually, but there are some exceptions to this general rule). Also, there is typically some degree of reflection of the waves off the beachface when they are forming. Most rip currents owe their formation to the occurrence of edge waves – standing waves with their crests normal to the beach, according to Bowen and Inman (1969) and Komar (1976). See Figure 38.

Ripples – Asymmetric, linear or cuspate features mostly composed of sand that are moved along the sediment surface by flowing water (or wind) with a spacing between their crests of less than 2 feet. On the beach, they usually occur in the intertidal troughs on the landward side of intertidal bars, where the water depths are greater than elsewhere on the beach. See Figures 52A and 55.

Riprap – Large pieces of rock used in shore-protection structures such as groins, revetments, and jetties.

River delta – A protruding, delta-shaped (triangular) mass of sediment that builds into a standing body of water off a river mouth.

Salt marsh – A wetland containing halophyte vegetation that is regularly flooded by marine waters during mid to high tides. Salinities range from 15-36 parts of "salts" per thousand parts of water.

Saltation – A mode of sand transport by the wind, whereby the sand grains bounce off the sand bed and "fly" for some distance at heights usually not more than a couple of feet. See Figure 71.

San Andreas Fault – A geological transform fault that runs a length of roughly 800 miles through Western California. Its motion is right lateral strike-slip (horizontal motion). It forms the tectonic boundary between the Pacific Plate and the North American Plate.

Sand – Sediment with a median diameter between 0.0625 and 2 mm. See Figure 6.

Sand bypassing system – Mechanical systems, such as land-based dredging plants and pumping systems, that move the sand from the updrift to the downdrift sides of jetties or man-made harbors that block the natural transport of sand along the shore.

Sand waves – Asymmetric, linear features (ripple shaped) moved along a sandy sediment surface by flowing water with a spacing between their crests of >18 feet. In coastal areas, they are mostly found in deep tidal channels or tidal inlets.

Sandstones – Sedimentary rocks made up mostly of sand-sized particles of minerals, most notably quartz and rock fragments.

Sea arch – A bedrock arch usually located offshore of a rocky coast in relatively shallow water. Many form in the situation where a narrow peninsula composed of layered rocks with a more resistant top layer projects offshore. Waves refracting around the headland focus the wave erosion such that two caves eroded into the opposite flanks of the peninsula meet, forming the arch. See Figures 47 and 94.

Sea cave – A natural chamber in a zone of relatively weak rocks that extends into the face of a wave-cut rock cliff. The cave is quarried out by wave action. See Figure 60.

Sea stacks – Isolated pinnacles of rock that are typically taller than they are wide with tops at a somewhat lower elevation than the top of the adjacent rock cliff (Davis and FitzGerald, 2004). The stacks are usually composed of rocks that are significantly more resistant to wave erosion than their neighboring rocks in the formerly existing wave-cut rock cliff. See Figure 46.

Seafloor spreading – A process first described by Hess (1962) after he observed that the floor of the Pacific Ocean was in motion and "expanding." This is happening in response to additional volcanic flows being added to the seafloor along mid-ocean ridges at the same time as similar volcanic materials are carried to great depths in subduction zones along the outer margin of the oceans. *A dramatic proof of sea-floor spreading was discovered in the mid 1960s when data revealed alternating stripes of magnetic orientation on the seafloor, parallel to the mid-ocean ridges and symmetric across them—that is, a thick or thin stripe on one side of the ridge is always matched by a similar stripe at a similar distance on the other side. This mirror-image magnetic orientation pattern is created by steady sea-floor spreading combined with recurrent reversals of Earth's magnetic field. Iron atoms in liquid rock welling up along a mid-ocean ridge align with Earth's magnetic field* (Enotes, 2009). See Figures 3 and 4B.

Seas – Waves in the area where the wind that forms them is blowing. Under sea conditions, the water surface is choppy and the waves are asymmetrical. Whitecaps are common. See Figure 32B.

Sedimentary rocks – Rocks most commonly formed by weathering and erosion of pre-existing rocks on the Earth's surface. This process creates sediment that can be transported by water, or wind in some cases, to be deposited in masses of significant size. Once deposited, these sediments may become buried where they are consolidated by chemical cementation and other processes to form solid rocks.

Seismic wave – Waves that pass through the Earth caused by the sudden release of energy into the surrounding Earth from the spot where an earthquake originates (the hypocenter or focus of the quake).

Seismic P wave – The primary seismic wave that moves through the Earth in a fashion similar to sound waves due to compression of the rock particles, with each particle moving backward and forward in a straight line from the origin of disturbance. They are faster moving than their associated seismic S waves (Collier, 1999).

Seismic S wave – The secondary seismic waves that are due to shearing or transverse motion that moves rock particles not forward and backward, but side to side or up and down the way ripples radiate out when a rock breaks the surface of a calm pond. They are the seismic waves that produce the rolling motion of the ground during an earthquake and are, therefore, more damaging to structures than P waves (Collier, 1999).

Seismograph – An instrument used to measure seismic waves, which runs continuously, recording all incoming waves. Because of their different speeds, the seismic P waves from a specific earthquake will arrive ahead of the seismic S waves. Although their rates of speed may vary significantly, depending upon what they are moving through, the ratio of the difference between the speeds of the two waves is quite constant; therefore, this difference can be used to determine how far away the hypocenter of the earthquake is from the seismograph. The closer together the arrival times are, the closer the hypocenter.

Serpentinite – A metamorphic rock composed of a number of complex, usually hydrated, ferromagnesium minerals. It usually forms at the crests of oceanic ridges, where seawater sinks into fractures that open as the two plates separate. The tremendous pressure of the depths forces the water down into the extremely hot mantle rocks below the ridge crest. The water reacts with hot peridotite to make serpentinite (Alt and Hyndman, 2000).

Setback lines – A "soft" solution for dealing with potential beach erosion in which a line is established landward the high-tide line, seaward of which building is prohibited. They work best and are easiest to implement in areas that have not been developed as yet. The criteria used to establish such lines include analysis of historical changes of the locations of the high-tide line, preservation of the line of foredunes, and defining areas of flooding and storm wave uprush.

Shale – Sedimentary rock made up mostly of silt and/or clay that is fissile (i.e., composed of very fine, multiple layers).

Shelf break – Point where the outer edge of the continental shelf joins a steeply descending planar surface called the continental slope. See Figure 83.

Significant wave height – A measure that portrays the average height of the waves that comprise the highest 33% of waves in a given sampling period.

Silica tetrahedron – The basic building block of all silicate minerals, which consists of a single silicon cation surrounded by four oxygen anions. Silicon cations, which have a charge of +4, are less than ⅓ the size of the oxygen anions, which have a charge of −2. As a result of these size differences, one silicon cation fits "snugly" between 4 oxygen anions, forming the tetrahedron.

Silicates – A group of minerals that are the fundamental building block of the Earth's crust, with the atoms of silicon and oxygen combined making up about 75% of it. The basic structure of the silicate minerals is the silica tetrahedron (see definition above). The tetrahedron is perfectly arranged geometrically, but the resultant ionic charge of −4 is far from balanced. To reduce the negative charge, the tetrahedra must bond with other positive ions, such as Mg^{2+} or link up with other tetrahedra in various ways, such as in sheets (e.g., mica;), chains (e.g., pyroxenes and micas), and so on. Because of its particular arrangement of bonding, called framework bonding, quartz (SiO_2) is the most stable of all the silicate minerals, both physically and chemically (Eggar, 2006).

Slip face – A steeply sloping face of a migrating wedge of sand, such as a bedform created by flowing water (e.g., ripples and sand waves or on the downwind side of a sand dune). The angle of slope of the slip face, which is essentially the angle of repose of the sand (dip angle achieved when a pile of sand is created by simply dumping the sand out of a container, such as a truck bed), averages around 22 degrees in water and 30 degrees in air. See Figures 52C and 53.

Spilling waves – Waves that start disintegrating into foaming lines fairly far offshore and continue to foam their way to shore, gradually decreasing in height as they go. See Figure 34B.

Spit – A linear projection of sediment across an embayment, stream outlet, and so on that builds in the direction that the littoral sediment transport system is moving the sediment. See Figure 110.

Spring tides – The maximum tides of the month that exist when the gravitational attraction of the Moon and Sun on the water bodies of the World Ocean are working together, during full and new moons (condition called syzygy). Minimum tides (called neap tides) occur during the first and third quarter of the Moon. See Figure 27.

Storm berm – A linear, pyramidal-shaped mound of cobbles and pebbles that typically occurs on the most landward portion of exposed gravel beaches. It is only activated during major storms and, unless the storm is unusually severe, it is built up even higher during the storm event and not eroded away. See Figure 58.

Storm profile – A term suggested by Paul Komar, who was working on the Oregon Coast, to indicate a flat beach profile eroded by storm waves. Hayes and Boothroyd (1969) used the term post-storm beach. See Figure 51B.

Strike-slip fault – A geologic fault in which the blocks of rock on either side of the fault slide horizontally past each other without any significant up or down motion. See Figure 13.

Subduction zone – A region where a portion of one of the Earth's tectonic plates is diving beneath a portion of another plate, in some instances resulting in the formation of an oceanic trench. Along ocean margins, an oceanic plate dives under a continental plate.

Submarine canyon – A deep canyon in the continental shelf and slope and extending about perpendicular to the shore.

Surf zone – The nearshore zone off the beachface where the waves usually break. See Figure 33B.

Surging waves – Waves that approach shore and wash directly up the beachface without breaking.

Swash zone – Area where the water released by a broken wave washes across the portion of the beach known as the beachface. The released water at first rushes up the beachface until it loses its momentum, at which point it stops momentarily before returning back down the beachface under the influence of gravity. Usually, some of this water is lost through percolation into the beachface sediments.

Swell – Waves that have passed beyond the area where they were formed. Swell waves are typically longer on the average than those in the area of formation. The swell surface is regular and more-or-less symmetrical (whitecaps are rare). See Figure 32B.

Swell profile – A term suggested by Paul Komar, who was working on the Oregon Coast, to indicate a beach with a well-developed depositional berm (formed between storms). Hayes and Boothroyd (1969) used the terms late accretional or mature profile. See Figure 51B.

Syncline – The condition where layers of rocks are folded into a form similar to what you would get if you pushed one edge of a rug along the floor, folding it into parallel bands. Where the folded layers bend down, the fold is called a syncline.

Syzygy – The times when the gravitational forces of the moon and sun work together with regard to generating tides in the World Ocean (full and new Moon). The largest tides of the month (spring tides) occur during syzygy. See Figure 27.

Talus – Debris piles of fragments of rocks up to boulders in size (and soil in some cases) at the base of cliffs.

Tectonic (lithospheric) plates – A dozen or so segments of the lithosphere that covers the Earth's outer part (Hyndman and Hyndman, 2006). The plates that comprise the continents (e.g., the North American Plate) are made up of relatively light material composed of silica, oxygen, and so on. The plates that constitute the ocean floors are composed of relatively heavy material that contains abundant iron, magnesium, and so on. See Figures 2A and 14B.

Tectonism – The forces that produce deformation of the Earth's crust, such as folding or faulting of the rocks, with the motion of tectonic plates commonly being the driving mechanism for this deformation.

Thrust, or reverse, fault – The type of fault that occurs when a rock or sediment mass located on the upper side of a sloping planar surface (a fracture in the rocks) slides up the fault relative to the lower side under the influence of compression. See Figure 13 (Middle).

Tidal flat – A flat intertidal surface covered and uncovered as the tides rise and fall. It is usually a depositional surface with the sediments supplied by tidal currents. In more exposed areas, where the waves are somewhat active and the tidal currents are strong, the flats are composed of sand. In more sheltered areas, they are composed mostly of mud.

Tidal inlet – Usually defined as major tidal channels that intersect barrier islands or spits, usually to depths of 10s of feet. See Figure 66.

Tidal range – The vertical distance between high and low tide.

Tide-dominated coasts – Term used by Hayes (1976) to define a specific type of geomorphic assemblage on depositional coasts. Tide-dominated coasts are characterized by such features as open mouthed, multi-lobate river deltas, abundant large estuarine complexes, and extensive tidal flats and salt marshes. Most tide-dominated coasts are macrotidal (tidal range >12 feet). They usually occur on coasts with a combination of relatively small waves and large tides.

Tide pool – Depressions in wave-cut rock platforms that hold water during low tide, noted for the wide diversity of plant and animal life that inhabits them.

Tombolo – A triangle-shaped body of sediment (usually sand) that forms in the lee of an offshore structure, such as a small island or ship wreck. Waves approaching shore bend around this offshore structure, with the result that two opposing littoral sediment transport directions are created that converge in the shelter of the offshore structure, resulting in the formation of the tombolo. See Figure 35.

Trailing-edge coasts – Coasts defined by Inman and Nordstrom (1971) as those that occur on the trailing edges of continental plates. They are typically characterized by wide coastal plains, numerous river deltas, and barrier islands. Rocky coasts are rare.

Transform fault – A form of a strike-slip fault that occurs between oceanic plates or within oceanic plates along the length of mid-ocean ridges in which the rocks on either side of the fault simply slide past each other in a horizontal plane without colliding or pulling apart. Defined and described by J. Tuzo Wilson in 1965, they provided an understandable mechanism whereby the new lava flows moving away from the mid-ocean ridges could fracture and fit onto the spherical surface of the Earth. See Figures 4A and 12.

Transverse sand dunes – A wind-formed, linear sand ridge that has a steep slip face on its downwind side. The slip face dips at the angle of repose of sand in air (30± degrees). See Figure 72.

Tsunami – A Japanese word translated as *harbor wave*. It is a large ocean wave usually generated by a submarine earthquake, although it may also be generated by volcanic eruptions, landslides (both those initiated on shore and submarine), and even by a comet or asteroid impact. *Seismic seawaves* is another term that scientists commonly use for this phenomenon. These are typically very long waves (up to several hundred miles) with relatively small wave heights (commonly less than 3 feet), and long periods (over an hour in some cases) that have sufficient energy to travel across entire oceans. They move very fast in the open water at speeds measured in hundreds of miles per hour, but they slow down before breaking in shallow water.

Turbidite – A coarse-grained sediment deposit formed by a swiftly moving, bottom-flowing current along the ocean floor called a turbidity current. See Figure 84.

Turbidity current – A rapidly flowing, bottom-hugging current that is composed of a dense turbulent plume that contains an unusually large amount of suspended sediment. This type of flow is commonly generated by a mechanism such as a landslide on a steep slope of the seafloor (there are several other possibilities). See Figure 84.

Unconformity – The general term for an erosional surface, usually near horizontal when it was formed, that represents a significant loss of time from the rock record. Therefore, it separates two rock masses, or strata, of significantly different ages.

Updrift – The direction from which the individual sand grains moving along the shore within the littoral sediment transport system have come.

Upper flow regime – See definition of flow regime.

Upwelling – An oceanographic process by which warm, less-dense surface water is drawn away from a shore by offshore currents and replaced by colder, denser, and more nutrient-rich water brought up from the depths.

Wave-cut rock platform – A flat, usually intertidal, rock platform left behind as an eroding, wave-cut rock cliff retreats. Some geologists refer to these features as shore platforms.

Wave-dominated coasts – Term used by Hayes (1976) to define a specific type of geomorphic assemblage on depositional coasts (not leading-edge coasts like the Pacific coast of North America). Wave-dominated coasts are characterized by such features as deltas with smooth outer margins and abundant long barrier islands. Most wave-dominated coasts are microtidal or low-mesotidal (tidal range less than 6 feet). They are usually found on coasts with a combination of relatively large waves and small tides.

Wave diffraction – The phenomenon by which energy is transmitted laterally along a wave crest. When part of a wave is interrupted by a barrier, the effect of diffraction is manifested by propagation of waves into the sheltered region within the barrier's geometric shadow. A good example would be waves passing between two offshore breakwaters and into the sheltered areas landward of the breakwaters.

Wave refraction – Any change in the direction of a wave resulting from the bottom contours. For example, a submerged hump would cause the two ends of the wave crest located away from the hump to bend toward each other. Such differences in the offshore topography can cause the energy of the waves to be focused in some areas along the shore and less energetic in others.

Wave uprush – The water that moves up the beachface under the impetus of a breaking wave.

Young mountain range coasts – Defined by Hayes (1965) as coasts that abut high mountains created by relatively recent orogenic (mountain building) activity. In most places, the land is continuing to rise. The bedrock is usually dominated by relatively young sedimentary and volcanic rocks. Mostly short, steep rivers empty into the sea along such coasts. They are most common on the leading edges of continental plates.

Zone of the turbidity maximum – A zone in the middle reaches of an estuary that contains an abundance of fine-grained sediment in suspension, because of the mixing of salt water with fresh water containing a profusion of suspended clay particles, which tend to flocculate. See Figure 78B.

REFERENCES CITED

Aguilar-Tunon, N.A. and P.D. Komar. 1978. The annual cycle of profile changes of two Oregon Beaches. Oregon Bin 40:25-39.

Allan, J.C., R. Geitgey, and R. Hart. 2005. Dynamic revetments for coastal erosion in Oregon. Final Report SPR 620. Oregon Department of Transportation Research Unit. Available at: http://www.oregon.gov/ODOT/TD/TP_RES/docs/reports/dynamicrevetments.pdf.

Allen, J.R.L. 1968. Current ripples – their patterns in relation to water and sediment motion. North-Holland Publ. Comp., Amsterdam, Holland. 433 pp.

Allan, J.C. and P.D. Komar. 2000. Are ocean wave heights increasing in the eastern North Pacific? EOS Trans. AGU 81(47):561-567.

Allan, J.C. and P.D. Komar. 2002. The wave climate of the Eastern North Pacific: Long-term trends and an *El Niño/La Niña* dependence. In: G. Gelfenbaum and G.M. Kaminsky, USGS Open File Report 02-229, Southwest Washington Coastal Erosion Workshop Report 2000, pp. 224-227.

Allan, J.C. and P.D. Komar. 2002. A dynamic revetment and artificial dune for shore protection. Proceedings of the 28th Conference on Coastal Engineering. Cardiff, Wales. ASCE. 2044-2056.

Allan, J.C. and P.D. Komar. 2004. Environmentally compatible cobble berm and artificial dune for shore protection. Shore & Beach 72:9-18.

Allan, J.C. and P.D. Komar. 2006. Climate controls on U.S. West Coast erosion processes. Journal of Coastal Research 22:511-529.

Allan, J.C., P.D. Komar, P. Ruggiero, and R. Witter. 2012. The March 2011 Tohoku tsunami and its impacts along the U.S. West Coast. Journal of Coastal Research 28:1142-1153.

Alt, D. 2001. Glacial Lake Missoula and its humongous floods. Mountain Press Pub. Co., Missoula, MT.199 pp.

Alt, D. and D.W. Hyndman. 1978. Roadside geology of Oregon. Mountain Press Pub. Co., Missoula, MT. 278 pp.

Alt, D. and D.W. Hyndman. 1984. Roadside geology of Washington. Mountain Press Pub. Co., Missoula, MT. 288 pp.

Alt, D. and D.W. Hyndman. 1989. Roadside geology of Idaho. Mountain Press Pub. Co., Missoula, MT. 394 pp.

Alt, D. and D.W. Hyndman. 2000. Roadside geology of Northern and Central California. Mountain Press Pub. Co., Missoula, MT. 369 pp.

Anderson, J.L., M.H. Beeson, R.D. Bentley, K.R. Fecht, P.R. Hooper, A.R. Niem, S.P. Reidel, D.A. Swanson, T.L. Tolan, and T.L. Wright. 1987. Distribution maps of stratigraphic units of the Columbia River Basalt Group, In: J.E. Schuster (ed.), Selected Papers on the Geology of Washington, Wash. Div. of Geol. and Earth Resources, Bull. 77, pp. 183-195.

Atwater, B.F., S. Musumi-Rokkaku, K. Satake, Y. Tsuji, and D.K. Yamaguchi. 2005. The Orphan Tsunami of 1700. U.S. Geological Survey Professional Paper 1707. 144 pp.

Audubon California. 2017. Western snowy pover population status and recovery. Available at: http://ca.audubon.org/birds-0/western-snowy-plover-population-status-and-recovery

Babcock, R.S., R.F. Burmester, D.C. Engebretson, A. Warnock, and K.P. Clark. 1992. A rifted margin origin for the

Crescent basalts and related rocks in the northern Coast Range Volcanic Province, Washington and British Columbia: Journal of Geophysical Research, 97(B5):6799-6821.

Bagnold, R.A. 1940. Beach formation by waves; some model experiments in a wave tank. Journal of the Institute of Civil Engineering 15:27-52.

Basan, P.B. and R.W. Frey. 1977. Actual-paleontology and neoichnology of salt marshes near Sapelo Island, Georgia. In: T.P. Crimes and J.C Harber (eds.), Trace fossils 2. Geog. Jour., Spec. Issue 9:41-70.

Bascom, W. 1954. Characteristics of natural beaches. In: Proc. 4th Conf. On Coastal Eng. Pp. 163-180.

Beachapedia. 2013a. State of Beach/State Reports/WA/Beach Fill. Available at: http://www.beachapedia.org/State_of_the_Beach/State_Reports/WA/Beach_Fill.

Beachapedia. 2013b. State of Beach/State Reports/OR/Beach Fill. Available at: http://www.beachapedia.org/State_of_the_Beach/State_Reports/OR/Beach_Fill\.

Beeson, M. 1979. The origin of the Miocene basalts of coastal Oregon and Washington: An alternative hypothesis. Oregon Geology 41, n.10.

Biggs, R.B. 1978. Coastal bays. In: R.A. Davis, Jr. (ed.), Coastal Sedimentary Environments. Springer Verlag, New York. Pp. 69-99.

Billings, M.P. 1954. Structural geology. 2nd edition, Prentice Hall. 514 pp.

Bird, E.C.F. 1996. Beach management. John Wiley and Sons, Chichester, New York. 292 pp.

Bird Research Northwest. 2016. Tern management on East Sand Island. Available at: http://www.birdresearchnw.org/feature-stories/2016_final_annual_report.

Bogdanov, N.A. and N.L. Dobretsov. 1987. The ophiolites of California and Oregon. Geotectonics 12:472-479.

Boothroyd, J. C. 1969. Hydraulic conditions controlling the formation of estuarine bedforms. In: Hayes, M. O. Coastal Environments – NE Massachusetts and New Hampshire: Field Trip Guidebook for Eastern Section of SEPM. Pp. 471-427.

Boothroyd, J.C. and D. Nummedal. 1978. Proglacial braided outwash: A model for humid alluvial-fan deposits. Can Soc. Petroleum Geologists Memoir, S, 641±668.

Boule, M.E., N. Olmstead, and T. Miller. 1983. Inventory of wetland resources and evaluation of wetland management in western Washington. Washington State Department of Ecology. Shapiro and Associates, Inc., Seattle, Washington.

Bowen, A.J. and D.L. Inman. 1969. Rip currents: Laboratory and field observations. Journal of Geophysical Research, 74:5479-5490.

Brown, D. 1970. Bury my heart at wounded knee: An Indian history of the American west. H. Holt and Co., New York. 487 pp.

Bues, S.S. and M. Morales. 1990. Grand Canyon geology. Oxford University Press, New York. 9 pp.

Bureau of Land Management (BLM), 2014. Oil and gas energy, Oregon and Washington. Available at: http://www.blm.gov/or/energy/oilandgas.php.

Burke Museum. 2014a. New lands along the old coast: Building the Pacific Northwest. Available at: http://www.burkemuseum.org/static/geo_history_wa/New%20Lands%20Along%20 an%20Old%20Coast%20v.2.7.htm.

Burke Museum. 2014b. The Cascade Episode. Available at: http://www.burkemuseum.org/static/geo_history_wa/Cascade%20Episode.htm

Byrne, J.V. 1962. Geomorphology of the Continental Terrace Off The Central Coast of Oregon. The ORE BIN 24: 65-80.

Cajete, G.A. 1995. Ensoulment of nature. In: A.B. Hirschfelder (ed.), Native Heritage: Personal Accounts by American Indians, 1790 to the Present. Macmillan Pub. Co., NY. 298 pp.

Carter, R.W.G. and C.D. Woodroffe (eds.). 1994. Coastal evolution. Cambridge University Press, Cambridge. 517 pp.

Cartwright, D.E. 1969. Deep sea tides. Science Journal 5:60-67.

Centala, M. 2013. Geology of the Seal Rock area. 94 pp. Available at: https://geologyofsealrock.files.wordpress.com/2013/11/final-geology-of-the-seal-rock-area2.pdf.

Chappell, J.M. 1983. A revised sea-level record for the last 300,000 years from Papua, New Guinea. Journal of Geophysical Research 14:99-101.

Clark, P. U. and Mix, A. C. 2002. Ice sheets and sea level of the Last Glacial Maximum. Quaternary Science Reviews 21:1-7.

Clemens, K.E. and P.D. Komar. 1988. Tracers of sand movement on the Oregon Coast. Coastal Engineering 1988, Chapter 100, pp. 1338-1351. Available at: http://ftp.soest.hawaii.edu/falinski/ORE664/Clemens%20and%20Komar.pdf

Coleman, J.M. and L.D. Wright. 1975. Modern river deltas: Variability of processes and sand bodies. In: M.L. Broussard (ed.), Deltas. 2nd ed., Houston Geological Society, Houston, TX. Pp. 99-150.

Collier, M. 1999. A land in motion: California's San Andreas Fault. University California Press, Berkeley. 118 pp.

Comet, P.A. 1996. Geological reasoning: Geology as an interpretive and historical science. Discussion. Geological Society of America Bulletin 108:1508-1510.

Connor, E. and D. O'Haire. 1988. Roadside geology of Alaska. Mountain Press Pub. Co., Missoula, MO., 251 pp.

Cook, J. 1842. The voyages of Captain James Cook, volume II. William Smith Publishers, London, 619 pp.

Cooper, W.S. 1958. Coastal sand dunes of Oregon and Washington. Geology Society of America Memoir 72. 169 pp.

Davies, J.L. 1973. Geographical variation in coastal development. Hafner Publication Co., New York. 204 pp.

Davis, M. 2001. Late victorian holocausts: *El Niño* famines and the making of the third world. Verso, London. 271 pp.

Davis, R.A., Jr. and M.O. Hayes. 1984. What is a wave-dominated coast? Marine Geology 60:313-329.

Davis, R.A., Jr. and D.M. FitzGerald. 2004. Beaches and coasts. Blackwell Publishing, Malden, MA. 419 pp.

Dawes, R.L. and C.D. Dawes. 2016. Geology of the Pacific Northwest: Virtual field site Rosario. Available at: https://commons.wvc.edu/rdawes/virtualfieldsites/Rosario/VFSRosario.html

Dietrich, G. 1963. General Oceanography. Wiley-Interscience, New York. 588 pp.

Dietz, R.S. and J.C. Holden. 1970. Reconstruction of Pangaea: Breakup and dispersion of continents, Permian to Present. Journal Geophysical Research 75:4939-4956.

Dingler, J.R. and H.E. Clifton. 1994. Chapter 4: Barrier systems, of California, Oregon, and Washington. In: R.A. Davis, Jr. (ed.), Barrier Islands. Springer Verlag, New York. Pp. 115-165.

Doughton, S. 2013. Full-Rip 9.0: The next big earthquake in the Pacific Northwest. Sasquatch Books, 288 pp.

Eggar, A.E. 2006. Minerals III: The Silicates. Available at: http://www.visionlearning.com/library/module_viewer.php?mid=140.

Emmett, R., R. Llanso, J. Newton, R. Thom, M. Hornberger, C. Morgan, C. Levings, A. Copping, and P. Fishman. 2000. Geographic signatures of North American West Coast estuaries. Estuaries 23(6):765-792.

Enotes. 2009. Sea-floor spreading. Available at: http//enotes.com/earth-science/sea-floorspreading.

Figge, J. 2009. Evolution of the Pacific Northwest. In: Chapter 3, The coast range episode: New lands along an evolving margin. Published by the Northwest Geological Institute, Seattle. 39 pp.

Fleming, K., Johnston, P., Zwartz, D., Yokoyama, Y., Lambeck, K. and Chappell, J. 1998. Refining the sea-level curve since the Last Glacial Maximum using far- and intermediate-field sites. Earth and Planetary Science Letters 163:1-4:327-342.

Folk, R.L. 1955. Student operator error in determination of roundness, sphericity, and grain size. Journal of Sedimentary Petrology 25:297-301.

Frey, R.W. and J.D. Howard. 1969. A profile of biogenic sedimentary structures in a Holocene barrier island-salt marsh complex. Trans. Gulf Coast Association of Geological Societies 19:427-444.

Gaines, E., D. Lauten, K. Castelein, D. Farrar, and A. Kotaich. 2016. Documenting causes of snowy plover nest failure with cameras. Portland State University, Portland, OR. Available at: http://inr.oregonstate.edu/sites/inr.oregonstate.edu/files/2016cameraposter101516.pdf

Gallup, C.D., H. Cheng, F.W. Taylor, and R.L. Edwards. 2002. Direct determination of the timing of sea level change during Termination II. Science, 295, 310–313.

García-Medina, G., H.T. Özkan-Haller, P. Ruggiero, R.A. Holman, and T. Nicolini. 2018. Analysis and catalogue of sneaker waves in the U.S. Pacific Northwest between 2005 and 2017. Natural Hazards, https://doi.org/10.1007/s11069-018-3403-z(01234567.

Geotimes, 2004. Roving Oregon's dunes. Geotimes. Available at: http://www.geotimes.org/aug04/Travels081204.html.

Gillis, J. 2013. Heat trapping gas passes milestone, raising fears. Article in the New York Times, published on May 10, 2013.

Geology News Blog. 2008. Weekend adventures – Capitola fossils. In: D. Schumaker, Geology News blog. Available at: http://geology.rockbandit.net/2008/04/27/weekend-adventures-capitola-fossils/.

Gerstel, W.J. and W.S. Lingley, Jr. 2000. Geology of the Forks 1:100,000 Quadrangle, Washington. Open File Report 2000-4. Washington Division of Geology and Earth Resources.

Glen, W. 1975. Continental drift and plate tectonics. C.E. Merrill Pub. Co., Columbus, OH. 188 pp.

Goldfinger, C., C.H. Nelson, A.E. Morey, J.E. Johnson, J.R. Patton, E. Karabanov, J. Gutiérrez-Pastor, A.T. Eriksson, E. Gràcia, G. Dunhill, R.J. Enkin, A. Dallimore, and T. Vallier. 2012. Earthquake hazards of the Pacific Northwest coastal and marine regions. U.S. Geological Survey Professional Paper 1661–F, 170 pp.

Graham, N.E. and H.F. Diaz. 2001. Evidence for intensification of North Pacific winter cyclones since 1948. Bulletin of the American Meteorological Society 82:1869-1893.

Griggs, G. 1986. Littoral cells and harbor dredging along the California Coast. Environmental Geology 10:7-20.

Griggs, G.B. and A.S. Trenhaile. 1994. Coastal cliffs and platforms. In: R.W.G. Carter and C.D. Woodroffe (eds.), Coastal Evolution, Cambridge University Press, Cambridge. Pp. 425- 450.

Griggs, G.B., K. Patsch, and L. Savoy. 2005. Living with the Changing California Coast. University of California Press, Berkeley, CA. 540 pp.

Guza, R.T. and D.L Inman. 1975. Edge waves and beach cusps. Journal of Geophysical Research 80:1285-1291.

Hales, T.C., D.L. Abt, E.D. Humphreys, and J.J. Roering. 2005. A lithospheric instability origin for Columbia River flood basalts and Wallowa Mountains uplift in northeast Oregon. Nature 438, 842-845 (8 December 2005) | doi:10.1038/nature04313

Hall, M.J. 1999. Federal On Scene Coordinator's Report and Assessment of M/V *New Carissa* Oil Spill Response. U.S. Coast Guard Marine Safety Office, Portland, OR.

Ham, Y.G. 2018. *El Niño* events will intensify under global warming. Nature 564:192-193.

Hand, B.M., J.M. Wessel, and M.O. Hayes. 1969. Antidunes in the Mt. Toby Conglomerate (Triassic), Massachusetts. Journal of Sedimentary Research 39:4:1310-1316.

Harari, Y.N. 2014. Sapiens: A Brief History of Humankind. HarperCollins, 443 pp.

Harden, D.R. 2004. California geology (second edition). Prentice Hall, Upper Saddle River, N.J. 552 pp.

Hartness, N.B. 2001. The 1964 Good Friday earthquake tsunami. Available at: http://www.geo.arizona.edu/~nhartnes/alaska/tsunam.html

Hayes, M.O. 1964. Lognormal distribution of inner continental shelf widths and slopes. Deep-Sea Research 11:53-78.

Hayes, M.O. 1965. Sedimentation on a semiarid, wave-dominated coast (south Texas); with emphasis on hurricane effects. Ph.D. Thesis, Dept. of Geol., University of Texas, Austin. 350 pp.

Hayes, M.O. 1967a. Relationship between coastal climate and bottom sediment type on the inner continental shelf.

Marine Geology 5:111-132.

Hayes, M.O. 1967b. Hurricanes as geological agents: Case studies of Hurricanes *Carla* (1961) and *Cindy* (1963). University of Texas, Austin, Bur. Econ. Geol. Rep. 61. 56 pp.

Hayes, M.O. (ed.) 1969. Coastal Environments – NE, Massachusetts and New Hampshire: Field trip guidebook for Eastern Section of SEPM, May 9-11. 462 pp.

Hayes, M.O. 1976. Lecture Notes. In: M.O. Hayes and T.W. Kana, (eds.), Terrigenous Clastic Depositional Environments. Tech Rep 11-CRD, Geol. Dept., University of South Carolina, Columbia. Pp. I-1 – I-131.

Hayes, M.O. 1979. Barrier island morphology as a function of tidal and wave regime. In: S. Leatherman (ed.), Barrier Islands, from the Gulf of St. Lawrence to the Gulf of Mexico, Academic Press, New York. Pp. 1-27.

Hayes, M.O. 1980. General morphology and sediment patterns in tidal inlets. Sedimentary Geology 26:139-156.

Hayes, M.O. 1999. Black tides. University of Texas Press, Austin. 287 pp.

Hayes, M.O. 2009. Barrier islands. In: R. Gillespie and D. Clague (eds.), Encyclopedia of Islands. University California Press, Berkeley, pp. 82-88.

Hayes, M.O. 2011. Coastal heroes. Pandion Books, Columbia, SC. 322 pp.

Hayes, M.O. and J.C. Boothroyd. 1969. Storms as modifying agents in the coastal environment. In: M.O. Hayes (ed.), Coastal Environments – NE Massachusetts and New Hampshire: Field Trip Guidebook for Eastern Section of SEPM, May 9-11. Pp. 290-315.

Hayes, M.O. and T.W. Kana (eds.). 1976. Terrigenous Clastic Depositional Environments. Tech Rep 11-CRD, Geol, Dept., University of South Carolina, Columbia. 315 pp.

Hayes, M.O. and C.H. Ruby. 1994 Chapter 10: Pacific Coast of Alaska. In: R.A. Davis, Jr. (ed.), Barrier Islands. Springer Verlag, New York. Pp. 395-433.

Hayes, M.O. and J. Michel. 2008. A coast for all seasons: A naturalist's guide to the coast of South Carolina. Pandion Books, Columbia, SC. 285 pp.

Hayes, M.O. and J. Michel. 2010. A coast to explore: Coastal geology and ecology of Central California. Pandion Books, Columbia, SC. 338 pp.

Hayes, M.O. and J. Michel. 2013. A tide-swept coast of sand and marsh: Coastal geology and ecology of Georgia. Pandion Books, Columbia, SC. 299 pp

Hayes, M.O. and J. Michel. 2017. A coast beyond compare: Coastal geology and ecology of Southern Alaska. Pandion Books, Columbia, SC.

Hayes, M.O., C.H Ruby, M.F. Stephen, and S.J. Wilson. 1976. Geomorphology of the southern coast of Alaska. In: Proc. Fifteenth Conference on Coastal Engineering, Honolulu, HI. Pp. 1992-2008.

Hayes, M.O., J. Michel, and F.J. Brown. 1977. Vulnerability of coastal environments of Lower Cook Inlet, Alaska to oil spill impact. In: Proc. 4th International Conference on Port and Ocean Engineering Under Arctic Conditions, 26-30 September 1977, Memorial University of Newfoundland. 12 pp.

Hess, H.H. 1962. History of ocean basins. In: A.E.J. Engel, H.L. James, and B.F. Leonard (eds.), Petrologic Studies: A Volume in Honor of A.F. Buddington, Geology Society of America. Pp. 599-620.

Highway Research Board. 1958. Landslides and engineering practice. Special Report 29. 544 pp.

Hooper, P.R. and D.A. Swanson. 1987. Evolution of the eastern part of the Columbia Plateau. In. Schuster, J. E. (ed.), Selected Papers on the Geology of Washington, Wash. Div. Of Geol. and Earth Resources, Bull. 77:7-121.

Hopson, C.A., J.M. Mattinson and E.A Pessagno, Jr. 1981. Coast Range Ophiolite, western California. In: W.G. Ernst (ed.), The Geotectonic Development of California. Prentice-Hall, Englewood Cliffs, NJ. Pp. 418-510.

Howard, J.D. and J. Dorjes. 1972. Animal-sediment relationships in two beach-related tidal flats, Sapelo Island, Georgia. Journal of Sedimentary Petrology 42:608-623.

Howard, K. 2002. The Milankovitch theory. Available at: http://academic.emporia.edu/aberjame/student/howard2/milan.htm.

Hutton, J. 1785. Theory of the Earth with proof and illustrations. Royal Society of Edinburgh.

Hyndman, D. and D. Hyndman. 2006. Natural Hazards and Disasters. Thomson Brooks/Cole, United States. 490 pp.

Imbrie, J., J.D. Hays, D.G. Martinson, A. McIntyre, A.C. Mix, J.J. Morley, N. Pisias, W.L. Prell, and N.J. Shackleton. 1984. The orbital theory of Pleistocene climate: Support from a revised chronology of the marine $\delta^{18}O$ record. In: Milankovitch and Climate: Understanding the Response to Astronomical Forcing, Proceedings of the NATO Advanced Research Workshop held 30 November - 4 December, 1982 in Palisades, NY. Edited by A. Berger, J. Imbrie, H. Hays, G. Kukla, and B. Saltzman. Dordrecht: D. Reidel Publishing, 1984., p.269. Inman, D.L. 2005. Littoral cells. In: M.L. Schwartz (ed.), Encyclopedia of Coastal Science. Springer, The Netherlands. Pp. 594-599.

Inman, D.L. and C.E. Nordstrom. 1971. On the tectonic and morphologic classification of coasts. Journal of Geology 79:1-21.

Inman, D.L. and P.M Masters. 1994. Status of research on the nearshore. Shore & Beach 62:11-20.

Inman, D.L. and S.A Jenkins. 2005a. Energy and sediment budgets of the global coastal zone. In: M.L. Schwartz (ed.), Encyclopedia of Coastal Science. Springer, The Netherlands. Pp. 408-415.

Inman, D.L. and S.A. Jenkins. 2005b. Climate patterns in the coastal zone. In: M.L. Schwartz (ed.), Encyclopedia of Coastal Science. Springer, The Netherlands, pp. 243-246.

International Commission on Stratigraphy. 2008. Revised geological time scale. Available at: http://www.stratigraphy.org/upload/ISChart2008.pdf

Intergovernmental Panel on Climate Change (IPCC). 2013: Climate change 2013: The physical science basis. In: T.F. Stocker, D. Qin, G.K. Plattner, M. Tignor, S.K. Allen, J. Boschung, A. Nauels, Y. Xia, V. Bex and P.M. Midgley (eds.), Contribution of Working Group I to the Fifth Assessment Report of the Intergovernmental Panel on Climate Change Cambridge University Press, Cambridge, United Kingdom and New York, NY, USA, 1535 pp. Inman, D.L. and T.K Chamberlain. 1960. Littoral sand budget along the southern California coast. In: Volume of Abstracts. Report of the 21st International Geological Congress, Copenhagen, Denmark. Pp. 245-246.

Jennings, R. and J. Shulmeister. 2002. A field based classification scheme for gravel beaches. Marine Geology 186:211-228

Kahle, S.C., R.R. Caldwell, and J.R. Bartolino. 2005. Compilation of geologic, hydrologic, and ground-water flow modeling information for the Spokane Valley—Rathdrum Prairie Aquifer, Spokane County, Washington, and Bonner and Kootenai Counties, Idaho: U.S. Geological Survey Scientific Investigations Report 2005-5227, 64 pp.

Kaminsky, G.M., M.C. Buijsman, and P. Ruggiero. 2001. Predicting shoreline change at decadal scale in the Pacific Northwest, USA. Proceedings of The International Conference on Coastal Engineering, ASCE, 2000.

Kaminsky, G.M. and M.A. Ferland. 2003. Assessing the connections between the inner shelf and the evolution of Pacific Northwest barriers through vibracoring. International Conference on Coastal Sediments 2003, ASCE.

Kaminsky, G.M., P. Ruggiero, M.C. Buijsman, D. McCandless, and G. Gelfenbaum. 2010. Historical evolution of the Columbia River littoral cell. Marine Geology, Volume 273, Issues 1–4, Pages 96-126,

Kaufman, Y. 2002. On the shoulders of giants-Milutin Milankovitch (Earth Observatory, NASA). Available at: http://earthobdservatory.nasagov.Library/Grants/Milankovitch.

Keeling, C.D., S.C. Piper, R. B. Bacastow, M. Wahlen, T. P. Whorf, M. Heimann, and H. A. Meijer. 2001. Exchanges of atmospheric CO_2 and $13CO_2$ with the terrestrial biosphere and oceans from 1978 to 2000. I. Global aspects, SIO Reference Series, No. 01-06, Scripps Institution of Oceanography, San Diego. 88 pp..

Kelsey, H.M., D.C. Engebretson, C.E. Mitchell, and R.L. Ticknor. 1994. Topographic form of the Coast Ranges of the Cascade Margin in relation to coastal uplift rates and plate subduction. Journal of Geophysical Research 99:12,245-12,255.

Komar, P.D. 1976. Beach processes and sedimentation. Prentice-Hall, Inc., Englewood Cliffs, NJ. 429 pp.

Komar, P.D. 1983. Coastal erosion in response to the construction of jetties and breakwaters. In: Handbook of Coastal Processes and Erosion, CRC Press, Boca Raton, FL.

Komar, P.D. 1992. Ocean processes and hazards along the Oregon coast. Oregon Geology 54:3-20.

Komar, P.D. 1998a. The Pacific Northwest coast: Living with the shores of Oregon and Washington. Duke University Press, 216 pp.

Komar, P.D. 1998b. The 1997-98 *El Niño* and erosion on the Oregon coast. Shore & Beach 66(3): 33-41.

Komar, P.D. 2004. Oregon's Coastal Cliffs: Processes and Erosion Impacts. In: M.A. Hampton and G.B. Griggs (eds.), Formation, Evolution, and Stability of Coastal Cliffs–Status and Trends. U.S. Geological Survey Professional Paper 1693. 123 pp.

Komar, P.D. 2005. Hawke's Bay, New Zealand: Environmental change, shoreline erosion and management issues. Unpublished report to the Hawke's Bay Regional Council.

Komar, P.D. 2007. The design of stable and aesthetic beach fills: Learning from Nature. Coastal Sediments '07, New Orleans, Amer. Soc. Civil Engrs., 10 pp.

Komar P.D. 2010. Oregon. In: E.C.F. Bird (ed.), Encyclopedia of the World's Coastal Landforms. Springer, Dordrecht.

Komar, P.D., J.R. Lizarraga-Arciniega and T.A. Terich. 1976. Oregon coast shoreline changes due to jetties: Jour. Waterways, Harbors and Coastal Engineering, ASCE, 102:13-30.

Komar, P.D. and C. Wang. 1984. Processes of selective grain transport and the formation of placers on beaches. Journal of Geology 92:637-655.

Komar, P.D. and Z. Li. 1988. Pivoting analyses of the selective entrainment of sediments by shape and size with application to gravel threshold. Sedimentology 33:425-436.

Komar, P.D., G.M. Diaz-Mendez, and J.J. Marra. 2001. Coastal processes and stability of the New River Spit, Oregon. Journal of Coastal Research 17:625-635.

Komar P.D. and J.C. Allan. 2002. Nearshore-process climates related to their potential for causing beach and property erosion. Shore & Beach 70:31-40.

Komar, P.D. and M. Styllas. 2006. Natural processes and human impacts as the causes of sediment accumulation in Tillamook Bay, Oregon: Shore & Beach. v. 74, n. 4, p. 3-15.

Komar, P.D., J.C. Allan and P. Ruggiero. 2013. U.S, Pacific Northwest Coastal Hazards: Tectonic and Climate Controls: In C.W. Finkl, (ed.), Coastal Hazards, Coastal Research Library 6, Springer Science+Business Media, pp. 587-673.

Komar, P.D., J.R. Lizarraga-Arciniega, and T.A. Terich. 1976. Oregon coast shoreline changes due to jetties. Journal of Waterways, Harbors and Coastal Engineering. ASCE 102 (WW1):13-30.

Komar, P.D., J.C. Allan and R. Winz. 2003. Cobble beaches – The "Design with Nature" approach for shore protection. Coastal Sediments '03, Amer. Soc. Civil Engrs., electronic publication.

Komar, P.D., J.C. Allan, and P. Ruggiero. 2011. Sea level variations along the U.S. Pacific Northwest coast: Tectonic and climate controls. Journal of Coastal Research 27(5), 808–823.

Komar, P.D., J.C. Allan, G. Dias-Mendez, J.J. Marra, and P. Ruggiero. 2000. *El Niño* and *La Niña* – erosion processes and impacts. Proceedings of the 27th International Conference on Coastal Engineering, ASCE, 2414-2427.

Komar, P.D., W.G. McDougal, J.J. Marra, and P. Ruggiero. 1999. The rational analysis of setback distances: Applications to the Oregon coast: Shore & Beach, 67:41-49.

Kozuka, K. and V. Chaidex. 2008. (Abstract) Strong population genetic structure and larval dispersal capability of the burrowing ghost shrimp (*Neotrypaea californiensis*). 31st Annual Meeting, Pacific Estuarine Research Society, Feb-Mar 2008, Newport, OR.

Kukla, G. J. 2000. The last interglacial. Science 287(5455): 987.

Jeffries, S., D. Lynch, and S. Thomas. 2017. Results of the 2016 survey of the reintroduced sea otter population in Washington State. Washington Department of Fish and Wildlife, Lakewood, WA.

Lahee, F.H. 1952. Field geology. McGraw-Hill Book Co., New York. 883 pp.

Lambeck, K. and J. Chappell. 2001. Sea level change through the last glacial cycle. Science 292:679-686.

Landry, M.R. and B.M. Hickey (eds.). 1989. Coastal oceanography of Washington and Oregon. Volume 47 of Elsevier Oceanography Series. Elsevier Science Publishers, Amsterdam. 606 pp.

Lasmanis, R. 1991. The geology of Washington. Journal of Rocks and Minerals 66(4):262-277.

Le Bars, D., S. Drijfhout, and H. de Vries. 2017. A high-end sea level rise probabilistic projection including rapid Antarctic ice sheet mass loss. Environmental Research Letters 12(4):1-9.

Li, M.Z. and P.D. Komar. 1992. Longshore grain sorting and beach placer formation adjacent to the Columbia River: Journal of Sedimentary Petrology 62:429-441.

Lillie, R.J. 2015. Beauty from the beast: Plate tectonics and the landscapes of the Pacific Northwest. Wells Creek Publishers, Philomath, OR, CreateSpace Independent Publishing Platform, 92 pp.

Lorang, M.S. 1991. An artificial perched-gravel beach as a shore protection structure. In: Proc. Coastal Sediments '91, American Society of Civil Engineers, New York. Pp. 1916-1925.

Lorang, M.S. 2000. Predicting threshold entrainment mass for a boulder beach. Journal of Coastal Research 16:432-445.

Lund, E.H. 1973. Landforms along the Coast of Southern Coos County, Oregon. The Ore Bin Volume 35, Number 12. Oregon Department of Geology and Mineral Industries, Portland, OR.

Lund, E.H. 1974. Coastal landforms between Roads End and Tillamook Bay, Oregon. The Ore Bin, Oregon Department of Geology and Mineral Industries, Portland, OR, 36(11).

Lund, E.H. 1975. Landforms along the Coast of Curry County, Oregon. The Ore Bin Volume 37, Number 4. Oregon Department of Geology and Mineral Industries, Portland, OR.

Milankovitch, M. 1941. Kanon der Erdbestrahlung und seine Andwendung auf das. Eiszeiten-problem. Royal Serbian Academy, Belgrade. English translation: 1998, Canon of Insolation and the Ice Age Problem. With introduction and biographical essay by Nikola Pantic. Alven Global, 636 pp.

Miller, C.D. 1990. Volcanic hazards in the Pacific Northwest. Geoscience Canada 17:183-187.

Miller, I.M., H. Morgan, G. Mauger, T. Newton, R. Weldon, D. Schmidt, M. Welch, and E. Grossman. 2018. Projected Sea Level Rise for Washington State – A 2018 Assessment. A collaboration of Washington Sea Grant, University of Washington Climate Impacts Group, Oregon State University, University of Washington, and US Geological Survey. Prepared for the Washington Coastal Resilience Project.

Monroe, J.S. and R. Wicander. 1998. Physical Geology (3rd edition). Wadsworth Publishing. Company, Belmont, CA. 646 pp.

Moores, E.M. and R.J. Twiss. 1995. Tectonics. W.H. Freeman, New York. 415 pp.

MTYcounty.com. 2002. Gray whale. Available at: http://www.mtycounty.com/pgs-animals/g-whale.html.

Muhs, D. R., Wehmiller, J. F., Simmons, K. R. and York, L. L. 2004. Quaternary sea-level history of the United States. Developments in Quaternary Sciences 1:147-183.

National Research Council. 2012. Sea-level rise for the coasts of California, Oregon, and Washington: Past Present and Future. National Academies Press, Washington, D.C.

Naughton, M.B., D.S. Pitkin, R.W. Lowe, K.J. So, and C.S. Strong. 2007. Catalog of Oregon seabird colonies. U.S. Fish and Wildlife Service, Biological Technical Publication BTP-R1009-2007. 487 pp.

National Marine Fisheries Service (NMFS). 1997. Draft environmental assessment: Use of Acoustic Pingers as a Management Measure in Commercial Fisheries to Reduce Marine Mammal Bycatch. NMFS, Office of Protected

Resources, Silver Spring, MD.

National Marine Fisheries Service (NMFS). 2014. Harbor Seal (*Phoca vitulina richardii*) Oregon/Washington Coast Stock. National Marine Fisheries Service. Available at: http://www.nmfs.noaa.gov/pr/sars/2013/po2013_harborseal-orwac.pdf.

National Marine Fisheries Service (NMFS). 2016. Steller Sea Lion (*Eumetopias jubatus*); Eastern Stock. National Marine Fisheries Service. Available at: http://www.nmfs.noaa.gov/pr/sars/pdf/stocks/alaska/2016/ak2016_ssl-eastern.pdf.

National Geophysical Data Center (NGDC). 2009. Age of the ocean floor. NGDC Data Announcement Number: 96-MGG-04NGDC, Report MGG-12. Available at: www.ngdc.noaa.gov/mgg/image/images/g01167-pos-a0001.pdf.

National Oceanic and Atmospheric Administration (NOAA). 2008. Northern elephant seal. Available at: http://www.afsc.noaa.gov/nmml/species/species_ele.php.

National Oceanic and Atmospheric Administration (NOAA). 2018. Global greenhouse gas reference network. Available at: https://www.esrl.noaa.gov/gmd/ccgg/.

National Oceanic and Atmospheric Administration (NOAA)/RPI. 2014. Environmental sensitivity index (ESI) maps of the shoreline of the Outer Coast of Washington and Oregon. By Research Planning, Inc. (RPI) under contract to the National Oceanic and Atmospheric Administration, NOS Data Explorer.

National Oceanic and Atmospheric Administration (NOAA)/RPI. 2015. Environmental sensitivity index (ESI) maps of the shoreline of the Pacific Northwest Coast (California/Oregon Border to Canadian Border). By Research Planning, Inc. (RPI) under contract to the National Oceanic and Atmospheric Administration, NOS Data Explorer.

NOVA Evolution. 2012. Extinction of the dinosaurs? Available at: http://www.pbs.org/wgbh/evolution/extinction/dinosaurs/asteroid.html.

Nummedal, D., J.J. Gonsiewski, and J.C. Boothroyd. 1976. Geological significance of large channels on Mars. Geol. Romana 15:407-418.

O'Brien, M.P. 1931. Estuary tidal prisms related to entrance areas. Jour. Civil Engineers 1:738-793.

Oregon's Adventure Coast. 2014. Oregon Dunes national recreation area. Available at: http://www.oregonsadventurecoast.com/trip-ideas/guide-to-the-oregon-dunes-national-recreation-area/.

Oregon Department of Fish and Wildlife. 2014. Razor clams. Available at: https://myodfw.com/crabbing-clamming/species/razor-clam.

Oregon Department of Fish and Wildlife. 2016. Salmon River Fall Chinook program. Hatchery and Genetic Management Plan. 46 pp.

Oregon Department of Geology and Mineral Industries. 2018. Oregon Tsunami Clearinghouse. Available at: https://www.oregongeology.org/tsuclearinghouse/pubs-evacbro.htm.

Oregon State Parks. 2014a. Cape Sebastian state scenic corridor. Available at: http://www.oregonstateparks.org/index.cfm?do=parkPage.dsp_parkPage&parkId=52

Oregon State Parks, 2014b. Arizona beach state recreation site. Available at: http://www.oregonstateparks.org/index.cfm?do=parkPage.dsp_parkPage&parkId=188.

Oregon State Parks, 2014c. Port Orford Heads state park. Available at: http://www.oregonstateparks.org/index.cfm?do=parkPage.dsp_parkPage&parkId=43.

Orr, E.L. and W.N. Orr. 2000. Geology of Oregon. Kendall/Hunt Pub. Co., 5th Edition, 254 pp.

Orr, W.N. and E.L. Orr. 2002. Geology of the Pacific Northwest. 2nd Edition. Waveland Press, Inc., 337 pp.

Packwood, A.R. 1983. The influence of beach porosity on wave uprush and backwash. Coastal Engineering 7:29-40.

Paine, R.T. 1966. Food web complexity and species diversity. American Naturalist, 100:65-75.

Paine, R.T. 1969. A note on trophic complexity and community stability. American Naturalist, 103:91-93.

Paine, R.T., J.T. Wootton, and P.D. Boersma. 1990. Direct and indirect effects of peregrine falcon predation on seabird abundance. The Auk 107:1-9.

Pararas-Carayannis, G. 2009a. The tsunami page. The March 27, 1964 Great Alaska Tsunami. Available at: http://www.drgeorgepc.com/Tsunami1964GreatGulf.html.

Pararas-Carayannis, G. 2009b. The tsunami page. Historical tsunamis in California, Available at: http://www.drgeorgepc.com/TsunamiCalifornia.html.

Patsch, K. and G. Griggs. 2006. Littoral cells, sand budgets, and sand beaches: Understanding California's Shoreline. Inst. of Marine Sciences, U.C. Santa Cruz, Cal. Sed. Management Workshop. 40 pp.

Peltier, W.R., 2002. On eustatic sea level history: Last Glacial Maximum to Holocene. Quat. Science Reviews 21:377-396

Peterson, C.D., G.W. Gleeson, G.W., and N. Wetzel. 1987. Stratigraphic development, mineral resources, and preservation of marine placers from Pleistocene terraces in southern Oregon, U.S.A. Sedimentary Geology 53:203-229.

Petit, J.-R., Jouzel, J., Raynaud, D., Barkov, N. I., Barnola, J.-M., Basile, I., Bender, M., Chappellaz, J., Davis, M. and Delaygue, G. 1999. Climate and atmospheric history of the past 420,000 years from the Vostok ice core, Antarctica. Nature 399:429-436.

Pilkey, O.H., R.S. Young, S.R. Riggs, A.W. Smith, H. Wu, and W.D. Pilkey. 1993. The concept of shoreface profile of equilibrium: A critical review. Jour. Coastal Research 9(1):255-278.

Pilkey, O.H. and M.E. Fraser. 2003. A celebration of the world's Barrier Islands. Columbia University Press, New York. 309 pp.

Pilkey, O.H. and L. Pilkey-Jarvis. 2006. Useless arithmetic: Why environmental scientists can't predict the future. Columbia University Press, New York. 230 pp.

Plafker, G. 1969. Tectonics of the 27 March 1964 Alaska Earthquake. U.S. Geological Survey Prof. Paper 543-1. 74 pp.

Playfair, J. 1802. Illustrations of the Huttonian theory of the earth. Edinburgh. 528 pp.

Postma, H. 1967. Sediment transport and sedimentation in the Estuarine Environment. Estuaries, 83:158-179.

Pritchard, D.W. 1967. What is an estuary? Physical viewpoint. In: G.W. Lauff (ed.), Estuaries, American Association for the Advancement of Science Publ. 83:3-5.

Rau, W.W. 1973. Geology of the Washington coast between point Grenville and the Hoh River. Washington Department of Natural Resources, Geology and Earth Resources Division Bulletin No. 66. 198 pp.

Rau, W.W. 1980. Washington coastal geology between the Hoh and Quillayute Rivers. Washington Department of Natural Resources, Geology and Earth Resources Division Bulletin No. 72.

Reimnitz, E. 1966. Late quaternary history and sedimentation of the Copper River Delta and vicinity, Alaska. Ph.D. Dissertation, University Cal., San Diego. 160 pp.

Ricketts, E.F. and J. Calvin. 1939. Between Pacific tides. Revised by D.W. Phillips, 1985. 5th/Revised edition, Stanford University Press, Stanford. 652 pp.

Ruggiero, P., G. Gelfenbaum, C.R. Sherwood, J. Lacy, and M.C. Buijsman. 2003. Linking nearshore processes and morphology measurements to understand large scale coastal change. Proc. of Coastal Sediments '03, ASCE.

Ruggiero, P., G.M. Kaminsky, G. Gelfenbaum, and B. Voigt. 2005. Seasonal to interannual morphodynamics along a high-energy dissipative littoral cell. Journal of Coastal Research 21:553-578.

Ruggiero, P., M. Buijsman, G.M. Kaminsky, and G. Gelfenbaum. 2006. Modeling the effect of wave climate and sediment supply variability on large scale shoreline change. Marine Geology. 63 p. Available at: http://www.geo.oregonstate.edu/files/geo/Ruggiero_etal_inpress_MG.pdf

Ruggiero, P., M.G. Kratzmann, E.A. Himmelstoss, D. Reid, J. Allan, and G. Kaminsky. 2013. National assessment of shoreline change - Historical shoreline change along the Pacific Northwest coast. U.S. Geological Survey Open-

File Report 2012–1007, 62 p.

Rust, B.R. 1978. Depositional models for braided alluvium. In: A.D. Miall (ed.), Fluvial Sedimentology. Can. Soc. Petrol. Geol. Memoir 5:605-625.

Satake, K., K. Shimazake, Y. Tsuji, and K. Ueda. 1996. Time and size of the giant earthquake in Cascadia inferred from Japanese tsunami records of January 1700. Nature 379:246-249.

Schuster, J.E. 2014. Northern Washington, general geology overview. Available at: http://www.westerngeohikes.0catch.com/Washington/NWGI.html

Shepard, F.P. 1950a. Bars and troughs. Tech. Memo 15, Beach Erosion Board. U.S. Army Corps of Engineers, Washington, DC. 38 pp.

Shepard, F.P. 1950b. Beach cycles in Southern California. Tech. Memo 20, Beach Erosion Board. U.S. Army Corps of Engineers, Washington, DC. 26 pp.

Shepard, F.P. and H.R. Wanless. 1971. Our changing coastlines. McGraw-Hill, New York. 579 pp.

Sherwood, C.R., D.A. Jay, R.B. Harvey, P. Hamilton, and C.A. Simenstad. 1990. Historical changes in the Columbia River estuary. Progress in Oceanography 25:271-297.

Shih, S.M. and P.D. Komar. 1994. Sediments, beach morphology and sea cliff erosion within an Oregon coast littoral cell. Journal of Coastal Research, v. 10, p. 144-157.

Silvester, R. 1974. Coastal Engineering, Vol. II. Elsevier Pub. Co. 338 pp.

Silvester, R. 1977. The role of wave reflection in coastal processes. In: Proc. Coastal Sediments '77, Amer. Soc. Civil Engineers, New York. Pp. 639-654.

Simons, D.B. and E.V. Richardson. 1962. Resistance to flow in alluvial channels. Am. Soc. Civil Eng. Transactions, 127:927-953.

Slingerland, R.L. 1977. The effects of entrainment on the hydraulic equivalence relationships of light and heavy minerals in sands. Journal of Sedimentary Petrology 47:753-770.

State of Oregon. 2014. Columbia River basalt: The Yellowstone hot spot arrives in a flood of fire. Oregon Dept. of Geology and Mineral Industries. Available at: http://www.oregongeology.org/sub/publications/ims/ims-028/unit08.htm

Stevens, A.W., G. Gelfenbaum, P. Ruggiero, and G.M. Kaminsky. 2012. Southwest Washington littoral drift restoration - Beach and nearshore morphological monitoring. U.S. Geological Survey Open-File Report 2012-1175, 67 p.

Stephenson, W. and R. Kirk. 2005. Shore platforms. In: M.L. Schwartz (ed.), Encyclopedia of Coastal Science. Springer, New Zealand. Pp. 873-875.

Sunamura, T. 1992. Geomorphology of rocky coasts. Wiley, New York. 314 pp.

Terich, T.A. and P.D. Komar. 1973. Development and erosion history of Bayocean Spit, Tillamook, Oregon. Oregon State University, School of Oceanography, Reference 73-16. 156 pp.

The Columbia River: A photographic journey, 2014. Cape Disappointment, Washington. Available at: http://www.columbiariverimages.com/Regions/Places/cape_disappointment.html.

The Natural History Museum. 2014. Mass extinctions. Available at: http://www.nhm.ac.uk/nature-online/life/dinosaurs-other-extinct-creatures/mass-extinctions/

The Oregon Encyclopedia. 2017a. Bayocean. The Oregon Encyclopedia, A project of the Oregon Historical Society. Available at: https://oregonencyclopedia.org/articles/bayocean/#.WoCOAWa-LGY

The Oregon Encyclopedia. 2017b. Tillamook Rock Lighthouse. The Oregon Encyclopedia, A project of the Oregon Historical Society. Available at: https://oregonencyclopedia.org/articles/tillamook_rock_lighthouse/#.WoHr_2a-LGY

Thoma, D.W. 1983. Changes in Columbia River estuary habitat types over the past century. Columbia River Estuary

Data Development Program, Columbia River Estuary Study Task-force, Astoria, Oregon. 102 pp.

Tillotson, K. and P.D. Komar. 1997. The wave climate of the Pacific Northwest (Oregon and Washington): A comparison of data sources. Journal of Coastal Research, 13(2): 440-452.

Trenhaile, A.S. 1987. The geomorphology of rock coasts. Oxford University Press, USA. 393 pp.

Twichell, D.C. and V.A. Cross. 2001. Holocene evolution of the southern Washington and northern Oregon shelf and coast: Geologic discussion and GIS data release. U.S. Geological Survey Open-File Report 01-076.

University of Nevada-Reno. 2009. Available at: http://quake.unr.edu/ftp/pub/louie/class/100/magnitude.html.

UCal Berkeley, 2004. Foraminifera. Museum of Paleontology. Available at: http://www.ucmp.berkeley.edu/index.html.

U.S. Army Corps of Engineers. 2002. Coastal engineering manual. Engineer Manual 1110-2-1100, U.S. Army Corps of Engineers, Washington, DC. Available at: http://bigfoot.wes.army.mil/cem026.html.

U.S. Fish & Wildlife Service. 2014. Nestucca Bay National Wildlife Refuge. U.S. Fish and Wildlife Service. Available at: http://www.fws.gov/refuges/profiles/index.cfm?id=13597

U.S. Fish & Wildlife Service. 2016. Rangewide status of Marbled Murrelets. 24 pp. Available at: https://www.fws.gov/wafwo/Documents/RangewideSOS/USFWS/Rangewude%20Status%20of%20Marbled%20Murrelets.pdf

USC Tsunami Research Group. 1964. 1964 Alaskan Tsunami. Available at: http://cwis.usc.edu/dept/tsunamis/alaska/1964/webpages/index.html.

U.S. Geological Survey (USGS). 1997. Geology of the Point Reyes national seashore and vicinity, Marin County, California. Open-File Report 97-456. Available at: http://pubs.usgs.gov/of/1997/of97-456/pr-geo.txt.

U.S. Geological Survey (USGS). 2005. Diagram of the interior of the earth. Available at: http://pubs.usgs.gov/gip/dynamic/inside.html

U.S. Geological Survey (USGS). 2009. Quaternary fault and fold database of the United States. Available at: http://earthquake.usgs.gov/regional/qfaults/

U.S. Geological Survey (USGS) Earthquake Hazard Program. 2016. Measuring the Size of an Earthquake. Available at: http://earthquake.usgs.gov/learn/topics/measure.php

U.S. Global Change Research Program (USGCRP). 2017. Climate Science Special Report: Fourth National Climate Assessment, Volume I. Wuebbles, D.J., D.W. Fahey, K.A. Hibbard, D.J. Dokken, B.C. Stewart, and T.K. Maycock (eds.), U.S. Global Change Research Program, Washington, DC, USA, 470 pp.

Van Straaten, L.M.J.U. 1950. Environment of formation and facies of the Wadden Sea sediments. Koninkl.Ned. Aardrijkskde Genoot. 67:94-108.

Walsh, T.J., M.A. Korosec, W.M. Phillips, R.L. Logan, and H.W. Schasse. 1987. Geologic map of Washington-Southwest quadrant. Washington Division of Geology and Earth Resources, Geologic Map GM-34.

Wang, P., B.A. Ebersole, and E.R. Smith. 2002. Sand transport – initial results from large-scale sediment transport facility. ERDC/CHL CHETN-II-46. Pp. 1-14.

Washington Department of Fish and Wildlife. 2017. Columbia River sea lion management. Available at: https://wdfw.wa.gov/conservation/sealions/questions.html

Washington State Department of Natural Resources. 2014. Geology of Washington. Available at: www.dnr.wa.gov/ResearchScience/Topics/GeologyofWashington/Pages/geolofwa.aspx

Wegener, A. 1912. Die Entstehung der Kontinente. Peterm. Mitt. 185–195, 253–256, 305–309.

Wentworth, C.K. 1922. A scale of grade and class terms for clastic sediments. Journal of Geology, 30:377-39.

Wiegel, R.L. 1964. Oceanographical engineering. Prentice-Hall, Englewood Cliffs, NJ. 532 pp.

Wikipedia, The Free Encyclopedia. 2014a. Columbia River Basalt Group. Available at: http://en.wikipedia.org/wiki/Columbia_River_Basalt_Group

Wikipedia, The Free Encyclopedia. 2014b. Geology of the Pacific Northwest. Available at: http://en.wikipedia.org/wiki/

Geology_of_the_Pacific_Northwest

Wikipedia, The Free Encyclopedia. 2014c. 1700 Cascadia Earthquake. Available at: http://en.wikipedia.org/wiki/1700_Cascadia_earthquake

Wikipedia, The Free Encyclopedia. 2014d. Gold mining in the United States. Available at: http://en.wikipedia.org/wiki/Gold_mining_in_the_United_States#Oregon

Wilson, D. 2008. Siletz Basin steelhead trapping and management activities. Oregon Department of Fish and Wildlife. 10 p. Available at: http://library.state.or.us/repository/2008/200805091147105/index.pdf.

Wilson, J.T. 1965. A new class of faults and their bearing on continental drift. Nature, 207:343-347.

Witter, R.C., H.M. Kelsey, and E. Hemphill-Haley. 2003. Great Cascadia earthquakes and tsunamis of the past 6700 years, Coquille River estuary, southern coastal Oregon: Geological Society of America Bulletin, 115:1289-1306.

Wood, F.J. 1982. Tides. In: M.L. Schwartz (ed.), The Encyclopedia of Beaches and Coastal Environments. Hutchinson Ross Pub. Co., Stroudsburg, PA. Pp. 826-837.

Woodroffe, C.D. 2002. Coasts. Cambridge University Press, Cambridge. 623 pp.

Yuill, B., D. Lavoie, and D.J. Reed. 2009. Understanding subsidence processes in Coastal Louisiana. Journal of Coastal Research: Special Issue 54:23-36.

Zingg, T.H. 1935. Beitrag zur Schotternalyse. Schweiz. Min. u. Pet. Mitt. 15:39-140.

INDEX

N

O

P

Q

R

S

ABOUT THE AUTHORS

Miles O. Hayes

Dr. Miles O. Hayes is a coastal geomorphologist and sedimentologist with over 50 years of research experience. He has authored over 250 articles and reports and seven books on numerous topics relating to tidal hydraulics, river morphology and processes, beach erosion, barrier island morphology, oil pollution, and petroleum exploration. Based on extensive field experience throughout the world, he has developed innovative techniques regarding environmental protection, oil-spill response, and shoreline processes. Three of the original concepts proposed and developed by him are: 1) Importance of hurricanes to barrier island and nearshore shelf sedimentation; 2) The effect of tides on shoreline morphology and sedimentation patterns; and 3) The environmental sensitivity index (ESI) for mapping shorelines (with co-author Michel), which has been applied worldwide. Hayes' teaching experience includes a range of both undergraduate and graduate courses while a Professor at the Universities of Massachusetts and South Carolina. Seventy-two graduate students received their degrees under his supervision, most of whom are now leaders in their respective academic, government and industry positions. He is presently Chairman of the Board of Research Planning, Inc. (RPI), a science technology company located in Columbia, SC.

Jacqueline Michel

Dr. Jacqueline Michel is an internationally recognized expert in oil and hazardous materials spill response and assessment with a primary focus in the areas of oil fates and effects, non-floating oils, shoreline cleanup, alternative response technologies, and natural resource damage assessment. As of this time, she has participated in research projects in 33 countries. Much of her expertise is derived from her role, since 1982, as part of the Scientific Support Team to the U.S. Coast Guard provided by the National Oceanic and Atmospheric Administration (NOAA). Under this role, she is on 24-hour call and provides technical support for an average of 50 spill events per year. She leads shoreline assessment teams and assists in selecting cleanup methods to minimize the environmental impacts of the spill. She has written over 150 manuals, reports, and scientific papers on coastal resource impacts, mapping, and protection. As a member of the Ocean Studies Board at the National Academy of Sciences for four years, she served on four National Research Council committees (chairing two), and is a Lifetime Associate of the National Academies. One of the original founders of RPI, which started in 1977, she now serves as the company President.